LÉON ROSENFELD

Physics, Philosophy, and Politics in the Twentieth Century

LÉON ROSENFELD

Physics, Philosophy, and Politics in the Twentieth Century

Anja Skaar Jacobsen

Niels Bohr Archive, Denmark

NEW JERSEY • LONDON • SINGAPORE • BEIJING • SHANGHAI • HONG KONG • TAIPEI • CHENNAI

Published by

World Scientific Publishing Co. Pte. Ltd.

5 Toh Tuck Link, Singapore 596224

USA office: 27 Warren Street, Suite 401-402, Hackensack, NJ 07601

UK office: 57 Shelton Street, Covent Garden, London WC2H 9HE

British Library Cataloguing-in-Publication Data
A catalogue record for this book is available from the British Library.

LÉON ROSENFELD
Physics, Philosophy, and Politics in the Twentieth Century

ISBN-13 978-981-4307-81-9
ISBN-10 981-4307-81-5

Typeset by Stallion Press
Email: enquiries@stallionpress.com

Printed in Singapore.

Preface

The incentive to write this book stems from a seminar I attended at the Niels Bohr Archive in Copenhagen in May 2002. Here, for the first time, I heard dialectical materialism mentioned in connection with Niels Bohr's complementarity interpretation of quantum mechanics. The talk was about the Russian theoretical physicist Vladimir A. Fock and it was given by one of his pupils, Yuri Demkov. Trained as a physicist at the University of Aarhus, Denmark, and with a master's thesis which was a comparative study of recent interpretations of quantum mechanics, I had never heard that link being made before.[1] Nor was I familiar with dialectical materialism. I asked the speaker about this link, and he sought approval for his claims among the audience, an assembly of distinguished physicists at the Niels Bohr Institute. Nobody objected and I went away with the feeling that here was something I would like to understand better, a story worth pursuing; but not for me since regrettably I do not read Russian. When I told Helge Kragh about my interest later that summer he suggested that I read Loren Graham's books about Russian science and, more importantly, that I looked at Léon Rosenfeld's Papers at the Niels Bohr Archive. I am deeply grateful to Kragh for that suggestion. Rosenfeld, who had been Kragh's supervisor in history of physics at one point, was known to be a disciple of Bohr and a Marxist. And so began my exciting journey. My original interest in the intersection between Marxist philosophy and physics developed into a broader curiosity about Rosenfeld, his extraordinary life in the *Age of Extremes*, as Eric Hobsbawm has called the 20th century; Rosenfeld's friendship

[1] Jacobsen (1995).

with Bohr; the history of Bohr's quantum philosophy; and the history of the political Left.

My first research proposal about Rosenfeld was supported by the Danish Research Council for the Humanities and took place at the Department of Mathematics and Physics, Roskilde University 2004–2006. I owe many thanks to colleagues at Roskilde University for their interest in, and for critical remarks and constructive comments on my research and especially to Tinne Hoff Kjeldsen for her untiring encouragement and friendship over the years. For further research with the purpose of writing a book manuscript, I received support from the Carlsberg Foundation and the research took place at the Niels Bohr Archive 2007–2010. I am deeply indebted to Director of the Niels Bohr Archive, Finn Aaserud, for his interest and his stimulating comments from the very beginning of the project. I am also deeply indebted to Felicity Pors whose assistance in gold digging in the treasure chambers of the Niels Bohr Archive has proved indispensable for the project. Pors had also assisted with the language revision of my manuscript. I also want to thank Lis Rasmussen, who has been tremendously helpful with all things practical related to the project.

It is a pleasure to thank Olival Freire Jr. for the interest he has taken in my work as well as his hints and generous sharing of ideas and perspectives on the quantum controversy and its political contexts. I also want to thank Arne Hessenbruch for putting me in contact with Robert S. Cohen, John J. Stachel and Silvan S. Schweber. Cohen generously provided me with a copy of Rosenfeld's *Selected Papers* which he and Stachel edited in collaboration with Rosenfeld, and Cohen introduced me to the relevant papers in his archive at the University of Boston.

As for Rosenfeld's role in the history of quantum physics I have been fortunate to be part of the international network of historians, philosophers, and physicists: *History and Foundations of Quantum Physics*, based at the Max Planck Institute for the History of Science in Berlin. This group is responsible for the sequence of History of Quantum Physics (HQ) conferences in which I took part in Berlin in 2007 and 2010. I owe special thanks to Christian Joas, Christoph

Lehner, and Jürgen Renn for giving me the opportunity to spend three months at the Max Planck Institute during the History of Science conferences in the winter 2008–2009. While there, I participated in the reading group pertaining to the quantum history project. Quite fittingly the group was discussing the history of early quantum field theory at the time and the perspective taken in the second chapter of my book was inspired by the discussion in that forum.

With respect to treating Rosenfeld's political commitment, I have been fortunate to be part of another fruitful international network of scholars interested in the political engagement of scientists in the twentieth century. This includes Vidar Enebakk, Patrick Petitjean, Martha Cecilia Bustamante, Anna K. Mayer, Chris Chilvers, Geert Somsen, Garry Werskey, and Everett Mendelsohn, among others. This network has been responsible for several well-attended conference sessions and symposia. Inspiration from the discussions in that forum looms large in the third and fifth chapters of my book.

The manuscript has improved considerably from comments on various parts by Finn Aaserud, Vidar Enebakk, Alexei Kojevnikov, Kenneth L. Caneva, and John J. Stachel. In connection with papers preceding the book and on which parts of the book are based, I owe thanks to Joan Bromberg, Jørgen Kalckar, Silvan S. Schweber, Skuli Sigurdsson, Don Salisbury, and Christoph Lehner for their criticism and comments.

Rosenfeld held chairs in Liège, Utrecht, Manchester, and Copenhagen and his correspondence and publications are written in French, German, Dutch, English, and Danish. I owe thanks to Leen Dorsman, Brigitte van Tiggelen, Geert Somsen, Leo Molenaar, David Baneke, Marijn Hollestelle, and Ernst Homburg for their valuable hints about the Dutch and Belgian contexts, source material, and archives. I am grateful to Albena Nielsen and Jens Gregersen for their help with translating Russian texts, and to Per Friedrichsen and Kurt Møller Pedersen for their help with translations from German and French. Naturally, however, I remain responsible for all translations. In addition, I would like to thank David Favrholdt, Anita Kildebæk Nielsen, Paul Josephson, Ivar Skaar Jacobsen, Cecile DeWitt-Morette, Christian Forstner, John Heilbron, Sabine Lee, Henrik

Knudsen, Roberto Salmeron, Gerald E. Brown, Simon Schaffer, Helmut Rechenberg, Kenneth Ford, Graham Farmelo, and Ole Knudsen.

For permission to consult and quote from the Rosenfeld Papers, I am indebted to Rosenfeld's children, Andrée and Jean L. J. Rosenfeld. (Sadly Andrée Rosenfeld did not live to see the book completed.) I would like to thank the Niels Bohr Archive for permission to quote Niels Bohr's letters; Sir Rudolf Peierls' daughter Mrs. Joanna Hookway for permission to quote from the Peierls Papers; Christian Møller's son Ole Møller for permission to quote from the Møller Papers; Georges Lochak for permission to quote Louis de Broglie's letters; University of Sussex, England, for permission to quote from J. G. Crowther's papers.

In addition, I want to thank archivists and librarians at the institutions for the sources of which I have made use of besides the Niels Bohr Archive. These include Lisbeth Dilling and Kader Rahman Ahmed at the library at the Niels Bohr Institute, Copenhagen; Albert Einstein Archives, Jewish National and University Library, Jerusalem; Arbejdermuseet and Arbejderbevægelsens Bibliotek og Arkiv, Copenhagen; Archive de L'Académie des Sciences, Institute de France and Fondation Louis de Broglie, Paris; Archives Joliot-Curie, Fonds Frédéric Joliot-Curie, Paris; J. G. Crowther Archive, University of Sussex Library, Special Collections, Brighton; Internationaal Instituut voor Sociale Geschiedenis, Amsterdam; Cambridge University Library, Cambridge; Paul A. M. Dirac Science Library, Florida State University, Tallahassee; Colin Harris at The Bodleian Library, University of Oxford, Oxford; Staatsbibliothek, Preussischer Kulturbesitz, Berlin; Special Collections Research Center, University of Chicago Library; The American Philosophical Society, Philadelphia; The Niels Bohr Library, AIP; U.S. Department of Justice, Federal Bureau of Investigation, Washington, D.C., USA; and Helmuth Rechenberg in charge of the Werner Heisenberg Nachlass, Werner-Heisenberg Institute, Munich.

Last but not least I want to thank my parents Vibeke Jacobsen and Ivar Skaar Jacobsen for many years of patience, support, and encouragement. I dedicate this book to my children Mai and Karl with all my love.

Contents

Preface v

List of Pictures xi

Introduction 1

Chapter 1. Physicist of the Second Quantum Generation 11

Chapter 2. Rosenfeld in Copenhagen 49

Chapter 3. Physics, Philosophy, and Politics in the 1930s 95

Chapter 4. Surviving the War in Utrecht 153

Chapter 5. Cold War and Political Commitment 185

Chapter 6. Bohr's Cold Warrior 255

Epilogue 303

Bibliography 315

Index 343

List of Pictures

Chapter 1

P19: Participants at the fifth Solvay Meeting 1927.
P28: Dirac and Rosenfeld in Göttingen in 1928.
P30: Rosenfeld 1930.
P31: Participants at the first Copenhagen Conference 1929.
P47: Picture of Gamow and Rosenfeld taken by Peierls at Piz da Daint where Gamow signed his note for *Nature*.

Chapter 2

P57: Bohr and Rosenfeld in Tisvilde 1931.
P66: Blackboard with Bohr's drawings from a visit by Bohr to Japan 1937.
P66: Gamow's Einstein box, a Christmas present to Bohr in 1930.
P70: A later reconstruction of Gamow's drawing of Landau gagged, pinioned to a chair and confronted by Bohr.

Chapter 3

P97: Piet Hein's drawing of the box with two dice meant to illustrate the complementarity principle.
P98: Participants in the Copenhagen Conference 1932.
P102: Leo Trotsky speaking in Copenhagen in 1932.
P102: Drawing of Trotsky by Piet Hein.
P103: Yvonne and Léon Rosenfeld in 1935.
P109: Rosenfeld, Niels, and Margrethe Bohr arriving in Moscow 1934.
P113: Bohr, Landau, Bronstein, and Rosenfeld discussing at the Kharkov Conference 1934.

P113: Participants in the Kharkov Conference 1934.
P118: Participants in the Copenhagen Conference 1933.
P130: Participants in the Second International Unity of Science Congress at the Carlsberg Mansion in June 1936.
P141: Walter Heitler and Rosenfeld at the Copenhagen Conference in 1937.
P141: Rosenfeld 1938.

Chapter 4

P158: Sketch of Rosenfeld by a Belgian student in 1940.

Chapter 5

P204: Sketch of Rosenfeld by a Dutch student in 1947.
P211: The board of WFSW in 1946.
P215: Personnel at the institute at Manchester in 1947, including Blackett and Rosenfeld.
P245: Léon and Yvonne in the French Alps in the early 1950s.

Chapter 6

P286: Pauli's nickname for Rosenfeld.

Introduction

One way to get insight into the fascinating history of twentieth century physics and its inextricable entanglement with general history is to look at the life stories of particular individuals.[1] This book is a social-intellectual biography of the Belgian theoretical physicist, Léon Rosenfeld (1904–1974). In the history of science literature Rosenfeld is largely a blank sheet, but he was at the centre of modern physics as one of the pioneers of quantum field theory and quantum electrodynamics in the late 1920s and the 1930s. Today he is best known for his publications with Niels Bohr in 1933 and 1950 on measurability in quantum electrodynamics and as the fierce spokesman of Bohr's complementarity interpretation of quantum mechanics after the Second World War.[2] Rosenfeld was, however, exceptionally erudite and made detailed and significant contributions across the whole range of theoretical physics, such as quantum field theory, nuclear physics, statistical physics, and astrophysics. In 1956, he founded the journal *Nuclear Physics*. Besides, he reflected deeply on the history and philosophy of science and its social role. During his career he held chairs at Liège, Utrecht, Manchester, and Copenhagen. He received the Francqui Prize in 1949, which is given exclusively to Belgian scientists and scholars, but in spite of its national character bestows an honour comparable to the Nobel Prize. In addition to his contributions to physics, Rosenfeld played a salient role as Bohr's close

[1] Söderqvist (2006), pp. 105–107. Hankins (1979).

[2] Rosenfeld's political engagement as regards quantum epistemology and history of science has been treated in Jacobsen (2007; 2008). The scientific context of the famous Bohr–Rosenfeld paper has been discussed in Jacobsen (2011). Rosenfeld's role in the quantum controversy has been treated in Freire Jr. (2001; 2005; 2007), and Rosenfeld's contribution to constrained Hamilton dynamics has been dealt with by Salisbury (2006; 2007).

collaborator as well as communicator of Bohr's quantum philosophy. In their collaboration Rosenfeld was in charge of the mathematical formulations of their theoretical considerations, but he also served as a sounding board for Bohr, that is, a critical mind on whom Bohr could try out his ideas. Such a sounding board was of crucial importance for Bohr's way of working.

Rosenfeld was a perceptive, polyglot cosmopolitan, whose life crossed those of many important people in many countries; he was part of a ramified international scientific and scholarly network in both the East and West. He possessed a strong, integrated personality and was capable of performing exotic calculations at one moment, disentangling subtle philosophical questions, or intervening in a political discussion, the next — all at the highest level. He was also an eminent linguist; he mastered eight or nine languages, speaking and writing at least five of them fluently. To some he was regarded as a sage.[3] He was kind, courteous, honest, decent, and upright, as well as shy, timid, and modest. He possessed a dry humour. Compared with the often boring and dull style of today's scientific publications which give the impression that science is an impersonal disembodied trade, Rosenfeld's publications are characterized by a rare erudition, originality, and personality. At the same time he insisted indefatigably and uncompromisingly on rigour, consistency, and logic in thought and argument. This strong commitment reveals temper and passion behind his shy appearance and it could make him rather biting and highly satirical in book reviews, polemical papers, not to mention private correspondence aimed at opponents of Bohr's complementarity interpretation or Marxists who blindly followed Stalin's decrees. Rosenfeld's vitriolic style was felt by many a "quantum dissident", that is, physicists who challenged the Copenhagen Interpretation of quantum mechanics from the 1950s onwards.[4]

As a Marxist, Rosenfeld fully recognized the political and social dimensions of scientific inquiry and took very seriously his

---

[3]Recollections of H. Levi (1993), NBA. Notes on L. Rosenfeld by S. Rozental, NBA. A. Bohr (1974). Møller (1974).

[4]Freire Jr. (2009).

responsibility as a scientist to speak out and challenge misinterpretations, as he saw them, of scientific concepts, in addition to working actively for socialism and peace. Rosenfeld was therefore actively involved in many controversies where scientific, philosophical, ideological, and political issues intersected. This makes the Rosenfeld Papers at the Niels Bohr Archive in Copenhagen a particularly valuable and rich source for exploring the development of modern physics in its ideological and political context. Rosenfeld clearly strove to combine his political interests with his views on science and this was relatively easy for him until the late 1940s. However, in the post-war years Rosenfeld's strong belief in a science driven by the researchers' curiosity rather than by state planning, and his conviction that philosophy should adapt to the novel achievements of modern science and not vice versa, could not be reconciled with the official communist conception of socialist scientific and scholarly work proper. He therefore sought to counter these new agendas. Rosenfeld's commitment seems to have been partly ignited and driven by the lessons he drew from his studies in the history of science. In the end, science will win over temporarily unfavorable receptions of new ideas. However, Rosenfeld was determined to defeat with stronger means than time the unfavorable reception of Bohr's complementarity among communist scientists and scholars and later among new generations of physicists.

The book is not a comprehensive biography of Rosenfeld. It does not treat Rosenfeld's research in its totality, and it deals only sporadically with his private life (for which there are only few sources). As reflected in the title, my motive has been to use the biography of Rosenfeld as a lens to study the intersection between physics, philosophy, and politics in the twentieth century. On the other hand, that focus makes sense *only* because it was such an integrate part of Rosenfeld's life. I have been driven by a strong curiosity and desire to understand Rosenfeld's world view, his activities and choices based on his convictions in physics and socialism, and to situate him within the historical, political, and scientific context, which *he* tried to understand and influence. The book therefore attempts to understand Rosenfeld and his activities in science and politics as a whole.[5] By following Rosenfeld's trajectory my book treats the intersection between Rosenfeld's science and politics in the different European cultures he lived in. Because of lack of sources documenting his specific activities and viewpoints at a given time, Rosenfeld's life sometimes takes the background in relation to events treated in the book rather than the focus. His views on scientific, philosophical, and political ideas come to the fore most clearly in the many controversies he engaged in. The study of these controversies provides new and relevant insights into, for example, the Cold War political debates and on the debates in the 1930s about quantum philosophy that have previously been overlooked or "black-boxed" for historical analysis or categorized in a stereotyped way.[6] My undertaking involves furthermore depicting the scientific and scholarly community within which Rosenfeld operated and describing how his professional practice influenced his world view and vice versa. It also involves portraying the personal relationships that mattered most to Rosenfeld, and it shows how these conditioned, or sometimes conflicted with, his scientific and political commitments.[7] In particular, the book revolves around the close friendship between Rosenfeld and Bohr. And since Rosenfeld was so strongly engaged in promoting Bohr's interpretation of quantum theory from the beginning of the 1930s, the book also constitutes a chapter of the story of the changing receptions of this influential, but also highly contested, point of view.

In the treatment of Rosenfeld and the controversies he took part in, scientific as well as political, I envision myself largely as a commentator illuminating and analyzing the issues at stake. My book is not a contribution to the defense of Bohr's interpretation of quantum mechanics, nor is it meant to defend Rosenfeld's support of Bohr

[5]To that extent the biography is existential in Thomas Söderqvist's meaning of the word. Söderqvist (1998), pp. 26–34. Among historical biographies of scientists employing a perspective similar to mine, see Cassidy (2005).

[6]In using a scientist less famous than, say, a Nobel laureate, as subject of a biography as well as analyzing scientific controversies with the aim of elucidating relevant themes, I have been inspired by the anthology Bertomeu-Sánchez and Nieto-Galan (2006). About scientific controversies, see for example Machamer *et al.* (2000).

[7]In writing about a leftist scientist, I have been inspired by Werskey (1988; 2007a).

and his criticism of quantum dissidents. On the other hand, I cannot claim to be completely neutral. Bohr used to say that "we are both onlookers and actors in the great drama of existence", and that applies at least as much to the historical practice as to physics.[8] Moreover, I fully sympathize with Rosenfeld in his criticism of the tendency, simply out of principle, to treat every scientific idea as equally true. Today there are many examples of how social, political, and ideological agendas, and the ever heavier political, industrial, and commercial focus on scientific products, interfere increasingly with sound scientific practice.[9] Therefore there may even be a lesson to be drawn from Rosenfeld's example with respect to the responsibility of scientists to speak out and challenge perverted scientific messages. Rosenfeld was intimately familiar with both Bohr's quantum philosophy and Marxism, and this made him take action in order to demonstrate that an interpretation was possible that combined complementarity and dialectical materialism in the period beginning in 1947 when the Communist Party intensified its ideological grip on culture and science. Knowledge entails responsibility!

A biography of a scientist should deal with the science itself. According to Thomas L. Hankins, quantum mechanics "contains a large philosophical content which makes it a good subject for biographical treatment".[10] At the same time, Hankins stressed that the biographer of a theoretical physicist or a mathematician "will have difficulty adhering to the biographical form if he includes too many technical details".[11] I have tried to keep this balance by giving a brief treatment of Rosenfeld's early contributions to quantum field theory that form the background of his important work with Bohr and which also serve to motivate a few considerations about Rosenfeld's style in physics. Bohr and Rosenfeld's joint work is treated in its context of the controversies in the small community of theoretical physicists around Bohr in the early 1930s. To a theoretical physicist

[8]Bohr (1934), p. 119. See also Bohr (1999b), p. 60[76].
[9]See for example Oreskes and Conway (2010).
[10]Hankins (1979), p. 13.
[11]*Ibid.*

my treatment may include too few technical details, whereas others may still find these sections difficult to read. I see no easy way around it. However, I hope the book conveys a lively, coherent, and faithful picture of the Rosenfeld I have come to know and have spent so much time with during the last seven years. The portrait reveals a complex life of a man who also experienced moral conflicts.

The book is divided into six chapters dealing chronologically with Rosenfeld's career, each representing a thematic episode and formed on the background of crucial events in world politics. The first chapter covers Rosenfeld's early career with a focus on the scholars and scientists he met, the physics that occupied him, and the scientific milieus he frequented. It includes briefly his education in Belgium and his subsequent route to becoming one of the foremost theoretical physicists at the time. This route went through Paris, where he was supervised by Louis de Broglie and Paul Langevin; to Göttingen where he was Max Born's assistant; to Zurich where around 1930 he matured as a physicist specializing in relativistic quantum mechanics under Wolfgang Pauli's guidance.

Chapter 2 deals with the early 1930s when Rosenfeld collaborated with Bohr in Copenhagen. Working part of each year in Copenhagen, Rosenfeld quickly adapted to the way physics was practiced at Bohr's institute. The collaboration between Bohr and Rosenfeld culminated in their paper of 1933, famous and notorious among physicists, on the measurability of quantum electrodynamics. The paper was an outcome of a controversy between mainly Bohr and Rosenfeld on the one hand and Lev Davidovich Landau and Rudolf Peierls on the other. However, other physicists such as Pauli also took part. The chapter illuminates the communication problems from which the physicists involved seem to have suffered. My aim is to provide the scientific and ideological context of its emergence and early reception. I hope I have succeeded in conveying the gist of the controversy of this highly academic subject. Bohr and Rosenfeld's collaboration took place parallel to Bohr's famous discussion with Albert Einstein about quantum epistemology, which culminated in 1935 when Einstein and his co-workers Boris Podolsky and Nathan Rosen confronted Bohr with the so-called EPR paradox in 1935. These two debates within the physics community proved important for the clarification of Bohr's ideas on how to understand quantum theory.

Chapter 3 is concerned with illuminating the formation of Rosenfeld's leftist position and how in the 1930s he related his political outlook to science, philosophy, and history of science. However, since Rosenfeld has left only sporadic testimonies from that decade, the chapter brings to the foreground mainly the social and political contexts as Rosenfeld may have experienced them in terms of the appearance of extremist movements and political instability in Europe. Since Rosenfeld collaborated very closely with Bohr in the period, Bohr's activities are also given much attention. We learn about Bohr and Rosenfeld's visit to Russia in 1934 and about the effect of Hitler coming to power in 1933 on the life and work at Bohr's institute in Copenhagen. In particular the chapter relates how the political climate influenced the philosophical debates about Bohr's complementarity interpretation of quantum mechanics, when Bohr and his disciples attempted to communicate the epistemological lesson of quantum theory to an intellectual lay audience.

Not only was Rosenfeld a remarkable, many-sided physicist, he also lived through a historical period in Europe of exceptional violence and danger. Chapter 4 treats Rosenfeld's stay in Utrecht during the difficult times of German occupation. As an outspoken socialist possessing a name of Jewish origin, Rosenfeld was extremely vulnerable in a country invaded by Germany during the Second World War and he wisely kept a low profile. The questions to be addressed in this chapter are why Rosenfeld took up a position in Utrecht in the first place, how he managed the difficult times, and what his working conditions were like. Rosenfeld's correspondence, much of it in Danish, reveals hitherto unknown contact between physicists across the occupied countries. The documentation includes Rosenfeld's correspondence with Bohr and Christian Møller in Copenhagen, with Werner Heisenberg in Berlin, and with Oskar Klein in Sweden. Rosenfeld tackled the difficult times in Utrecht during German occupation by means of quiet courage and sanity, continuous contact in writing with his colleagues in Copenhagen, and by burying himself in work.

Chapter 5 covers the period 1945–1958, during which Rosenfeld spent first in Utrecht and from 1947 as a professor of theoretical physics in Manchester. It deals with the ethical and political dilemmas in which physicists found themselves after the Manhattan Project and the emerging Cold War atmosphere where the social responsibility of the scientist was highlighted. The chapter focuses on the basic concepts that Rosenfeld deployed in contests over the ideology and cultural values embodied by science in the early Cold War as well as on his political activism. As soon as the Second World War was over, he appeared optimistic that the time had come for uniting the Left in the creation of a better world based on reason, critical reflection, and socialism. He was a driving force in establishing the Dutch *Federation of Scientific Researchers* (VWO) and an active board member of the communist-led *World Federation of Scientific Workers*. He associated with, but was less active in, the *World Peace Congress*. Rosenfeld's alliance with the Left meant that he did not fully accept Bohr's idea that it would take an Open World, in which all information on technology, science, and social conditions was shared between nations, to prevent a catastrophic war with atomic weapons. Thus, Rosenfeld was simultaneously Bohr's strongest supporter in physics and his critic in politics. Even so, Rosenfeld's admiration for Bohr was undiminished. Bohr reciprocated the feelings and paved the way for Rosenfeld to take up a position as NORDITA professor in Copenhagen in 1958, at a time after Stalin's death when there was an opening for renewed contact and collaboration with Russian physicists.

The culmination of my treatment of the intersection between science and politics as seen through Rosenfeld is found in the final Chapter 6. It focuses on Rosenfeld's role as Bohr's spokesman and defender beginning in the late 1940s, when Cold War tensions framed controversies over the interpretation of quantum physics both in the East and in the West. Rosenfeld's defense of Bohr involved an outspoken critique of the turn in Russian domestic policy in matters of science and culture. I analyze Rosenfeld's view on the relation between science and philosophy and how it was in opposition to Marxism–Leninism. The chapter also treats Rosenfeld's bitter controversy with the American physicist, David Bohm, and a group of young communist physicists in Paris, as well as the opposition his blending of dialectical materialism and complementarity met from the creators of quantum mechanics, Pauli, Born, and Heisenberg.

A brief epilogue about Rosenfeld's continued engagement in defending Bohr in the later years concludes the book. The style and attitude Rosenfeld adopted in this regard, was not always conducive to his evident intentions, *viz.*, to promote and elucidate Bohr's subtle views.

Chapter 1
Physicist of the Second Quantum Generation

The first chapter takes us through Rosenfeld's education and follows his tracks around Europe as assistant professor and postdoc at some of the centers of theoretical physics in the late 1920s. The scientific context of Rosenfeld's early career was primarily the development and consolidation of quantum physics and its combination with relativity theory beginning in 1926. In 1929, he began making his own important contributions to this research area. The pioneers of quantum theory included Max Planck, Albert Einstein, Niels Bohr, Arnold Sommerfeld, followed by Werner Heisenberg, Wolfgang Pauli, Pascual Jordan, P. A. M. Dirac, Erwin Schrödinger, Louis de Broglie, Oskar Klein, Hendrik Anthony Kramers, among others. Rosenfeld can be characterized as a physicist of the second quantum generation including Rudolf Peierls, Lev Davidovich Landau, Enrico Fermi, Christian Møller, Jacques Solomon, Matvei Bronstein, George Gamow, Hans Bethe, among others.[1] Generation here does not refer to age in a strict sense, but rather to the timing of the research of the physicists in question. Other characteristics can be used to divide the generations of theoretical physicists. Suman Seth has recently argued that during the early twentieth century, theoretical physics emancipated itself from experimental physics; a new generation of theoretical physicists, including Heisenberg, Pauli, Dirac, Klein and Kramers, comprised theoreticians who did not have a solid training

[1] Kragh (1999).

in laboratory practice like the older generation.[2] Rosenfeld can be described as a physicist of the second quantum generation in both senses.

"You see, I was Belgian"

Léon Jacques Henri Constant Rosenfeld was born in the mining town of Charleroi in the Walloon part of Belgium on 14 August, 1904, and grew up as an only child. Belgium is a country with two major peoples, the Dutch-speaking Flemish in the north and the French-speaking Walloons in the south-east, sharply separated by both language and culture. It was one of the first countries to be industrialized on the European continent with metallurgy and coal mining primarily in the Walloon part; the Flemish part was predominantly agrarian. During the industrialization, the bourgeoisie gained social dominance and shaped the social, political, and economic life, mixing with Belgium's traditional aristocracy. French was the official language even in Flemish areas and dominated higher education. In the early twentieth century, during the so-called second industrial period, Belgium was ruled by the Catholic Party and during its reign a strong tradition in engineering was cultivated.[3] Rosenfeld's father, Léon Rosenfeld (1872–1918), was a Jewish Russian immigrant from St. Petersburg who married a Belgian, Jeanne Marie-Laure Mathilde Pierre (1880–1955). He was an electrical engineer and inventor working for an electrical company. There seems to have been no contact between him and his family in Russia. However, his father's Russian background may have initiated Rosenfeld Jr.'s curiosity and interest in Russia and the Russian language. His father's background as a migrating Jew may also have affected Rosenfeld Jr. His father distanced himself from the Jewish community; their home was secular and Rosenfeld Jr.'s position became atheistic. His father's example may have prepared Rosenfeld Jr. for not finding a migrating identity alien; he does not seem to have been particularly tied to Belgium.

[2]Seth (2010), p. 182.
[3]Cook (2004), p. 79. Baudet (2007). Halleux *et al.* (2001), pp. 58–59.

Rosenfeld's father died in an accident at the factory during the First World War when Rosenfeld Jr. was only 14. Naturally, this traumatic incident deeply and emphatically affected Rosenfeld Jr.'s life. The premature death of his father had a marked impact on his immediate choices and dispositions in life. His primary intellectual interests were initially history, Greek, Latin, and natural history, but his father's premature death made him turn to study natural science and mathematics, "to be like his father", despite the poor prospects in that field in Belgium at the time and his mother's opposition to this choice.[4]

However, Rosenfeld's background and interest in the human sciences came to mark his activities for the rest of his life, also as a physicist. He stood out from most of his later physicist colleagues as one who sought the philosophical meaning on a deeper level and traced the conceptual roots of physics. Already during his student years, Rosenfeld published numerous small pieces about history of science in *Bulletin Scientifique de l'Association des Elèves des Ecoles Spéciales* at the University of Liège.[5] Wherever he worked, whether in Liège, Utrecht, Manchester, or Copenhagen, Rosenfeld engaged with philosophers and historians of science as well as with physicists and astrophysicists. Early on in the twentieth century, other scholars in Belgium also had strong interest in the history of science, notably George Sarton, the later founder of the American History of Science Society and the leading journal in the field, *Isis*. Sarton was a progressive, secular humanist who regarded history of science as the human activity demonstrating the progress of mankind.[6] Besides Sarton, Rosenfeld was well-acquainted with the mathematician and historian of science Jean Pelseneer, whose approach to the history of science was rather conservative. Both Sarton and Pelseneer were involved with the institutionalization of the history of science in Belgium and internationally.[7] Already in 1927, Rosenfeld published two papers in *Isis*.

[4]Kuhn and Heilbron (1963a), pp. 1–3, 9. A. Rosenfeld (Rosenfeld's daughter), in email correspondence with A. S. Jacobsen, 15 May 2004.
[5]Kuhn and Heilbron (1963a), p. 15. Rosenfeld (1979c), p. 911.
[6]Christie (1990), pp. 16–17.
[7]Baudet (2007), pp. 212–213.

Apart from the great loss of his father, Rosenfeld seems to have grown up in "a close and warm family life, not particularly rich, but not in poverty either".[8] According to Rosenfeld's daughter Andrée, "after his father's early death his maternal uncle took care of the well being of him and his mother and played an important role as a father figure in his later childhood".[9] Rosenfeld was quite close to his mother and assumed a great measure of responsibility for her. He brought his mother along with him (and later with him and his wife, Yvonne) to Copenhagen when he was living there at frequent intervals in the 1930s. His mother suffered from grave illness in the 1930s and during the Second World War, when Rosenfeld went to a great deal of trouble to visit her in Belgium from the Netherlands. After the war, she came to live with the Rosenfelds in Manchester.[10]

Rosenfeld began his mathematics and physics studies in 1922 at the University of Liège, which was in fact more of an engineering school, but a good one, and he graduated in 1926 with great distinction.[11] At this time he began his lasting friendship with another student, the later professor of spectroscopy and astrophysics at the same university, Polydore Swings.[12] Belgium had a strong tradition in engineering, but the country was rather peripheral with respect to modern physics. This fact lay behind the first sentence in Rosenfeld's story about his life as a physicist when interviewed by Thomas S. Kuhn and John Heilbron in 1963, "You see I was Belgian";[13] he did not have the best starting point for a career in theoretical physics. According to Rosenfeld, the teaching he received was at a very low level. The students were introduced to neither relativity theory nor quantum theory, and Rosenfeld studied these topics on his own during his last year at university.[14] For a comparison, it may

---

[8] A. Rosenfeld, in email correspondence with A. S. Jacobsen, 15 May 2004.
[9] *Ibid.*
[10] RP, copy of letter from L. Rosenfeld to S. Chandrasekhar, 19 Nov 1933 and 11 Nov 1936. Original in CP. BSC, L. Rosenfeld to N. Bohr, 4 Jun 1955. Bohr to Rosenfeld, 16 Jul 1955.
[11] RP, "Liège (1922–1940): Charleroi and Liège (1922–1926)". Halleux *et al.* (2001), Vol. 1, pp. 76–77.
[12] Swings (1964), pp. 301–302.
[13] Kuhn and Heilbron (1963a), p. 1.
[14] *Ibid.*, pp. 5–8.

be mentioned that not even at the Institute for Theoretical Physics in Copenhagen was quantum mechanics part of the curriculum until 1928. However, in Copenhagen graduate students could follow lectures and seminars and discuss the recent developments in quantum mechanics with the many visiting physicists.[15]

During their studies, the professor of topography at the University of Liège, Marcel Dehalu, played an important role as patron for Rosenfeld and Swings. When he received the Francqui Prize in 1949, Rosenfeld described how Dehalu never stopped encouraging, guiding, and giving him support and sympathy and, at the same time, never tried to restrict Rosenfeld's activities or influence the orientation of his work. Rosenfeld further expressed gratitude to all those who had made it possible for him to follow the studies of his own choice.[16] After his graduation Rosenfeld obtained scholarships from the Belgian government, the University Foundation, and the patrimony of the University of Liège to continue his studies in physics at the prestigious École Normale Superieure in Paris.[17]

Paris 1926–1927: Physics and Socialism

In Paris Rosenfeld took courses and was supervised by Paul Langevin, Léon Brillouin, and Louis de Broglie. Besides physics and mathematics he was introduced to the strong tradition of the history of science and philosophy at the Sorbonne. Here Rosenfeld attended Abel Rey's lectures "La Science Grecque".[18] However, Rosenfeld's main interest lay at the very frontier of theoretical physics, the combination of quantum theory and relativity, and for a while he worked with de Broglie. According to Rosenfeld, de Broglie, who came from

[15]Robertson (1979), p. 135. C. Møller heard Heisenberg's lectures and also a lecture by Schrödinger in 1926. Kuhn (1963a), pp. 3–4. Kragh (1992), p. 301.
[16]RP, "Manchester (1947–1958): Prix Francqui (1949)", Typewritten speech of thanks. Halleux *et al.* (2001), Vol. 2, p. 139.
[17]Swings (1974), p. 656.
[18]RP, Rosenfeld notebooks. At the Sorbonne he also attended lectures by E. Vessiot on "Invariants differentials", L. de Broglie on "Méchanique ondulatoire", and E. J. Cartan's on "Espaces de Riemann". In 1927, he attended Langevin's lecture "Atomes et Etoiles" at the Conservatoire des Arts et Métiers, his lectures "Structure de la Lumiére" at Collège de France, and E. Picard's on "Equations functionelles" at the Sorbonne. For Rey and history of science and philosophy, see Chimisso (2008), pp. 6, 72–73, 86, 93–100.

a noble family, was extremely shy and isolated even from other physicists in Paris, but the equally shy Rosenfeld still managed to establish relatively good contact with him. De Broglie, who was a pioneer in quantum physics, was the first with whom Rosenfeld discussed how to make sense of the new quantum theory. By drawing an analogy to the duality of light, which Einstein had suggested in 1905 could be described as both waves and particles, de Broglie proposed in 1923 that material particles like the electron possess a similar dual nature. He attributed to such particles a wavelength, since called the de Broglie wavelength. It was on this basis that the theory of wave mechanics that united the physics of light and matter was created a few years later by Erwin Schrödinger.[19] Under de Broglie's supervision Rosenfeld worked on combining relativity with wave mechanics, developing the wave equation in five dimensions, a topic which the Swedish physicist Oskar Klein was also working on independently at the time.

Besides being introduced to quantum theory, it was also in Paris that Rosenfeld's political and social awakening began. "In Paris I got the first glimpse of social problems and international politics... It was only in Paris by listening to the animated discussions of the people that I realized that there was a problem there that was worth thinking about".[20] At Langevin's lectures on statistical mechanics at the Collège de France, Rosenfeld met another physics student, Jacques Solomon, and the two soon developed a close friendship. They shared an interest in the synthesis of quantum theory and relativity, including quantum gravity, and published two joint papers on the quantum theory of radiation.[21] Solomon seems to have been the catalyst for Rosenfeld's interest in the combination of science and socialism and for Rosenfeld's later occupation with Marxist thought. Solomon was a militant socialist; he became a member of the French Communist Party in 1933. With the French philosopher and Marxist theoretician, George Politzer, he translated Friedrich Engels' oeuvre

[19]Darrigol (1986).
[20]Kuhn and Heilbron (1963a), p. 16.
[21]Rosenfeld and Solomon (1931a; 1931b).

into French.[22] Solomon introduced Rosenfeld to the group of leading scientists, which was also dominantly left-wing, in Paris, including Paul Langevin, Marie Curie, Iréne and Frédéric Joliot-Curie, and Jean Perrin, among others.[23] Rosenfeld, on the other hand, seems to have mediated a contact between Solomon and de Broglie and was also instrumental in arranging Solomon's visit at Bohr's institute in 1931–1932.[24] Solomon's Doctoral Dissertation (1931), which was partly based on the work he did in Copenhagen, was dedicated to his wife and parents-in-law, but also to Rosenfeld.[25] Tragically, however, Solomon was executed by the Gestapo in 1942 because of his activities in the French Resistance.[26]

Rosenfeld also got acquainted with the physicists Edmund Bauer and Alexandre Proca as well as the mathematician Jean Leray.[27] Apart from these young friends, the example of Langevin's personality, kindness, humanism, and political engagement made a deep impression on Rosenfeld as it did on many young scholarly and scientific aspirants at the time.[28] When Rosenfeld returned to Paris in early 1931 to teach a course in quantum electrodynamics at the newly established Institut Henri Poincaré, he stayed at Langevin's home. By this time Solomon had married Langevin's daughter Hélène and the couple lived in Langevin's apartment. All in all, Rosenfeld's stays in Paris during this time of social, political, and intellectual fermentation seem to have given him the impetus that soon made him engage in socialism and explore Marxism. Rosenfeld later saw a

---

[22]Bustamante (1997), p. 55.

[23]Rosenfeld (1979f). Bustamante (1997). Nye (1975). Weart (1979).

[24]RP, "Correspondance particulière", L. de Broglie to L. Rosenfeld, 6 Jan 1928. RP, "Copenhague: Copenhague (1931–1939)", J. Solomon to L. Rosenfeld, 16 Nov 1931, 19 Dec 1933, 8 Feb (no year given, but probably 1934). BSC, correspondence between N. Bohr and J. Solomon.

[25]Solomon (1931). See also footnote on p. 3 where Solomon expresses his gratitude to his "friend L. Rosenfeld".

[26]As a tribute to his deceased friend Rosenfeld dedicated his book *Nuclear Forces* (Amsterdam, 1948) to Solomon.

[27]RP, "Liège (1922–1940): Paris 1927", J. Leray to L. Rosenfeld, 12 Jan 1930, and 29 Feb 1931. RP, "Correspondance particulière: Proca" and "Bauer".

[28]Rosenfeld (1937a). RP, Rosenfeld, "Souvenir" de Langevin, manuscript, p. 1. Probably written in connection with Langevin's death in 1946. Biquard (1965), pp. 25–28, 58.

possibility for making himself useful as an instrument for propagating Bohr's views in a French context.

Göttingen 1927–1929: The "Indeterminist" School

With regard to physics, however, Langevin warned Rosenfeld that he would be isolated in Paris. Rosenfeld therefore looked for an opportunity to go to Göttingen, where much of the new formalism and mathematical technique of quantum mechanics was cast, to continue his research. Rosenfeld's work on the five-dimensional wave function had brought him into contact with Brussels physicist and mathematician Théophile de Donder, who was an expert on the theory of relativity and gravitation. In October 1927 Rosenfeld went to work with him. De Donder took Rosenfeld along with him to the later famous fifth Solvay Congress in Brussels to enable his friend to ask Max Born if he could come to Göttingen.[29] Among other philanthropic initiatives the illustrious Belgian industrial magnate Ernest Solvay founded the Conseil Solvay in 1911, as a meeting assembling some of the most brilliant scientists in the world, taking place in Brussels every three years.[30] Rosenfeld was not among the invitees to this exclusive, elitist meeting. However, his work on the five-dimensional wave function was presented by de Donder during the general discussion.[31] In the meantime Rosenfeld waited outside the lecture hall to approach Born when an opportunity arose. As it happened, Born immediately accepted the proposed arrangement, and during the winter semester 1927 Rosenfeld could attend lectures at the Georgia-Augusta University by David Hilbert on the foundations of mathematics, by Born on "atomic mechanics", by Pascual Jordan on quantum statistics, and by the historian of science Edmund Hoppe on the fundamental concepts of physics.[32]

---

[29] Kuhn and Heilbron (1963a), pp. 11–14.

[30] Halleux *et al.* (2001), Vol. 1, pp. 78–80, 197–198, Vol. 2, pp. 109–121.

[31] Bacciagaluppi and Valentini (2008), p. 500.

[32] Kuhn and Heilbron (1963a), p. 14. RP, "Liège (1922–1940): Göttingen (1928–1929)". Hoppe died while Rosenfeld was still in Göttingen and he wrote an obituary of him to *Isis*, Rosenfeld (1929). Rosenfeld kept meticulous notebooks over the lectures he attended. RP, Rosenfeld notebooks, box 3.

A. PICCARD E. HENRIOT P. EHRENFEST Ed. HERZEN Th. DE DONDER E. SCHRÖDINGER E. VERSCHAFFELT W. PAULI W. HEISENBERG R.H. FOWLER L. BRILLOUIN

P. DEBYE M. KNUDSEN W.L. BRAGG H.A. KRAMERS P.A.M. DIRAC A.H. COMPTON L. de BROGLIE M. BORN N. BOHR

I. LANGMUIR M. PLANCK Mme CURIE H.A. LORENTZ A. EINSTEIN P. LANGEVIN Ch.E. GUYE C.T.R. WILSON O.W. RICHARDSON

Absents : Sir W.H. BRAGG, H. DESLANDRES et E. VAN AUBEL

Participants at the fifth Solvay Meeting in 1927. Courtesy of The Niels Bohr Archive.

Rosenfeld landed in this stronghold of mathematics and theoretical physics just at the time when intense discussion over the meaning of quantum mechanics was at its highest, following the Solvay Conference and the preceding congress of physics in Como commemorating the centenary of Alessandro Volta's death. Bohr introduced his idea of complementarity in Como. In Brussels, Heisenberg and Born presented the new matrix mechanics, its paradoxes and probabilistic interpretation, while Schrödinger presented his wave mechanics.[33] Bohr presented his idea of complementarity during the general discussion. In between presentations at the Solvay meeting, Bohr and Einstein began their famous discussion about the interpretation of quantum mechanics which was to continue at the Solvay Congress in 1930, and culminated in 1935 with the publication of the EPR-paper (see Chapter 2).

Overwhelmed and bewildered by all the radical new ideas in quantum theory, Rosenfeld continued for a while to discuss the central ideas of quantum mechanics through correspondence with de Broglie. De Broglie was happy to learn that Rosenfeld was "satisfied with your stay in Göttingen where you can thoroughly bridle the "indeterminist" school".[34] As it was, during this period de Broglie struggled with coming to terms with Heisenberg and Bohr's indeterministic ideas and Born's probabilistic interpretation of the wave function. At the Solvay meeting, de Broglie had presented his realistic interpretation of Schrödinger's wave function known as the "pilot wave theory" or the "theory of the double solution".[35] In order to account for the wave–particle duality of microscopic phenomena, de Broglie suggested that Schrödinger's wave function played a twofold role as, on the one hand, determining the motion of a particle if its initial position was known, and if not, one could calculate from the wave function, in complete accordance with Born's probabilistic interpretation, the probability density of the particle's presence at

[33]Bacciagaluppi and Valentini (2008).
[34]RP, "Correspondance particulière", L. de Broglie to L. Rosenfeld, 14 Dec 1927. Originally in French.
[35]Bacciagaluppi and Valentini (2008).

a given position in space. The wave function was to be conceived not only as a probability wave, but also as what he termed a "pilot wave" that determined the trajectory of the particle in space. He furthermore suggested that the discrete structure of matter and radiation, that is the corpuscles, whether electrons or photons, manifested themselves formally as mobile singularities in the interior of the propagating waves. The continuous solutions to the wave equations, he maintained, provided only statistical information.

De Broglie was eager to learn what Rosenfeld made of the indeterminist ideas, which he could not reconcile himself with: "It would be paradoxical everywhere in these formulas to reach the conclusion that the corpuscles do not have a determinate position and that the elementary processes are not determinable at all!" he wrote to Rosenfeld.[36] It was only natural that Rosenfeld was influenced by de Broglie's ideas since it was through conversations with him that he first became acquainted with the topic. Retrospectively, in the light of how his viewpoints were soon to change, however, Rosenfeld saw his arrival in Göttingen in the following way: "I was still a bit heretical and I still insisted on the possibility of giving more reality to the waves and so on. Then Jordan took the trouble to explain the details to me, or rather, to point out the arguments; that contributed a great deal to make me abandon them [the realistic waves] completely".[37]

Jordan's lectures on quantum statistics and on Bohr's correspondence principle linking the old quantum theory with quantum mechanics completely captured Rosenfeld.[38] On the very last day of 1927 Jordan, who was then in Hannover, sent a long letter to Rosenfeld clarifying in detail some questions Rosenfeld had apparently asked him about the role of causality and Heisenberg's idealized localization experiment by means of a gamma ray microscope.[39] The analysis of Heisenberg's gamma ray microscope experiment was to become a key element in the new indeterministic interpretation

---

[36]RP, "Correspondance particulière", L. de Broglie to L. Rosenfeld, 14 Dec 1927. Originally in French.
[37]Kuhn and Heilbron (1963a), p. 18.
[38]*Ibid.*, p. 14.
[39]RP, "Liège (1922–40): Göttingen (1928–29)", P. Jordan to L. Rosenfeld, 31 Dec 1927.

of quantum mechanics. The thought experiment involved measuring the position of an electron by detecting a high energy photon scattered by it. It was designed by Heisenberg to give an "anschaulich" interpretation of the uncertainty relations of position and momentum for a material particle like the electron. Heisenberg reasoned that in order to detect the position of an electron in an atom for example, with as small an uncertainty as possible, we have to use high energy radiation with a small wavelength compared to the radius of the orbiting electron. The accuracy in the measurement of the position of the electron was governed by the wavelength of light. However, from Compton scattering it is known that the electron will be diffracted by high energy radiation; it will be excited. Therefore, the measurement with high energy radiation which would ensure precision in position measurement in fact "disturbs" the particle in such a way that its momentum becomes uncertain. In this way, Heisenberg elucidated the reciprocal relationship between the uncertainty in momentum and position.

Meanwhile, Bohr found that Heisenberg's analysis had not fully exhausted the thought experiment's potential for elucidating how the uncertainty relation between position and momentum of an electron appears as a result of the limitation built into the very description of both "the agency of measurement and of the object".[40] Bohr did not agree that the accuracy of the measurement of position was governed merely by the wavelength of light employed. According to Bohr, the conditions for the measurement have to be taken into account in order to get a deeper understanding of quantum mechanical effects such as the uncertainty relation — in this case how microscopes work. It is known from optics that such apparatus has a limited resolving power. Thus there is an uncertainty of the position determination already built into the apparatus. And because of the properties of optical lenses, it also cannot be known exactly which direction the scattered photon takes when it enters the focusing lens after being

[40]Bohr (1985c), pp. 582–584. Heisenberg (1927), pp. 197–198. Rosenfeld (1979g), pp. 291–292. For a thorough exposition of the thought experiment and Bohr and Heisenberg's different conceptions of it, see also Darrigol (1991), pp. 139–149.

diffracted by the electron. It can only be known that it entered the lens. Therefore, due to the limits of the measurement apparatus in question, an uncertainty in the x-component of the momentum of the photon arises and therefore also in the x-component of the position of the electron. Bohr wanted Heisenberg to revise his paper, which, however, was already in press, and Heisenberg refused to change it. In the end, however, Heisenberg added a postscript in the proof of the article which acknowledged Bohr's criticism of parts of it.[41]

In addition to clarifying Bohr and Heisenberg's discussion of the gamma ray microscope experiment, Jordan listed three scenarios regarding causality and determinism for Rosenfeld:

> (α) Quantum mechanics is correct; and there is no causality beyond quantum mechanics.
> (β) Quantum mechanics is correct; but there is nevertheless complete determinism, which, however, we are still ignorant of.
> (γ) Quantum mechanics is wrong; and there is complete determinism.
>
> Supposition α corresponds to my own interpretation. Supposition γ I consider unlikely; but of course in principle one should always regard γ as possible since in science one can never pass definite judgments.
> I assert however that β is logically impossible, the question "determinism or not?" will not be left open by quantum mechanics, but will be decided unequivocally in the negative; this decision can then only be wrong if quantum mechanics is wrong.[42]

If Rosenfeld confronted Jordan with de Broglie's ideas and questioned indeterminism, Jordan seems to have conceived it merely as if Rosenfeld did not understand the subtleties. Thus, Jordan ended his letter with the following remark, probably intended to comfort Rosenfeld: "Strictly speaking, these things are all so difficult, that

[41] Cassidy (1992), pp. 240–246. Camilleri (2009b), p. 77.

[42] RP, "Liège (1922–1940): Göttingen (1928–1929)", P. Jordan to L. Rosenfeld, 31 Dec 1927. Emphasis in original. Originally in German.

one can never completely grasp them; one has always to be satisfied if one has reached clarity approximately at least".[43]

Rosenfeld may have forwarded the message in Jordan's letter to de Broglie who wrote to Rosenfeld on 6 January 1928: "to correspond with you is certainly not only a very pleasant but also very instructive occupation".[44] De Broglie could not understand that when referring to Heisenberg's localization experiment, Jordan talked about measuring the properties of *individual* electrons by detection of *individually* scattered photons. Instead, de Broglie maintained that in order to make any sense of Schrödinger's wave function it had to be understood statistically. "As for me", he wrote to Rosenfeld, "I am profoundly convinced that under this statistical appearance a much more profound reality hides itself, or that one will recover the real determinacy of individual phenomena".[45] Thus, de Broglie favored the second or third of the three scenarios outlined by Jordan. In March the same year, de Broglie wrote to Rosenfeld who had informed him about Dirac's, Jordan's, and Klein's recent publications: "I have noticed, as you have pointed out to me by the way, that these works have really been quite far from my own conceptions".[46] By referring in particular to Dirac's quantum algebra, de Broglie continued:

> While carefully reading these memoirs, I have felt quite a vivid admiration for the logical expressions and ingenuity of the authors. Dirac, in particular, goes straight ahead in developing with perseverance this idea that every time one finds two canonically conjugated variables, you must replace them with q-numbers and since the energy and the phase are conjugated for a stable cavity of an electromagnetic wave, the energy and the phase are consequently q-numbers! It is very well, but I fear that these completely formal... theories do not have a

[43] *Ibid.* Emphasis in original.
[44] RP, "Correspondance particulière", L. de Broglie to L. Rosenfeld, 6 Jan 1928. Originally in French.
[45] *Ibid.*
[46] RP, "Correspondance particulière", L. de Broglie to L. Rosenfeld, 25 Mar 1928. Originally in French.

> true physical meaning. When I was in Brussels in October, Mr. Einstein told me: "The desolating thing is that one can make theories which are completely satisfactory from the formal point of view and which probably have no connection with reality!" That is precisely the impression I also get by studying the recent developments of quantum mechanics.[47]

Dirac's publications were devoid of the deeper epistemological considerations in connection with the symbols that entered the equations that were typically found in Bohr's writings.[48]

In a letter to Rosenfeld in April 1928, de Broglie continued to ponder over the interpretation of the wave function but maintained that "Schrödinger's waves are mainly only a statistical representation of the distribution of an ensemble of atoms".[49] However, at about this time de Broglie buried his dislike for what became the orthodox indeterministic interpretation of quantum mechanics and surrendered to the conglomerate of ideas later referred to as the Copenhagen Interpretation. When de Broglie revived his realistic interpretation of the wave function in the early 1950s, Rosenfeld became one of his fiercest critics (see Chapter 6).

Rosenfeld Finds his Style in Physics

In the spring of 1928, Rosenfeld applied for a fellowship from the International Education Board and wrote to Bohr and Einstein asking if he could work with them. Einstein responded positively: "I am glad that you will work with me on the topic you mention", which was "the relation between quantum mechanics and the theory of relativity".[50] At this time, however, Einstein's growing opposition

---

[47] *Ibid.*

[48] For an exposition of Dirac's quantum mechanics, see for example Moyer (1981a). Kragh (2005), p. 122. Darrigol (1992), p. 327. See also Bromberg (1977).

[49] RP, "Correspondance particulière", L. de Broglie to L. Rosenfeld, 9 Apr 1928. Originally in French.

[50] Albert Einstein Archives. Jewish National and University Library, Jerusalem, L. Rosenfeld to A. Einstein, 25 Apr 1928. A. Einstein to Rosenfeld, 3 May 1928. Originally in German.

towards quantum theory was becoming well-known. Since 1926 he had put his efforts into searching for a unified field theory which would at the same time serve as a deeper foundation for quantum mechanics. One of the sharpest critics of this research program was Pauli, who initiated a research program with a different approach to field theory, *viz.*, one that took its starting point in quantum theory.[51] Rosenfeld ended up contributing to Pauli's research program (see the next section) and working with Einstein was never again an issue for Rosenfeld. After he became Bohr's collaborator in the 1930s, Rosenfeld was a close witness to the culmination of Bohr and Einstein's discussion on the epistemology of quantum physics with the publication of the EPR paper in 1935 (see Chapter 2). In later years, Rosenfeld studied Einstein's epistemology more closely, in particular his view on Ernst Mach's philosophy.[52]

Bohr welcomed a stay by Rosenfeld in Copenhagen in 1928. However, in the end Rosenfeld postponed going to Copenhagen and instead accepted a post as assistant to Born for six months from May 1928, which was extended the following September until April 1929. Rosenfeld's work consisted of assisting Born with the textbook he co-authored with Jordan, *Elementare Quantenmechanik* (1930). This work "really gave me [Rosenfeld] an opportunity to get deep into questions and also into history; it was then that I learned about the preceding period".[53] As mentioned above, Rosenfeld's strong interest in the historical development of science made him attend lectures in history of science in both Paris and Göttingen. Tracing and analyzing the conceptual roots of physical theory was a means for him to acquire a better understanding of contemporary scientific concepts. Thus, while struggling with matrix mechanics, etc., he published a historical paper in *Isis* about the controversies about

[51]Howard (1990), pp. 66, 79–80, 98. Pais (1982), pp. 329, 346–347. See also Bustamante (1997), pp. 63–64 about the two programs in field theory.

[52]Rosenfeld (1979u), p. 518. Rosenfeld discussed these issues at length in the decade 1958–1968 in his correspondence with the East German historian of physics and Einstein specialist, Friedrich Herneck. RP, "History of science 7: Herneck".

[53]RP, "Correspondance particulière", M. Born to L. Rosenfeld, 13 Jan 1930. BSC 9 and 15, N. Bohr to M. Born, 15 Feb 1929. Bohr to Rosenfeld, 2 Mar and 16 Aug 1929. Kuhn and Heilbron (1963a), pp. 14, 22. Kuhn and Heilbron (1963b), pp. 5–6.

the wave–particle nature of light in the seventeenth and eighteenth centuries.[54]

When learning about Rosenfeld's plans for leaving Göttingen for Copenhagen, Born indirectly blamed Bohr for stealing his assistant.[55] After this misunderstanding was cleared up, however, Born warmly and wholeheartedly recommended Rosenfeld to Bohr as "widely read", as well as "a nice, diligent, conscientious, and clever man".[56] Born and Rosenfeld continued to hold each other in great esteem and their friendship was marked by respect and warm affection until Born died in 1970. In the 1950s, their correspondence involved extensive and heated discussions over political and ideological issues, as we shall see in Chapter 6.

While in Göttingen, Rosenfeld befriended other visiting physicists, among them the German Walter Heitler, the Russians George Gamow and Igor E. Tamm, Dirac, and the Dutch–Austrian–German Fritz G. Houtermans, the latter "a man with pronounced leftwing sympathies".[57] Rosenfeld stayed at the same pension as Dirac and Tamm during their visits: Geismar Ldstr. 1. He later remembered that they had "plenty of discussions during the meals, and we also took together a long excursion to the Harz Mountains, most of it on foot".[58] Dirac and Tamm were very close friends and were enthusiastic mountain climbers. Tamm was a socialist and had been politically active during the Russian Revolution as a radical leftist Menshevik. Dirac had close connections with the Russian physicists and frequently visited the Soviet Union.[59] Among the mathematicians, Rosenfeld was acquainted with Paul Ehrenfest's daughter, Tatyana Pavlovna Ehrenfest, who studied mathematics and physics in Göttingen in 1928 and

---

[54]Rosenfeld (1928).
[55]BSC 9, M. Born to N. Bohr, 11 Feb 1929.
[56]BSC 9, M. Born to N. Bohr, 16 Feb and 5 Mar 1929.
[57]Casimir (1983), pp. 220–223. Rosenfeld (1979k), pp. 335–337, Rosenfeld (1971), Khriplovich (1992), Kuhn and Heilbron (1963a), pp. 14, 24.
[58]RP, Supplement 1971–1974, Copy of letter from L. Rosenfeld to E. L. Feinberg, 14 Jan 1972, in connection with the death of I. E. Tamm. The address in Göttingen appears on Rosenfeld's correspondence with Einstein. Albert Einstein Archives. Jewish National and University Library, Jerusalem. Rosenfeld to Einstein, 25 Apr 1928.
[59]Kojevnikov (1996). Kojevnikov (2004), pp. 64–65. Hall (2008), on p. 256. Kragh (2005), pp. 68–69, 153–155.

attended some of Rosenfeld's lectures when he deputized for Born.[60] Some of the mathematicians in Göttingen, notably Emmy Noether and a group around her, held radical leftwing and pro-Soviet convictions. Noether spent half a year in the Soviet Union while Rosenfeld was in Göttingen.[61] Whether or not Rosenfeld interacted with Noether and her group is not clear, but he was certainly aware of Noether's mathematics at a time when it did not otherwise seem to be recognized as important among physicists (see below).

Dirac and Rosenfeld in Göttingen in 1928. From Rosenfeld album. Courtesy of The Niels Bohr Archive.

[60] RP, "Supplement: History of Quantum Theory", L. Rosenfeld, manuscript Ehrenfest as I saw him (An autobiographical chapter), May 1971. Kuhn and Heilbron (1963b), p. 2.
[61] McLarty (2005).

During the evening meals, Rosenfeld also discussed physics with E. E. Witmer and the two became involved with calculations related to some of Emil Rupp's experiments on electron diffraction on passage through thin metal foils, which Rupp carried out in Robert Pohl's laboratory at the time. Rosenfeld and Witmer did not question the data of this prominent and well-reputed experimentalist, who was even a close collaborator of Einstein, but a few years later it was discovered that Rupp had systematically forged his results probably over a period of ten years. And so it happened that Rosenfeld and Witmer became victims of Rupp's systematic scientific fraud.[62]

In general, close cooperation with experimentalists did not appeal much to Rosenfeld at the time. When much later he recollected his times in Göttingen, he stated "I was always inclined to emphasize a general aspect, general formulations, whereas the others [Gamow for example] were anxious to solve practical problems".[63] According to Rosenfeld, his colleagues would say "Well, you have a beautiful set of Hamiltonian equations and a beautiful formalism of perturbation theory, but try to give the answer to calculated transition probabilities with your formalism".[64] Typically for Rosenfeld's apparent lack of self-importance, he would later characterize his working style as "pedantically accurate on the formal side, and rather short-sighted about the physical aspects".[65] Rosenfeld was attracted to the mathematical foundations of physics, which, of course, is always on the edge of what is immediately useful. In this way he shared the interest and style of his close friend Solomon and his teacher Jordan.[66] During these years (1927–1929) Rosenfeld had a thorough introduction to the Göttingen culture of the marriage between mathematical innovation and the new

[62] RP, "Liège: Göttingen 1928–1929", E. Rupp to L. Rosenfeld 16, 20, 27 Nov, and 10, 17 Dec 1928. RP, "Supplement 1971–1974", copy of letter from L. Rosenfeld to B. H. Muller, 1 Sep 1972, with enclosed notes from an interview with Rosenfeld at the Varenna Summer School 12 August 1971, about Rupp and Leo Szilard. Kuhn and Heilbron (1963a), p. 22. Rosenfeld and Witmer (1928a; 1928b; 1928c; 1928d). Van Dongen (2007). French (1999).
[63] Kuhn and Heilbron (1963a), p. 20.
[64] *Ibid.*
[65] RP, "Supplement: History of Quantum Theory", L. Rosenfeld, manuscript "Ehrenfest as I saw him" (an autobiographical chapter), May 1971.
[66] Bustamante (1997), p. 86.

physics. He worked hard to catch up with the recent theoretical developments which he had missed in Liège and Paris. For example, he studied what he later called "Neumanistics", that is the Hungarian mathematician John von Neumann's axiomatic foundation of quantum mechanics.[67] From another Hungarian, the physicist and mathematician Eugene Wigner, Rosenfeld learned group theory. In the process Rosenfeld matured as a highly sophisticated mathematical physicist.[68]

Rosenfeld 1930. Courtesy of The Niels Bohr Archive.

[67]Von Neumann (1927a). Von Neumann (1927b). RP, "Supplement: History of Quantum Theory", L. Rosenfeld, manuscript Ehrenfest as I saw him (An autobiographical chapter), May 1971. Kuhn and Heilbron (1963b), pp. 1, 3. As regards to von Neumann's work in context with Hilbert's mathematical program, see Lacki (2000). See also Rédei and Stöltzner (2001).

[68]For a comparison with other styles in theoretical physics, see for example Grandin (2008), p. 194 about the Swedish physicist I. Waller. A. Warwick discusses local traditions and practices in theoretical physics in Warwick (1992). Seth (2010).

Participants at the first Copenhagen Conference 1929.
Courtesy of The Niels Bohr Archive.

Apart from his exchange of letters with de Broglie and Jordan mentioned in the previous section, there is no indication that Rosenfeld occupied himself seriously with the epistemology of quantum physics in the late 1920s. In early 1929, Rosenfeld again attempted to arrange going to Copenhagen. Bohr was then too busy to take him, but he invited Rosenfeld to participate in what was later known as the first Copenhagen Conference at Easter 1929. This conference was an informal gathering of about thirty physicists from Denmark and abroad, at which the participants briefly presented their field of interest and discussed recent developments in physics. The meeting proved such a success that a tradition was established and a series of conferences were held during the 1930s. Bohr arranged these conferences every year, usually around Easter. They constituted a venue much more informal than the Solvay meetings; no program had been planned in advance, for example. At these meetings, new theories were presented and discussed by physicists who were or had previously worked in Copenhagen or who were Bohr's friends.[69] As such,

[69]Robertson (1979), pp. 136–138.

these meetings took place in an atmosphere in favor of the latest developments in quantum theory. Rosenfeld went along to the first of these conferences with his friend Walter Heitler and met Bohr in person for the first time.[70] Here Rosenfeld was introduced to Bohr's quantum philosophy by Bohr himself, but that seems not to have helped Rosenfeld to get a firmer grasp of it. He later recalled that he felt completely mystified during the conversation.[71] Apart from attending Bohr's conference, Rosenfeld received his first ever invitation to give a colloquium in September 1929, namely at Paul Ehrenfest's colloquium in Leyden. The meeting with the friendly but direct, satirical, and bombastic Ehrenfest made a great impression on the young, shy physicist.[72]

The drive to seek the very roots of ideas came to mark all of Rosenfeld's intellectual activities. After his interest in politics and socialism had been raised in Paris, and as soon as his settlement in Göttingen allowed him a brief moment, he began studying Karl Marx's *Das Kapital*, which would also enable him to practice the German language: "When I opened the book and read the first chapter, it struck me that I had [here] a man of genius, just as when you open Darwin and read it, you feel there is something there which is a really superior grasp of the questions".[73]

Two Approaches to Quantum Electrodynamics: Jordan versus Dirac

In order to understand the significance of Rosenfeld's early contributions to physics, it is necessary for a brief introduction to the early development of quantum electrodynamics. Quantum electrodynamics is the branch of quantum field theory which accounts for the

[70]BSC 15, N. Bohr to L. Rosenfeld, 2 Mar 1929. BSC, Supplement 1910–1962, Rosenfeld to Bohr, 12 Mar 1929. Rosenfeld (1979i; 1971), p. 304. Kuhn and Heilbron (1963b), pp. 5–6. Robertson (1979), pp. 136–138. Hoffmann (1988).

[71]Rosenfeld (1979i; 1971).

[72]RP, "Supplement: History of Quantum Theory", L. Rosenfeld, manuscript Ehrenfest as I saw him (An autobiographical chapter), May 1971.

[73]Kuhn and Heilbron (1963a), p. 17. He expressed a similar view in a letter to Pauli, 6 Apr 1952, von Meyenn (1996), Vol. 4, Part 1, pp. 598–600.

interaction between charged particles like the electron and electromagnetic radiation; it accounts for such phenomena as, in Dirac's formulation, "a system in which the forces are propagated with the velocity of light instead of instantaneously, of the production of an electromagnetic field by a moving electron, and of the reaction of this field on the electron".[74]

After their victorious route to non-relativistic quantum mechanics in 1925–1927, so rich in discoveries and with such in the end simple, though paradoxical, outcomes, the quantum physicists were excited and optimistic about the next challenge, *viz.*, to combine relativity with quantum mechanics and create a quantum field theory and quantum electrodynamics along similar lines. However, new and seemingly insurmountable paradoxes appeared from this work which led to a new crisis in quantum theory in the period 1927–1933.[75]

The foundations of quantum field theory and quantum electrodynamics were laid in Jordan's pioneering approach for treating radiation according to quantum theory in his publications with Pauli and Heisenberg in the period 1925–1927, and Dirac's quantum theory of radiation from 1927 and relativistic treatment of the electron from 1928.[76] In constructing their theories, however, Dirac and Jordan drew quite different insights from the wave–particle duality of light and matter. Jordan broke radically with classical physics in his approach. He took his starting point in the wave nature of radiation and matter, treated both as *fields* and intended to deduce the corpuscular properties of both through the quantization conditions imposed on wave-propagating quantum fields. He suggested in 1927:

> [O]ne may construct a quantum wave theory in which electrons are represented by means of quantum waves in three-dimensional space... the basic fact of electronic theory, the existence of discrete particles of electricity, is explained as a

[74]Dirac (1927), p. 243.
[75]Kragh (1999), p. 196. Pais (1991), p. 364.
[76]Kragh (2005), pp. 127, 132–132. Schweber (1994). Darrigol (1986). See also Miller (1994). Pais (1986).

> characteristic quantum effect, or, in other words, it means that matter waves appear only in discrete quantum states.[77]

Particles were conceived as the quanta of fields. Jordan's method was later called "second quantization".[78] Jordan considered Schrödinger's wave function as a *field* amplitude in ordinary three-dimensional space analogous to the field quantities of the electric and magnetic fields or the vector potential. The field was then quantized, that is it was considered along with its conjugates as operators that were subjected to commutation relations the way observables would normally be presented in quantum theory. The *particle* nature of electrons then arises from this process of quantization.[79]

Dirac, on the other hand, sought a quantum electrodynamic theory in close analogy with classical Hamiltonian theory and in which the corpuscular nature of matter was preserved as a primary property in no need of explanation. Only radiation was treated as a field which was conceived as a macroscopic coherent state. Dirac's relativistic wave equation for the free electron seemed at first to be the relativistic analogue to Schrödinger's non-relativistic equation.[80] However, like all relativistic theories, Dirac's equation had solutions representing particles with both positive and negative energy, and even if the latter seemed absurd it could not simply be ignored. Dirac put forward the following hypothesis as possible explanation of "the negative energy paradox":

> There are so many electrons in the world that all the states of negative energy are filled up, and that there are some electrons left over, which are obliged by the Pauli principle always to have positive energy. This will mean an infinite density of electrons with negative energy, but since their distribution is quite uniform they will not be observable. If there is a state of negative energy that is not occupied, the resulting "hole"

[77]Jordan (1927), p. 473. English quotation in Kragh (2005), p. 130.
[78]Jordan and Klein (1927). Jordan and Wigner (1928).
[79]Kragh (2005), pp. 119–122. See also Heitler (1954), p. 114. Schweber (1994), p. 33.
[80]Dirac (1928), p. 613. Kragh (2005), p. 130. Schweber (1994), p. xxii. Bromberg (1977).

> will have a positive energy, and will also move in an external field as though it has a positive charge. We can therefore interpret this hole as a proton.[81]

The idea of an "infinite sea" of negative energy electrons was met with much skepticism.[82] The vacant negative energy states Dirac proposed to be "holes" which obeyed Pauli's exclusion principle and were thought to be occupied by either protons or positively charged electrons. The proton hypothesis was considered unlikely, however, because the mass of electrons and protons differ so much.[83] Positively charged electrons had never been observed so that suggestion was considered unlikely too. Dirac, nevertheless, boldly proposed the existence of the positive electron in 1931. In 1932, Carl D. Anderson discovered this very anti-particle which was given the name positron. The discovery was immediately confirmed by P. M. S. Blackett and G. P. S. Occhialini. It became clear later that out of the combination of quantum theory and relativity theory there emerges automatically a description of the creation and annihilation of particles and anti-particles.[84]

Zurich 1929–1930: The Pauli–Heisenberg Quantum Field Theory

While awaiting the opportunity to go to Copenhagen, Rosenfeld joined Pauli's group in Zurich for the academic year 1929–1930, again supported by Belgian scholarships, probably this time by the Fond National de la Recherche Scientifique (FNRS), which had just been established in 1928.[85] When Rosenfeld arrived in Zurich, Heisenberg and Pauli had just published their first joint paper, "Zur Quantendynamik der Wellenfelder" (On the quantum dynamics of wave fields),

[81]Grandin (2008), p. 206, quoting Dirac in a letter to I. Waller. Dirac gives pretty much the same explanation in a letter to Bohr, 26 Nov 1929, see Moyer (1981b), p. 1057.
[82]Kragh (2005), pp. 87–111.
[83]See for example Rosenfeld (1932a), on p. 59.
[84]Schweber (1994), pp. xii, 5, 77. Moyer (1981c).
[85]Kuhn and Heilbron (1963b), p. 5. Baudet (2007), p. 202. Halleux *et al.* (2001), Vol. 2, pp. 35, 57–61.

which introduced a comprehensive quantum field theory based on Jordan's idea of the quantization of waves.[86] With this pioneering relativistic treatment of quantum fields, Heisenberg and Pauli hoped to create a general field theory which would incorporate the important results found by Dirac in his one-particle relativistic theory of the electron. In 1930, Heisenberg and Pauli's second paper "Zur Quantenmechanik der Wellenfelder II" (On the quantum mechanics of wave fields II), with some improvements of their initial approach, appeared in print.[87] Pauli "was eager to have people brush up the details and explore the consequences" of his and Heisenberg's theory.[88] Rosenfeld was enrolled in this ambitious project, which also included the American physicist Robert Oppenheimer and the Swedish physicist Ivar Waller.[89]

Heisenberg and Pauli took their starting point for treating quantum fields in an analogy with the Lagrangian and Hamiltonian methods of classical analytical point mechanics. They constructed a field theory from the Lagrangian density depending on the field amplitudes at each point of space and their derivatives with respect to the space and time coordinates. From the Lagrangian, they found the equations of motion for the field variables by the variation principle for the action of the field system. The quantization was effected by requiring that the conjugate field variables obey commutation relations.[90] Heisenberg and Pauli had proved that their theory was Lorentz-invariant. The Lagrangian and the derived commutation relations could be proved to be Lorentz-invariant if the action

---

[86]Heisenberg and Pauli (1929). Later the same year, they submitted their second paper, which was published in Heisenberg and Pauli (1930).

[87]Heisenberg and Pauli (1930). In between Heisenberg and Pauli's two publications, the Italian physicist Enrico Fermi published an alternative approach to field quantization: Fermi (1929). Through his review article Fermi (1932), Fermi's theory later became quite influential among the new generation of physicists who considered Fermi's theory more easily comprehensible than Heisenberg and Pauli's. See Schweber (2003), pp. 194–195 and Schweber (2002). See also Darrigol (1986), pp. 239–240.

[88]Kuhn and Heilbron (1963b), p. 5.

[89]About Waller, see Grandin (2008). About Oppenheimer in Zurich, see for example Cassidy (2005), pp. 126–129.

[90]Heisenberg and Pauli (1929), pp. 11–12. Schweber (1994), pp. 39–41.

function and hence the Lagrangian density was relativistically invariant. However, the proof was so complicated that, according to Gregor Wentzel, Pauli is quoted as having said "I forewarn the curious".[91] Rosenfeld's first contribution to this research program was to correct an error in that proof.[92]

In accordance with Jordan's second quantization scheme, both radiation and the electron were treated as *fields*. Matter was described by Dirac's relativistic wave equation for the electron from 1928. Electromagnetic radiation was described by the quantized Maxwell field. When Heisenberg and Pauli applied their general theory to electrodynamics, they encountered several problems. The quantization procedure failed because the conjugate momenta to the fourth component of the four-potential turned out to be zero. The commutation relations could therefore not be applied.[93] Heisenberg found a technique, which today is called gauge-fixing, to avoid this problem. This trick, of course, should not violate the other properties of the theory such as relativistic invariance and gauge invariance, but Heisenberg's trick unfortunately disturbed the gauge invariance of the Lagrange function. In the sequel to their first paper, Heisenberg and Pauli found a way to treat the problem without this trick by making use of gauge invariance. However, this method destroyed the Lorentz-invariance of the theory. A third method to get around the problem proposed by the Italian physicist Enrico Fermi resulted in the sacrifice of gauge invariance.[94]

In the hope to get around these apparently more or less *ad hoc* adjustments to the theory, Rosenfeld made a successful attempt at a more general mathematical framework of the theory that took symmetry properties of the fields into account. This work resulted in a paper published in 1930, "Zur Quantelung der Wellenfelder" (On the Quantization of Wave Fields) in *Annalen der Physik*. Rosenfeld

[91] Wentzel (1973), p. 382.
[92] Rosenfeld (1930a; 1968), p. 231. Kuhn and Heilbron (1963b), p. 9.
[93] Heisenberg and Pauli (1929), pp. 24–27.
[94] Heisenberg and Pauli (1930). Fermi (1929).

acknowledged his debt to Pauli in the introduction: "As I was investigating these relations in the especially instructive example of gravitation theory, Professor Pauli helpfully indicated to me the principles of a simpler and more natural manner of applying the Hamiltonian procedure in the presence of identities".[95] In practice, Rosenfeld unified the Heisenberg–Pauli quantization of the electromagnetic field with recent attempts by Hermann Weyl and Vladimir A. Fock to treat the coupling between Dirac's electron field and gravitation fields. Rosenfeld's results in this paper anticipate what is nowadays referred to as "constrained" Hamiltonian dynamics, reinvented by Dirac and Peter Bergman independently in the early 1950s. Constrained Hamilton dynamics is a more general technique of presenting quantum fields than the usual Lagrangian or Hamiltonian formalisms, in which the fundamental relations between symmetries and invariances of the Lagrangian and conserved quantities are taken into account. The procedure constitutes the basis of all canonical treatments of local gauge theories today.[96]

Rosenfeld based his procedure on Emmy Noether's theorem from 1918 which gives the general relationship between symmetries and conservation laws in classical field theory. Noether's theorem states that every conservation law associated with a system arising from a variational principle is connected with a symmetry property. Thus to every symmetry there corresponds a conserved quantity and vice versa. It means that if a dynamical model possesses symmetry under a transformation involving arbitrary functions, a specific linear combination of equations of motions will vanish accordingly. In classical dynamics for example, if the variational principle leading to the so-called Euler–Lagrange equations is invariant under translations of the time variable, there is conservation of momentum.[97] Rosenfeld extended such considerations to the Heisenberg–Pauli relativistic quantum field theory. He applied his method to all known fields at the

[95]Rosenfeld (1930b), pp. 113–114. For an English translation and a commentary of this paper, see Salisbury (2009).
[96]Salisbury (2006). Salisbury (2007). Kragh (2005), pp. 197–199.
[97]O'Raifeartaigh (1997), p. 20. Salisbury (2006). Gray (1994), Vol. 1, pp. 473–474.

time, the electromagnetic field, the Dirac matter field, and the gravitational field and included their mutual interactions. In sum, Rosenfeld showed that symmetries lead to non-unique evolution in time, in which there exist constrained relations among the variables in question. In this way the specific choices made by Pauli, Heisenberg, and Fermi to get around the problem with the vanishing momentum in quantum electrodynamics were shown to be special cases of more general considerations. Rosenfeld presented his results in a more condensed form in a comprehensive review article in *Annales de l'Institut Henri Poincaré* (1932).[98]

Rosenfeld did not receive much recognition for this work. In fact it seems that hardly anyone but Pauli and Klein were aware of it. Pauli considered Rosenfeld's accomplishment as mainly securing the ground for the previous more specific assumptions in quantum electrodynamics done by himself, Heisenberg, and Fermi. Their techniques could now be used with an easy mind after Rosenfeld's demonstration of their general validity. Rosenfeld's work was referred to in Pauli's influential 1933 *Handbuch der Physik* article in connection with the following comment:

> There are already general systematic investigations on the quantisation of arbitrary classical field equations and the possibility of deriving the C.R. [commutation relations] from a canonical scheme. We need not, therefore, go into them, nor even into their connection with the quantisation of matter waves..., but refer the reader to the literature.[99]

In fact, Pauli's own papers with Heisenberg were relegated to the same footnote because at this time Pauli doubted that his and Heisenberg's theory was fruitful (see also the next chapter). As for "second quantization", Pauli merely mentioned this "peculiar" technique in passing and suggested that one had to be cautious with the concept of matter waves which differed essentially from light

[98] Rosenfeld (1932a).
[99] Pauli (1980), p. 196.

waves "because the functions ψ and ψ^* are symbolic quantities, not themselves directly observable, and they contain the quantum of action".[100] Pauli seems to have considered Rosenfeld's work too long and complicated and containing some questionable assumptions. In 1955, Pauli drew Klein's attention to Rosenfeld's paper:

> I would like to bring to your attention the long work by Rosenfeld, *Annalen der Physik* (1930), Vol. 5, Part 4, p. 113. He worked it out during his time with me in Zurich and was known here accordingly as the "man who quantized the Vierbein" (sounds like the title of a Grimm's fairy tale doesn't it?) ... I still remember that not all aspects of Rosenfeld's work were satisfactory since he had to introduce certain additional assumptions which no-one could really understand.[101]

To which Klein responded:

> I am particularly grateful that you reminded me of Rosenfeld's publication from 1930. I knew it fairly well and also thought about it in this connection at some point. However, I fear that I would have forgotten it without your reminder.[102]

Besides inventing a more general formal treatment of fields, Rosenfeld also made investigations related to another difficulty with the Heisenberg–Pauli theory, *viz.*, the prediction of an infinite value for the self-energy of the electron. The self-energy is the total energy of the particle in free space when isolated from other particles or light quanta. In other words, it is the energy of the interaction of a particle with the field produced by itself. There was a problem with the infinite self-energy of the electron already in the classical theory where the divergence appeared in the description of the electron as

---

[100]Pauli (1980), pp. 127–129. Darrigol (1986), pp. 247–248. Kragh (2005), pp. 137–138.
[101]W. Pauli to O. Klein, 25 Jan 1955, von Meyenn (2001), Vol. 4, Part 3, pp. 63–64. Originally in German.
[102]O. Klein to W. Pauli, 11 Feb 1955, von Meyenn (2001), Vol. 4, Part 3, p. 98. Originally in German.

a mass point, that is, where its assigned radius becomes zero. The classical Coulomb self-energy, the energy of the electrostatic field of the electron $\frac{e^2}{r}$ diverges for an infinitely small radius r of the electron. Since the point model of the electron was maintained in quantum theory, it was to be expected that the singularities would still exist. Indeed, the process of quantization did not rid the theory of this paradox, but it changed the nature of the singularities.[103] In 1929, Pauli complained to Bohr: "In particular the self-energy of the electron makes much bigger difficulties than Heisenberg had thought at the beginning. Also the *new* results to which our theory leads to are very suspect and the risk is very great that the entire affair loses touch with physics and degenerates into pure mathematics".[104] Oppenheimer found an additional self-energy for the bound electron, and he and Waller pointed out that the divergent self-energy was not directly connected with the negative energy states.[105] Rosenfeld calculated the self-energy of the light quantum arising from its gravitation field to see if it was infinite like the self-energy of the electron. Unfortunately Rosenfeld's calculations showed that analogously to the self-energy of the electron, the gravitation energy of the light quantum became infinitely large. Hence, this was seen as a further difficulty for the Heisenberg–Pauli quantum field theory.[106] The problems with infinities in quantum electrodynamics were only solved with the introduction of renormalization procedures introduced after the Second World War.[107]

As the focus shifted from field theory to quantum electrodynamics, the problems with the theory seemed to block further development. Besides, the two papers by Heisenberg and Pauli were considered terribly tedious and mathematically indigestible by the

[103]Weisskopf found that the divergence reduced to a logarithmic expression, Weisskopf (1939). Pais (1972), p. 80. Kuhn and Heilbron (1963b), pp. 8–9.
[104]W. Pauli to N. Bohr, 17 Jul 1929, Hermann *et al.* (1979), Vol. 1, pp. 512–514. English translation quoted from Schweber (1994), p. 83.
[105]Miller (1994), pp. 34–35.
[106]Rosenfeld (1930c). Kuhn and Heilbron (1963b), pp. 8–9.
[107]See Schweber (1994).

majority of physicists and only few followed their lead into the conceptually abstract Jordan–Wigner "second quantization".[108] Nowadays, quantized matter waves constitute a central part of quantum field theory, but at the time, Dirac, among others, criticized Heisenberg and Pauli's theory for being unphysical because it included quantities that did not represent observables. In Dirac's view, only the corspuscular nature of *radiation* should be derived as a second quantization, such as he had done in 1927.[109] As Heisenberg and Pauli's theory also seemed to have exhausted the possibilities of Jordan's second quantization, several of the physicists involved with quantum electrodynamics and quantum field theory thought the problems of the theory could not be solved without the introduction of some revolutionary new ideas. As a result, Jordan stopped working on quantum electrodynamics. Dirac did not contribute anything to quantum electrodynamics in the four-year period 1928–1932, and even Heisenberg and Pauli decided to leave their theory behind and start all over again by taking correspondence arguments into account.[110] The crisis in quantum theory 1927–1933 and the controversies it gave rise to in the small theoretical physics community will be dealt with in more detail in the next chapter.

It was Dirac's and Fermi's electrodynamics that became most influential among physicists working on quantum electrodynamics later in the 1930s, and the two's approaches also served as the starting point for the further development of quantum electrodynamics after the war. Jordan's second quantization and the entire research program commenced by Pauli and Heisenberg, including Rosenfeld's contributions, seem to have been largely ignored, only to be reinvented later in the century.[111] Walter Heitler's textbook, *The Quantum Theory of Radiation*, first published in 1936, was immensely influential for the education of the next generation of physicists in quantum electrodynamics.

[108] Darrigol (1986), pp. 244, 246. Kragh (2005), p. 131.

[109] Dirac (1927). Kragh (2005), p. 131. Darrigol (1986), p. 239.

[110] Cini (1982), pp. 237–239. Darrigol (1986), p. 200. Enrico Fermi also lost interest in quantum electrodynamics, see Schweber (2003).

[111] Schweber (1994), pp. 12, 56, 128–129. Cini (1982), p. 250. Fermi (1932). Schweber (2003), pp. 194–195. Schweber (2002). See also Darrigol (1986), pp. 239–241.

Its various editions may serve as an indicator of how quantum electrodynamics developed over the years. In the first editions, Heitler presented the quantum electrodynamics from Dirac's and Fermi's point of view mentioning Pauli and Heisenberg's approach only in a footnote.[112] However, a later edition seems to reflect how the tide eventually turned in favor of Jordan's second quantization. Thus, following a section on Dirac's "hole" theory in the third edition of the book in 1954, Heitler introduced a section, "Second quantization of the electron field", as an alternative method for treating a system of positrons and electrons "which introduces the properties of the positive electrons from the start... This method has the advantage that no 'infinite sea' of negative energy electrons occurs and only physically significant features are introduced. Moreover, it allows in a natural way for the possibility of creation and annihilation of particles".[113] Leonard I. Schiff's textbook *Quantum Mechanics*, first published in 1949, has a chapter on the quantization of wave fields taking as its starting point Pauli and Heisenberg's papers twenty years earlier. For the quantum–electrodynamical case, Rosenfeld's paper from 1932 in *Annales de l'Institut Henri Poincaré* is cited alongside Fermi's 1932 paper as well as Heitler's and Dirac's textbooks.[114] In his *Selected Papers on Quantum Electrodynamics* (1958), Julian Schwinger took as his starting point Dirac's and Fermi's pioneering papers, while Pauli and Heisenberg's papers were omitted because of their considerable length.[115]

In his historical publications from the early 1930s, Rosenfeld was occupied with discoverers versus their precursors in history of science.[116] He wanted to understand and explain why some discoveries are not assimilated into the scientific community straight away. He suggested that it is not possible to understand and appreciate the position forerunners occupy in the evolution of scientific ideas unless one takes into account their social interaction with the scientific community. Scientific work is never the act of an isolated individual but

[112]Heitler (1936), p. 58. Schweber (1994), pp. 82, 129.
[113]Heitler (1954), p. 114.
[114]Schiff (1949), pp. 330, 361.
[115]Schwinger (1958).
[116]Jacobsen (2008).

is always a more or less fruitful interaction between the personal contributions of individuals and the social milieu in which they move as well as the accepted ideological and experimental or mathematical techniques within the scientific community. Rosenfeld believed that the established scientific community is never prepared to accept a new idea without resistance and that introduction of a new idea in science will happen only as a result of a veritable fight between the pioneer who made the discovery and the conservative scientific tradition of the scientific community which he confronts. In such a fight, as in all social phenomena, two factors would determine the outcome, in Rosenfeld's view. On the one hand, there is opposition from the established community to a new idea, on the other, there may be limitations in human character or social abilities which may prevent the proponent of this idea from convincing the scientific community in question about its truth. If the pioneer is not able to convey his ideas, for example if he does not make himself comprehensible, he may not succeed, and if the pioneer loses the fight he remains a forerunner. It is then necessary to wait, sometimes a very long time, for the community and the science in question, to arrive at a state more favorable to the acquisition of the idea, or that a researcher better fitted puts forward the problem and its definitive solution anew. Favorable circumstances may hasten the birth of a new idea, but if time is not ripe for the assimilation of a discovery it will have to be rediscovered at a later stage.[117]

If we apply this standpoint to Rosenfeld's mathematical innovation for treating quantum fields, Rosenfeld's contribution might be seen in the broader context of the reception of Heisenberg and Pauli's quantum field theory and Jordan's second quantization. In light of the later development of quantum field theory in which the mathematical technique that Rosenfeld developed was successfully reinvented, it may be argued that his invention was done at a time

[117]Rosenfeld (1938), p. 74. See also Rosenfeld's much later paper, Rosenfeld (1979r), p. 903. Prompted by G. S. Stent's concept of prematurity in scientific discovery from the early 1970s, a remarkably similar historiographical approach has recently been taken up anew. Hook (2002).

when the physics community was not yet ripe for it. This explanation may be supported by the observation that gauge symmetry was not yet considered the fundamental principle it is today, and Noether's theorem had also not yet acquired the fundamental status in quantum field theory that it gained later.[118] We should also take into account Rosenfeld's shy nature, lack of self-assertion, and susceptibility to opinions of authorities, particularly his loud mentor Pauli's critical mind and quick judgement of others' works. Still, it has puzzled physicist–historians why Rosenfeld did not include the 1930 *Annalen der Physik* paper among his *Selected Papers* which he himself selected just before he died.[119]

Friendships

Among the young physicists in Zurich at the time were besides Gamow another Russian, Lev Davidovich Landau, and the German Rudolf Peierls. Peierls had been a student of Arnold Sommerfeld in Munich. He had worked with Heisenberg in Leipzig and was now Pauli's assistant.[120] Peierls became Rosenfeld's close friend, "we were always together" Rosenfeld said later.[121] Peierls later described his impression of Rosenfeld as "a stubby young man with a serious round face. He was given to philosophical contemplation. He had a powerful command of mathematics without losing sight of the physical facts. Pauli liked to discuss the finer points of field theory and relativity with him".[122] This characterizes very well Rosenfeld and his contributions to physics in this period. Rosenfeld also possessed a sense of dry humour; Peierls recalled "one occasion when Rosenfeld and I had coffee together, and I offered him cream and sugar. 'No, thank you,'

[118] O'Raifeartaigh (1997), pp. 4–7, 22.
[119] When Pol Swings summarized Rosenfeld's most important works in order to promote him as a candidate for the Francqui Prize in 1949, he did not mention Rosenfeld's 1930 study, and the version from 1932 was counted as one of Rosenfeld's didactic contributions. RP, "Manchester (1947–1958) 2, Prix Francqui (1949)", enclosed document "Travaux de L. Rosenfeld" in P. Swings to L. Rosenfeld, 17 Nov 1948.
[120] Rosenfeld (1979k), pp. 337–340.
[121] Kuhn and Heilbron (1963b).
[122] Peierls (1985a), p. 60.

he said, 'I drink my coffee black and without sugar.' He then reflected a little and said, 'I got this habit during the war' (the First World War, of course), 'when we had no cream and no sugar in Belgium.' He reflected a little more. 'It is true we had no coffee, either'".[123]

Already at the age of 21, Landau had obtained a Rockefeller fellowship to visit the European centres of modern physics during the period 1929–1931. He has been described as having a very quick and brilliant mind and possessing a reckless, teasing, and biting character.[124] Landau and Peierls became very close friends and published a couple of papers together. The second of their papers, on measurability of field quantities in relativistic quantum mechanics, gave rise to a heated dispute with Bohr, a dispute in which Rosenfeld also became intimately involved as we shall see in the next chapter.

Gamow was famous among physicists for his playful and witty intellect and for his many crazy ideas. While in Zurich in 1930, Gamow, Rosenfeld, and Peierls climbed the peak called Piz da Daint. Gamow had just finished a manuscript for *Nature* which outlined the explanation of the fine structure of the alpha rays emitted in spontaneous radioactivity which had been discovered by the experimental physicist in Paris, Salomon Rosenblum. Gamow decided that he wanted to sign his paper at the top of this snowy mountain and he wanted to have his picture taken while doing so.[125] The weather turned nasty and the three arrived at the top in a snow storm. Allegedly, Rosenfeld suggested that instead of ruining his original manuscript Gamow should just take some odd piece of paper, nobody would notice that it was not the original on the picture, which Peierls was ordered to take. However, Gamow insisted on taking the original.[126] It was a trip of several hardships as reported by Peierls to his future wife, Genia: "Gamow had difficulties with his foot (he had some accident in England before) and Rosenfeld's shoes went entirely to pieces and we got into a fog and so on. But Piz Da

---

[123] *Ibid.*, p. 60.
[124] Kojevnikov (2004), pp. 85–88.
[125] See also G. Kanegisser to R. Peierls, 10 Oct 1930, and Peierls to Kanegisser, 16 Oct 1930, in Lee (2007), pp. 136, 145–146.
[126] Rosenfeld (1979k), pp. 339–340.

Daint was also very beautiful".[127] These recollections may reflect the humoristic characters of Gamow, Rosenfeld, and Peierls more than they accurately describe the fine climb they had, which is documented in Rosenfeld's photo album at the Niels Bohr Archive.

Picture of Gamow and Rosenfeld taken by Peierls at Piz da Daint where Gamow signed his note for *Nature*. Rosenfeld Album. Courtesy of The Niels Bohr Archive.

It is very likely that this group of young physicists in Zurich discussed politics and Marxism, besides physics. Rosenfeld later commemorated that "he found the way to Marx and Lenin" in "a club of Marxist students" in Zurich. [128] Many of these young colleagues shared an insatiable curiosity about the grand social experiment that took place in the Soviet Union during the First Five Year Plan and the Cultural Revolution 1929–1932, "The Great Break" as Stalin called it. One can imagine that the European physicists would have had many questions for their Russian colleagues. Landau was a juvenile radical; he was proud of the Russian Revolution and enthusiastic about the on-going social and political developments and parts

[127] R. Peierls to G. Kanegisser, 16 Oct 1930, in Lee (2007), Vol. 1, p. 146.
[128] RP, Supplement "Science and society", L. Rosenfeld, De socialistische student in de maatschappij, manuscript.

of Marxist thought.[129] In Russia, Gamow and Landau were members of the so-called Jazz Band, a small group of progressive physicists also including Matvei Petrovich Bronstein, Dmitry Ivanenko, and Genia Kanegisser (who married Peierls in 1931). They shared a reckless and intense attitude to life in this period where this new progressive music could still be enjoyed. However, a few years later, travel abroad for Soviet scientists was severely restricted. Gamow attempted to escape the Soviet Union with his wife in a kayak across the Black Sea, but failed. He chose not to return to the Soviet Union after he was allowed by the Russian authorities to attend the Solvay Conference in Brussels in 1933. Landau was not permitted to travel abroad after 1934.

[129] Gorelik and Frenkel (1994), pp. 20–22. Hall (2008), pp. 250–254. Kojevnikov (2004), pp. 76, 91. Casimir (1983), pp. 104–116. Interview with Landau about the conditions in the Soviet Union, *Studenten*, 12 March, No. 22 (1931). Provided in English translation in Casimir (1983), pp. 111–114. Gorelik (1997).

Chapter 2
Rosenfeld in Copenhagen

This chapter is about Rosenfeld's time in Copenhagen in the early 1930s. It focuses on the milieu Rosenfeld encountered at Bohr's Institute for Theoretical Physics and on the physics Rosenfeld was involved with in close collaboration with Bohr. Throughout the 1930s, Rosenfeld worked on quantum field theory. His main contributions in this area were his general method of representing quantized fields taking explicit account of symmetry properties, a discussion of the implications of quantization for the gravitational field, and a proof that a new formulation of quantum electrodynamics proposed by Dirac was not an alternative to the original Heisenberg–Pauli theory, but simply an equivalent representation of the latter. In 1940, Rosenfeld published a general method for constructing the energy–momentum tensor of any field.[1] In addition, Rosenfeld published several papers on meson theory with Christian Møller. Bohr and Rosenfeld's collaboration culminated with their joint paper published 1933, "On the Question of the Measurability of Electromagnetic Field Quantities", famous among theoretical physicists and no doubt Bohr's most subtle work.[2] Rosenfeld's role in their collaboration was contributing with his expertise in quantum field theory. In other words Rosenfeld was responsible for the formal side of their work while Bohr took care of the more principal side. In the process of writing the paper with Bohr and witnessing Bohr's controversy

[1]Rosenfeld (1979n). RP, "Supplement 1971–1974", scientific autobiography.
[2]Bohr and Rosenfeld (1979a).

with Landau and Peierls which led up to it, Rosenfeld was thoroughly initiated into Bohr's subtle quantum epistemology. Rosenfeld further assisted Bohr with his reply to the EPR-paper in 1935.[3]

The Institute for Theoretical Physics

In 1930, Rosenfeld obtained a lectureship in theoretical physics at his alma mater, the University of Liège. With his students Jean Serpe and Jean Humblet, he inaugurated theoretical physics at this institution, and in 1937 he became professor. He was in charge of a course on the physics of radiation and from 1933 an optional course on statistical mechanics, which he later stated "remained one of my favorite courses when I moved to Utrecht".[4] In 1930, he also began collaborating with Bohr in Copenhagen. Rosenfeld arranged his lectures so that he could work in Copenhagen for several months each year, and this commuting continued for the next ten years, also after he married and had children, until the invasion by Germany stopped travel between the occupied countries.[5]

When Rosenfeld arrived in Copenhagen, Bohr's institute was world-famous for the original ideas in quantum theory fostered here that had opened a wholly new (and controversial) perspective on physics. With Rosenfeld's words, the first Copenhagen Conference in 1929 marked "the completion of a heroic period in the life of the Institute".[6] In his lectures in Chicago in 1929, Heisenberg talked about the "'*Kopenhagener Geist der Quantentheorie*'... which has directed the entire development of modern atomic physics".[7] Central in this direction of the research was Bohr's correspondence principle, which had so successfully guided the construction of the theory in

[3]Bohr (1996a).
[4]RP, "Supplement: History of Quantum Theory", L. Rosenfeld, Ehrenfest as I saw him (An autobiographical chapter), May 1971, manuscript. Swings (1974), p. 656. Serpe (1980), pp. 390–391.
[5]Kuhn and Heilbron (1963a), pp. 10–11. Kuhn and Heilbron (1963b), pp. 13–14. Miss Have's guestbook, NBA.
[6]Rosenfeld (1971).
[7]Heisenberg (1930), preface.

its early phases, although apparently exclusively in a Copenhagen setting.[8] More radical perhaps was the interpretation of quantum phenomena as indeterminate, a-causal, discontinuous, indivisible, probabilistic, and complementary, which caused disapproval among several physicists as already noted in the previous chapter. The so-called Copenhagen spirit also referred to the informal social working atmosphere prevailing at the institute, as testified to by the majority of its visitors. Rather than formal seminars, Bohr favored personal conversations, dialogues, which could take place on walks, while biking, on sailing trips, at his home — from 1932 the honorary residence at Carlsberg — or at his summer house in Tisvilde north of Copenhagen.[9]

The informal tone at Bohr's institute is exemplified by the now-famous Faust parody performed in connection with the Copenhagen Conference at Easter in 1932,[10] on the hundredth anniversary of Goethe's death. Max Delbrück changed Goethe's Faust into a parody about the current crisis in quantum theory, with the main figures being Bohr as the Lord, Pauli as Mephistopheles, and Ehrenfest portraying the troubled, suffering Faust torn between the Lord and Mephisto. Bohr and Pauli were performed by Felix Bloch and Rosenfeld respectively, but it is unclear who played Ehrenfest.[11] According to Rosenfeld, "Delbrück's enthusiasm was contagious and a number of us soon found ourselves, after each strenuous day of serious discussions of unsolved problems, involved in the composition and rehearsal of the scenes evoking them with gentle irony".[12] According to Rosenfeld, the parody greatly appealed to Ehrenfest. "I shall never forget the expression of childish delight in his face as he keenly watched the play... I could see that my impersonation

[8]See for example Jammer (1966), p. 116.
[9]Aaserud (1990). Robertson (1979), p. 135.
[10]Segré (2007). *Faust and Journal of Jocular Physics Volumes I, II, and III.* Reprinted on the Occasion of Niels Bohr's Centenary, 7 October 1985, NBA.
[11]von Weizsäcker (1985), p. 190.
[12]RP, "Supplement: History of Quantum Theory", L. Rosenfeld, Ehrenfest as I saw him (An autobiographical chapter), May 1971, manuscript.

of Pauli [as Mephistopheles] in his diabolic mood pleased him particularly".[13]

With his institute Bohr had created a Mecca for theoretical physics, and he served as a charismatic and enterprising leader. According to Rosenfeld, the unique position Bohr enjoyed among physicists was due to his urge "to look for the deeper logical aspects of the problems presented by the analysis of the physical phenomena, as much as his uncanny intuition for their essential features and his supreme ability to trace at one glance their widest implications", and he added that Bohr "was quite conscious of his powers".[14] It is evident from the correspondence between the theoretical physicists in Bohr's group at the time, that this trait was as central in the further development of quantum electrodynamics in the early 1930s as it had been in the late 1920s during the construction of quantum mechanics.

For the young postdocs, a stay in Copenhagen involved, if the physicists were receptive to it (which far from all of them were), training in Bohr's philosophical thought. In contrasting Sommerfeld's problem-solving approach to Bohr's and Einstein's philosophical approaches to physics, Born suggested in 1928 that if "several of [Sommerfeld's] most recent and most significant students have achieved great things in the study of foundations, this must... be due to later influences, particularly to contact with Bohr".[15] In addition to Bohr and Born, also Ehrenfest and Pauli were convinced that such training was important for the quantum physicist. When Pauli learned that Heisenberg was invited to Copenhagen in 1924, he told Bohr that "[h]opefully then Heisenberg, too, will return home with a philosophical orientation to his thinking".[16] Ehrenfest sometimes complained that the new generation of theoretical physicists lacked interest in fundamentals. However, he recognized in Landau a "Talmudist", a metaphor for a physicist who took an interest in

[13] *Ibid.*
[14] Rosenfeld (1967), p. 118.
[15] Born (1928), p. 1036. English translation in Seth (2010), p. 184.
[16] W. Pauli to N. Bohr, 11 Feb 1924, Hermann *et al.* (1979), Vol. 1, pp. 143–144.

foundations.[17] Likely prompted by his stay in Copenhagen, Landau was eager to make an impact in quantum foundations, as we shall see. It is plausible that Rosenfeld's original motivation for working with Bohr was to become further initiated in Bohr's quantum epistemology. Rosenfeld did not simply become a Bohrian after his first meeting with Bohr or the day he arrived in Copenhagen; it was a gradual process. In Rosenfeld, however, Bohr eventually found more resonance than in most physicists with regards to his philosophical ideas.

In spite of the word "theoretical" in the institute's name, Bohr was from the beginning convinced that physicists should "have the opportunity to carry out and guide scientific experiments in direct connection with the theoretical investigations". He had made this clear already in his original application for building the institute.[18] Indeed, fruitful collaboration between experimentalists and theoreticians took place at Bohr's institute; the discovery of Hafnium in 1922 and nuclear research in the 1930s are just two examples of this. Experiments at Bohr's institute were generally seen as being directed by theory among physicists and philosophers of science alike in a great part of the twentieth century. However, for Bohr there was another aspect to this relation between theory and experiment, as Rosenfeld often repeated, *viz.*, that theory needed to be solidly anchored in reality. Purely formal exercises did not enjoy Bohr's blessing.[19] In practice, however, and this cannot come as a surprise for anyone today, the experimental focus at the institute was often directed by potential funding possibilities.[20]

---

[17]P. Ehrenfest to A. F. Joffe, 6 Jan 1933: "One should admit that something of a Talmudist is present in his [Landau's] style of thinking (I can say the same about Einstein and myself). In any case, it is more prominent in his talks than in his thinking. A couple of heated discussions about his unfounded paradoxical judgments convinced me that he is a *clear* and *graphic* thinker — especially in classical physics". Quoted from Gorelik and Frenkel (1994), p. 57. Einstein called Bohr "the Talmudist". A. Einstein to E. Schrödinger, 19 June 1935, quoted from Howard (1990), p. 105.

[18]Bohr to the Science Faculty at University of Copenhagen, 18 Apr 1917, quoted from Aaserud (1990), p. 17.

[19]Rosenfeld (1967), pp. 114–115.

[20]Aaserud (1990).

Besides Gamow, Landau, Peierls, and Delbrück, Rosenfeld also made friends with the Indian astrophysicist Subrahmanyan Chandrasekhar, and the quantum physicists Victor Weisskopf, George Placzek, and Hendrik B. G. Casimir in the early 1930s. They stayed at Miss Have's pension in the vicinity of the institute where they had conversations over dinner about physics and the world situation.[21] Due to Gamow's initiative, Gamow, Delbrück, Casimir, and Rosenfeld addressed each other with "du" at a time when the German language and the term "Sie" were otherwise the norm. Bohr used "du" only to his nearest collaborators Hendrik Kramers, Heisenberg, Klein, Pauli, and Rosenfeld.[22] The Danish physicist Fritz Kalckar jokingly referred to Rosenfeld as "the man who is 'Dus' with Bohr and Picard", where Picard may be a reference to Rosenfeld's historical and literary orientation and French accent. Kalckar also joked about "the small kind professor's" balloon-like shape.[23] The shy and melancholic Chandrasekhar, who felt like an outsider among the quantum physicists, became an especially close friend of Rosenfeld. In 1936, Chandrasekhar left Europe for a position in the US. This planted in Rosenfeld "the melancholy feeling of initiating a wider separation from you. I tell you this without any sentimentality, just as an expression of the truth, namely that I have no other friend like you".[24]

Besides befriending these visitors from abroad, Rosenfeld established a close relationship with the Danish physicist Møller, who was the same age as himself. During the late 1930s, Rosenfeld collaborated with Møller on meson theory and they even continued their collaboration by mail during the war.[25] Bohr compared the two physicists suggesting that just as Rosenfeld was "the only real representative of modern theoretical physics in Belgium... Møller

[21]Miss Have's guestbook, NBA. Rosenfeld (1962b). Weart (1977). RP, copy of letter from L. Rosenfeld to S. Chandrasekhar, 30 Dec 1936. Original in CP.
[22]RP, "Supplement 1971–1974", L. Rosenfeld to D. Danin, 20 Jul 1973.
[23]Kalckar (1935).
[24]RP, copy of letter from L. Rosenfeld to S. Chandrasekhar, 30 Dec 1936. Original in CP.
[25]See Chapter 4.

represented the younger generation in this field in such an excellent way in Denmark".[26] Møller was appointed scientific assistant at Bohr's institute in 1931 and two years later he became associate professor. However, in contrast to Rosenfeld, philosophical aspects of physics had absolutely no appeal to Møller. As he expressed it in 1963 in an interview conducted by Thomas Kuhn:

> Although we listened to hundreds and hundreds of talks about these things, and we were interested in it, I don't think, except Rosenfeld perhaps, that any of us were spending so much time with this thing... When you are young it is more interesting to attack definite problems. I mean this was so general, nearly philosophical.[27]

Assisting Bohr

As mentioned, Rosenfeld played a prominent part as Pauli in the "Blegdamsvej Faust" in 1932. However, according to the recollections of Hilde Levi, one of the Jewish–German physicists to find refuge at Bohr's institute after Hitler took power in Germany, Rosenfeld worked intensely with Bohr during his frequent visits at the institute and rarely took part in the social life.[28] Rosenfeld's first task in Copenhagen was to assist Bohr with completing for publication the long manuscript from his Faraday Lecture of 8 May 1930 in London, "Chemistry and the Quantum Theory of Atomic Constitution".[29] During 1930 and 1931, Bohr was very busy with lectures around Europe, and Rosenfeld was invited to Copenhagen again in the late summer of 1931 to assist with Bohr's contribution, "Atomic Stability and Conservation Laws", for the congress in Rome in October arranged by the Fondazione Alessandro

[26]BSC, N. Bohr to H. A. Kramers, 25 Apr 1938. Originally in Danish.
[27]Kuhn (1963a), p. 21.
[28]Recollections of H. Levi (1993), NBA.
[29]Bohr (1985a). Bohr thanked Rosenfeld for his help in RP, "Copenhague 2: Bohr (1931–1938)", Bohr to Rosenfeld, 29 Jan and 21 Mar 1932.

Volta.[30] Besides assisting Bohr with his papers (and writing his own), Rosenfeld prepared a French translation (published in 1932) of Bohr's popular papers from the 1929 *Festschrift* for the University of Copenhagen.[31]

Bohr had his own very characteristic working method. Especially in later years, he hardly ever wrote his own papers (or letters) but dictated the text to someone who either typed it or wrote it down by hand while at the same time discussing the ideas and formulations with Bohr.[32] This role had been played by his younger brother, the mathematician Harald Bohr, and his mother, Ellen Adler, during his dissertation work. His wife, Margrethe, put his early ideas on the atomic theory down on paper, and later it was the young physicists who spent time in Copenhagen who worked with him.[33] Rosenfeld's predecessor in this role as sounding board was the Swedish physicist Oskar Klein who returned to Sweden in January 1931 in order to take up a professorship in Stockholm after four years in Copenhagen. Prior to Klein, Kramers had served in this role for a period of ten years.[34] According to Rosenfeld, when he worked in Copenhagen, Bohr still valued highly the opinions of these former colleagues; and if Pauli or Heisenberg were visiting, others had a very hard time getting a moment with Bohr, "but he weighted with the same scrupulousness the modest observations of his younger and very much less experienced helpers of those years, Casimir and myself".[35]

---

[30]RP, "Copenhague 2: Bohr (1931–38)", M. Bohr to L. Rosenfeld, 8 Aug 1931. Bohr (1986).

[31]Bohr (1929). Bohr (1932). See also *BCW*, Vol. 10, pp. xxvi–xxvii. RP, "Copenhague 2: Bohr (1931–1938)", N. Bohr to L. Rosenfeld, 29 Jan 1932. The publication also appeared in German (1931) and English (1934) translations. All translated versions of this publication included a fourth paper, a lecture on the general discussion of complementarity that Bohr gave at the meeting of Scandinavian Scientists in Copenhagen in 1929.

[32]Rosenfeld (1967), pp. 119–120. Kuhn and Heilbron (1963c), p. 3.

[33]Pais (1991), p. 102.

[34]Robertson (1979), p. 157.

[35]Rosenfeld (1967), p. 118. RP, "Correspondance particulière", L. Rosenfeld to M. Strauss, 27 Oct 1935.

Bohr and Rosenfeld in Tisvilde 1931. Courtesy of The Niels Bohr Archive.

"After Klein's departure I have in Rosenfeld acquired an excellent and understanding collaborator", Bohr wrote to Pauli in March 1931.[36] Bohr and Rosenfeld developed a close and intense collaboration in which their quite different approaches and strengths complemented each other harmoniously and efficiently.[37] It was a collaboration that was to prove important to both men. Bohr admired Rosenfeld's shrewd thinking, his language skill, and capacity for formulation. In fact, given Bohr's well-known fussy, elaborate, and laborious style in speech and writing, Rosenfeld was Bohr's complete antithesis. Whereas Bohr sought perfection through numerous proof-readings of everything he wrote, whether papers, talks or letters, at the same time striving not to offend anybody, Rosenfeld had obvious literary talent as well as exquisite elegance and clarity of expression. In polemical papers and in book reviews he could

[36]Bohr to Pauli, 21 Mar 1931, in Hermann *et al.* (1985), Vol. 2, pp. 68–69. Originally in Danish.
[37]Bohr, Aa. (1974). Møller (1974), p. 3.

appear quite blunt and biting, and many people took exception to this. Rosenfeld's collaborator in later years, the Belgian chemist Ilya Prigogine therefore called Rosenfeld a "paper tiger".[38] Contrary to Bohr, whose papers always went through numerous proofs, Rosenfeld seldom revised a paper after he had submitted it.[39]

Rosenfeld possessed a vast knowledge in most areas of physics, including its history and philosophy, and an eager ability for rapidly mastering new problems. He was fast in picking up Bohr's quantum epistemology and with time Rosenfeld became Bohr's most outspoken disciple. Bohr could explain his thoughts of a more philosophical nature to Rosenfeld with the assurance of an intelligent response. The enthusiasm for Bohr's ideas which Rosenfeld developed during the 1930s may also have encouraged Bohr. He would often, even after someone else had assisted him with a manuscript, await Rosenfeld's perusal before he considered it completed and ready for publication.[40] For example when writing an entry for the first volume of the *International Encyclopedia of Unified Science* (1938), edited by Otto Neurath, Jørgen Jørgensen, and Rudolf Carnap, Bohr seems to have relied heavily on Rosenfeld's philosophical expertise. From his letters to Neurath, one may get the impression that Bohr would *only* work on it when Rosenfeld was around.[41] When Bohr worked on a popular lecture for the Newton celebration in Cambridge in 1946, assisted by Abraham Pais, both felt "what a great help it would be if we could question you [Rosenfeld] closely about the historical connection".[42] Sometimes Rosenfeld would simply supply Bohr with a historical essay about a particular scientist who Bohr was going to talk about. In 1958, in connection with the preparation

---

[38]Prigogine (1974), p. 845.

[39]See for example Rosenfeld's correspondence with the editor of *Physics Today*. "Physics Today Archives" Box 19, corresp. 48–70, Centre for History of Physics, AIP, L. Rosenfeld to R. H. Ellis Jr., 29 Jul and 8 Aug 1969. Another example is the correspondence with the editor of *Dædalus: Journal of the American Academy of Arts and Sciences*, RP, "Supplement D", S. R. Graubard in May–June 1970.

[40]Kuhn and Heilbron (1963c), p. 4.

[41]Bohr (1999a). BGC, copies of letters from N. Bohr to O. Neurath, 18, 29 Jan and 14 Feb 1938. See also RP, N. Bohr to L. Rosenfeld, 2 Sep 1949 and RP, "Copenhague: Bohr (1940–1962)", Bohr to Rosenfeld, 31 Mar 1954 and Bohr to Rosenfeld, 22 Jan 1955.

[42]RP, "Copenhague: Bohr (1940–1948)", N. Bohr to L. Rosenfeld, 7 Jul 1946. Originally in Danish. Bohr (2007b).

of a speech during Bohr's visit to Zagreb, Yugoslavia (now Croatia), Rosenfeld sent him a brief biography he had written of the Jesuit scientist Ruggiero Boscovich.[43] Particularly during the 1930s, Bohr and Rosenfeld were practically inseparable. John Wheeler's wife, Janette, experienced it in this way: "you could not invite the one without inviting the other".[44]

Whereas others among Bohr's assistants, for example Weisskopf, have jokingly expressed some qualms about the task as Bohr's helper, because it tended to drain a person of professional originality and creativity, this seems not to have bothered Rosenfeld.[45] Rosenfeld was always willing to postpone his other duties, for example his lectures in Liège, if Bohr needed him.[46] In addition, although it may be true that Bohr had a tremendous working capacity for hours on end, which could totally exhaust his collaborators, Bohr *also* needed breaks, and, according to Margrethe, in such a situation Bohr would worry about what to do with the person in question, for example when work took place at the Carlsberg Mansion relatively far away from the institute where there would be nothing relevant for him to do. However, Bohr did not need to entertain Rosenfeld, for he would simply read a book he had brought with him from his growing collection of historical books in science.[47]

In their collaboration on measurability in quantum electrodynamics (see below) Bohr depended on Rosenfeld's mathematical expertise. However, with respect to Rosenfeld's predominantly formal style in physics when he first arrived in Copenhagen, Bohr

> used to tease me by proposing for my consideration some simple physical process involving electromagnetic radiation and — in the sweetest tone of voice but with an unmistakably malicious twinkle in his eye — asking me how I would handle this case with the learned methods of quantum electrodynamics. Usually the

---

[43]BGC, Rosenfeld to Bohr, 14 Oct 1958.
[44]The author's private conversation with Janette Wheeler, Princeton, 30 Mar 2006.
[45]Aaserud (1990), p. 13.
[46]See for example RP, "Copenhague 2: Bohr (1931–1938)", N. Bohr to L. Rosenfeld, 14 and 21 Nov 1938. See also BSC, Rosenfeld to Bohr, 16 Oct 1949.
[47]Kuhn (1963b), p. 14. Aaserud (1990), p. 12.

> answer came only after much fumbling and looked terribly complicated. This was the result he expected, and which gave him the opportunity of impressing upon me the danger of extrapolating formal procedures beyond the domain of experience for the description of which they had originally been devised, and the necessity of seeking a solid foundation for our lofty abstractions in some simple concrete aspect of the phenomena immediately accessible to observation.[48]

This "solid foundation" was concrete empirical lessons from experiments or logical considerations drawn from thought experiments. Ehrenfest shared Bohr's skeptical attitude towards a highly abstract mathematical approach and preferred to have the point of a physical problem spelled out in a few words or by means of a simplified model.[49] At the other end of the spectrum, there were physicists such as Dirac and Heisenberg who would hold mathematics more important than empirical evidence in theory construction. Rosenfeld later criticized others' "formalistic tendencies" and spoke, as exemplified here, with self-irony about his own position.[50]

Regarding Bohr as a father figure was by no means uncommon among the younger physicists who spent time in Copenhagen. Rosenfeld who lost his father as a young boy may have been especially susceptible to the intense interest and attention Bohr showed everyone working with him and his ability to advise, guide, comfort, and support his co-workers in all circumstances of life. When a postdoc arrived in Copenhagen, "truly he feels himself received into a spiritual family, strongly united under Niels Bohr's paternal aegis", according to Rosenfeld.[51] In 1949, in his speech of thanks upon receiving the Francqui Prize, Rosenfeld listed the various influences on him and the support he had enjoyed since his student years.

---

[48]Rosenfeld (1967), pp. 114–115.

[49]Seth (2010), pp. 185–187. The discussion of the role of mathematics in physics is of course as old as the science of physics. For the Danish natural philosopher H. C. Ørsted's views on the relation between mathematics and physics in the early 19th century, see Lynning and Jacobsen (2011), pp. 47–50.

[50]See also Rosenfeld (1979l), p. 442.

[51]Rosenfeld (1979j), p. 313. See also Kalckar (1967) and Heilbron (1985).

He naturally mentioned Niels Bohr, "to whom I am connected in a quasi-filial friendship".[52]

Crisis in Quantum Theory

The scientific context of Rosenfeld's time in Copenhagen in the early 1930s was the crisis in quantum theory noted in the previous chapter. This crisis led to great confusion and pessimism among the theoretical physicists involved with quantum electrodynamics, leading to heated controversies in their small community. One such controversy was about the competing research programs represented by Dirac, on the one hand, and Jordan, Pauli, and Heisenberg, on the other, which was briefly touched upon in the previous chapter. Another controversy was about the interpretation of measurement, pitting Bohr and Rosenfeld against Landau, Peierls, and Pauli. These controversies constitute the key to understanding Rosenfeld's publications and activities in Copenhagen during these years and in particular the important Bohr–Rosenfeld paper. Besides providing a picture of Rosenfeld in Copenhagen, the study of these controversies reveals that contrary to how it is often presented (for example by the protagonists involved), the interpretation of quantum mechanics as seen by Bohr, Heisenberg, Born, etc. was *not* finished with the Solvay Congress in 1927. Bohr spent several years clarifying, consolidating, and communicating his radical new quantum epistemology after his famous lectures on complementarity in Como in 1927 (see next chapter).[53] And a closer look reveals that there was no uniform "Copenhagen" view on all aspects of the interpretation of quantum theory in the late 1920s and the early 1930s.[54] Indeed, discussions on the meaning of quantum theory, and in particular measuring problems, continued well into the 1930s, now in the context of relativistic quantum mechanics and quantum field theory. The publication of

[52]RP, "Manchester (1947–1958): Prix Francqui (1949)", typewritten thanks speech. Originally in French. C. Møller also characterized Rosenfeld's filial relationship with Bohr in Møller (1974), p. 3.
[53]The ideas in the Como lecture were later published as Bohr (1985c).
[54]See for example Camilleri (2009b).

Bohr and Rosenfeld's paper in 1933 was the culmination of these discussions.[55]

As mentioned in the previous chapter, Pauli and Heisenberg were the first to recognize the problems with their own quantum field theory. After the lack of success with the second quantization procedure as a mathematical foundation of quantum electrodynamics, Heisenberg and Pauli shelved their quantum field theory, and Heisenberg took up a pre-Dirac approach to radiation problems drawing inspiration from Klein's 1927 publication "Electrodynamics and wave mechanics from the point of view of the correspondence principle".[56] In the end, Heisenberg's new theory proved equivalent to the Heisenberg–Pauli quantum electrodynamics. However, he introduced a new way of presenting quantum electrodynamics in terms of what is now known as the "Heisenberg picture" in contrast to Schrödinger's representation. In the Schrödinger presentation the state vectors move, that is, they change in time, whereas the operators are fixed. In the Heisenberg picture, the state vectors are fixed and the operators change in time.[57]

Rosenfeld followed suit. Quite fittingly, coinciding with his arrival in Copenhagen, he began to invoke the correspondence principle in his publications. He took his starting point in Heisenberg's new "refined correspondence procedure" for dealing with radiation problems, and treated spectroscopic line widths. He showed that to a first approximation the intensity of a transition line predicted by Heisenberg's new formulation would correspond to the spectral distribution of a classical dipole. In another paper taking the same starting point, Rosenfeld extended Møller's relativistic treatment of collisions between two free particles to arbitrarily many-body collisions. However, the correspondence approach proved of limited success.[58]

In the fall of 1930, Landau visited Bohr's institute for the second time that year as part of his tour of the European centres of modern

---

[55]The following is an extension of Jacobsen (2011).
[56]Klein (1927).
[57]Heisenberg (1931). Darrigol (1986), pp. 247–248.
[58]Rosenfeld (1932a; 1931b; 1931c), pp. 131–135. Kuhn and Heilbron (1963b), p. 13.

physics. Upon Bohr's return from the Solvay meeting in October Bohr and Landau discussed field measurements.[59] The topic of the Solvay meeting in 1930 was the magnetic properties of matter, but the problems of quantum electrodynamics were discussed between sessions. In his correspondence from these years Bohr worried about the strange physical predictions of the theory.[60]

Ever since he published his atomic model in 1913, Bohr had analyzed idealized measurements with the purpose of exploring the meaning of concepts of a strictly quantum character such as *stationary state*, *uncertainty relation*, *matter wave*, *photon*, and *spin* that would vanish from the theory in the classical limit. He wanted to reach an understanding of these phenomena, so unfamiliar from a classical physics perspective, by investigating how they could be logically justified in an imaginary measurement situation. According to Bohr, the description of the measurement results presupposed ordinary language supplemented with the terminology of classical physics: "We must, in fact, realise that the unambiguous interpretation of any measurement must be essentially framed in terms of the classical physical theories, and we may say that in this sense the language of Newton and Maxwell will remain the language of physicists for all time".[61] Therefore, despite the apparent symmetry in the complementarity principle with respect to wave–particle duality, Bohr attributed greater epistemological status to the wave picture of light and the particle picture of matter than to photons and matter waves. In his Maxwell Lecture in Cambridge in October 1931, he reflected on the classical concepts of field and particle:

> When one hears physicists talk nowadays about "electron waves" and "photons", it might perhaps appear that we have completely left the ground on which Newton and Maxwell

---

[59]Landau visited Copenhagen 8 Apr to 3 May and from 20 Sep to 22 Nov 1930, and again from 25 Feb to 19 Mar 1931. Aaserud (1990), p. 61 note 49. Pais (1991), p. 359. The guestbook of the institute, NBA. BSC, N. Bohr to W. Heisenberg, 19 Nov 1930. This letter seems not to have been sent. See also Darrigol (1991), p. 154.

[60]BSC, N. Bohr to P. A. M. Dirac, 23 Dec 1929. See also BSC, N. Bohr to W. Heisenberg, 18 Mar 1930.

[61]Bohr (1985b), p. 360.

> built; but we all agree, I think, that such concepts, however fruitful, can never be more than a convenient means of stating characteristic consequences of the quantum theory which cannot be visualized in the ordinary sense. It must not be forgotten that only the classical ideas of material particles and electromagnetic waves have a field of unambiguous application, whereas the concepts of photons and electron waves have not. Their applicability is essentially limited to cases in which, on account of the existence of the quantum of action, it is not possible to consider the phenomena observed as independent of the apparatus utilised for their observation.[62]

The wave concept has in the case of light an immediate correspondence with classical electromagnetic theory whereas the particle concept of matter has a direct classical correspondence. However fruitful the concepts of photons and electron waves may be, "it is a purely formal matter".[63]

To Bohr, it was irrelevant whether his idealized experiments could be realized in practice. The purpose of his analyses was to pair the conceptual *definitions* derived from the quantum formalism with the possibility of *measuring* these properties in an idealized experimental setup in order to arrive at an interpretation of them. In this way, his sole purpose when examining measurement problems in quantum theory was to give meaning to the formalism.[64] This, however, was never a trivial point in Bohr's papers and has led to many misunderstandings.[65] Even among Bohr's collaborators there was confusion about this point, as we shall see. No wonder it was difficult to grasp for those who were not part of Bohr's inner circle. Among

---

[62] *Ibid.*, p. 359.
[63] *Ibid.*, p. 360.
[64] Bohr MSS, "Solvay Conference" Oct 1930 (should be 1933) Morning October 28, microfilm 12, AHQP. Bohr and Rosenfeld (1979a). Rosenfeld (1979m), p. 424. See also Bohr (1996a), p. 294[698] footnote. Bohr (1996b), p. 362. RP, "Epistemology 1959–1964", Rosenfeld, Report on Louis de Broglie's *La théorie de la mesure en mécanique ondulatoire*, requested by Pergamon Press Limited, enclosed in correspondence between Miss S. Stratford-Lawrence and L. Rosenfeld, 26 Jan and 3 Feb 1959. See also Bokulich and Bokulich (2005), in which these issues are presented in a very clear way.
[65] See for example Bohr (1985c), pp. 150–152 [582–584]. Bohr and Rosenfeld (1979a).

other things, de Broglie, Einstein, and Schrödinger questioned Bohr's understanding of the behavior of *individual* quantum systems in his idealized measurements.[66] Other physicists had difficulty understanding the use or purpose of Bohr's thought experiments, no matter how many times Bohr explained himself, whether in writing or in the lecture room.[67] Rosenfeld later emphasized that Bohr's aim was never to provide a "theory of measurement" like von Neumann's. Von Neumann gave an axiomatic elucidation of quantum measurements in his book *Mathematische Grundlagen der Quantenmechanik* (1932).[68] His mathematical exposition of the so-called measurement problem in the same book has been taken as the starting point for all the later discussions on this topic (see epilogue).

Thought experiments that particularly appealed to Bohr for elucidating the meaning of quantum mechanics were the double-slit experiment, Heisenberg's gamma ray microscope experiment from the spring of 1927, and Einstein's photon box experiment, discussed by Einstein and Bohr in 1930. The double-slit experiment demonstrates the inherent indeterminacy of quantum phenomena, Heisenberg's uncertainty relation for position and momentum, the wave–particle duality, and the correlation between a quantum system and the measurement apparatus. The gamma ray microscope experiment was intended to make the wave–particle duality of light and matter and the uncertainty relation for position and momentum "anschaulich" (see Chapter 1). The photon box experiment was, in Bohr's interpretation of it, about the use of a weighing mechanism for keeping track of the energy exchange between a photon and the measurement apparatus as well as measuring the exact energy of the photon. It was therefore meant as a challenge both to the quantum correlation between object and measurement apparatus and to Heisenberg's uncertainty relation between energy and the time parameter of the photon. Bohr's analysis of it, however, bringing in arguments based on Einstein's own theory of general relativity, demonstrated the consistency of the quantum theory.

[66]RP, "Correspondance particulière", L. De Broglie to Rosenfeld, 6 Jan and 25 Mar 1928. Forman (1984), pp. 340–343.

[67]"Experience shows that he is very little understood", Kuhn and Heilbron (1963c), p. 4.

[68]von Neumann (1932). Jammer (1974), p. 487.

Blackboard with Bohr's drawings from a visit by Bohr to Japan 1937. Courtesy of The Niels Bohr Archive.

Gamow's Einstein box, a Christmas present to Bohr in 1930. The drawing of Einstein is made by Piet Hein. Courtesy of The Niels Bohr Archive.

With respect to quantum electrodynamics, Bohr was concerned with the use of the classical concept of a *field* in quantum theory. He hoped that an investigation of the measurability of the field components would enable physicists to get a better grasp of the idea of a quantized field and how it differed from the classical conception of fields.[69] The basis for this undertaking was Heisenberg's Chicago lectures in 1929 in which Heisenberg had, as the first, derived uncertainty relations for the electromagnetic field components (later to be corrected by Bohr and Rosenfeld). He traced the origin of these uncertainties in an imaginary measurement, as well as pondered over the meaning of field components as functions of time and space.[70]

A measurement of electromagnetic field quantities is done by measuring the change in momentum of electrically charged test bodies placed in the field. Bohr suggested using test bodies, designed to probe the electromagnetic field in a measurement situation, much bigger than the electron, in order to avoid the infinities introduced by the treatment of point charges.[71] In a series of letters (of which only a couple were actually sent) Bohr briefed Heisenberg about how his discussions with Landau developed. In late November Landau returned to Zurich where he involved Peierls in the considerations about measurability in a relativistic quantum frame.[72] Peierls later remembered: "[i]n our discussions we came back again and again to the unsolved problems of the quantum theory of fields, of which the infinite self-energy of the electron was a symptom".[73] Encouraged by Pauli, the two young physicists decided to publish

[69]Bohr, Field Measurements, 1930–1931, handwritten manuscript, NBA, AHQP. Bohr (1985b). BSC, N. Bohr to W. Heisenberg, 25 Dec 1930. Rosenfeld (1971). See also Darrigol (1991), pp. 158–159.
[70]Heisenberg (1930), pp. 47–54.
[71]BSC, N. Bohr to W. Heisenberg, 8 and 25 Dec 1930.
[72]Miss M. Have's guestbook and the institute's guestbook suggest that Landau stayed until 22 November 1930. See also Peierls Papers, Department of Western Manuscripts, Bodleian Library, Oxford University, ref. no. CSAC 52.6.77/c.179, L. D. Landau to R. Peierls, 9 Aug 1930. The Russian physicist V. A. Fock was also originally involved in this work with Landau and Peierls. See Lee (2007), Vol. 1, pp. 137, 167–68, 192, 206–09, 224. Fock published his own results with Jordan, Fock and Jordan (1931).
[73]Peierls (1985a), p. 66. See also Peierls (1980), p. viii.

their ideas and were soon ready to circulate a paper to Bohr and Heisenberg dealing with these issues.[74] They introduced their paper, later published with the title "Erweiterung des Unbestimmtheitsprinzips für die relativistische Quantentheorie" (Extension of the Uncertainty Principle to Relativistic Quantum Theory), with some general reflections on what they regarded as reasonable premises of measurements in terms of repeatability and predictability. They then drew attention to the important fact that during the very act of measurement, while the charge of the test body is accelerated by the external field it is supposed to measure, it gives rise to an additional field superposed onto the very field to be measured.[75] Hence, in addition to the uncertainty of field components due to their non-commuting properties, there will be an uncertainty attributed to *each single* field component, on account of an uncontrollable momentum transfer when the test body is accelerated. In sum, Landau and Peierls' analysis of the measurability of the field components suggested that fields could not be measured as accurately as the theory stated, from which they concluded that fields were not observable quantities in the theory. Paired with the problem of singularities and the prediction of negative energy states in Dirac's electron theory — Landau and Peierls firmly believed that all these problems were closely interrelated and were consequences of an inconsistent mathematical formalism — they were led to the dramatic conclusion that "it would be surprising if the formulation of quantum electrodynamics bore any resemblance to reality".[76] As a result, they opted for overthrowing the very foundations of the theory all together.

[74]Unfortunately, the manuscript is not in the Niels Bohr Archive. W. Pauli wrote to O. Klein, 12 Dec 1930: "Landau had interessante Ungleichungen und das Versagen vieler Begriffe aus Kopenhagen hierhergebracht, die mich wirklich interessiert haben", in Hermann *et al.* (1985), Vol. 2, pp. 43–46, on p. 46.

[75]Landau and Peierls (1983). BSC, N. Bohr to W. Heisenberg, 19 Nov and 8 Dec 1930. The last letter seems not to have been sent. Bohr to Heisenberg, 25 Dec 1930.

[76]Landau and Peierls (1983), p. 475. For a closer analysis of the equations in Landau and Peirls' paper as well as in the Bohr–Rosenfeld paper, see Darrigol (1991), pp. 157–173 and Kalckar (1996).

Peierls informed Rosenfeld about his and Landau's dramatic conclusion in a letter of 15 January 1931:

> Due to Landau'ian considerations we are... now of the opinion that the whole theory [quantum electrodynamics] is even more false than it would seem, since all the infinities are only expressions to be expected from the fact that the theory widely exceeds its range of validity, and constantly operates with unobservable quantities. In truth, according to relativity, all quantities occurring in the theory, i.e. fields, positions and momenta of electrons and light quanta, etc., are not measurable quantities,... so the theory is not at all consistent.[77]

Peierls ended his letter to Rosenfeld using a metaphor for the rebellious young physicists' uprising against Bohr and his collaborators, Klein, Heisenberg, and Pauli — in this matter, asking Rosenfeld to greet "den gesamten Sowjet" (the group of young postdocs in Copenhagen at the time, Gamow and Rosenfeld) from himself and Landau.[78]

Updated by Peierls' report, Rosenfeld went off to Paris where in February 1931 he gave a series of lectures on the new quantum field theory and its problems at the Institut Henri Poincaré. Based on these lectures, he wrote the review paper "La théorie quantique des champs".[79] Upon his review of the latest development of quantum field theory and quantum electrodynamics, Rosenfeld listed the difficulties with field quantization, such as the infinite self-energy in the field description of the point electron and the "unacceptable results" of the radiation field in terms of the infinite energy of the gravitation field produced by one photon. Because of these difficulties, Rosenfeld claimed, in keeping with Pauli and Heisenberg and with Landau and Peierls' conclusion, that it was necessary to abandon quantum electrodynamics, and analyze the situation again

[77]RP, "Correspondance particulière", R. Peierls to L. Rosenfeld, 15 Jan 1931. Originally in German.
[78]*Ibid.*, The guest book of the institute, NBA.
[79]Rosenfeld (1932a).

from a purely physical point of view, *viz.*, going back to scratch and invoking the correspondence principle anew. Rosenfeld then noted how the correspondence principle had laid the basis for Heisenberg's matrix mechanics, and how Klein had shown the relation between Schrödinger's formalism and the correspondence principle, and how Heisenberg's recent extension of Klein's method demonstrated the relation between the correspondence principle and the method of field quantization.[80]

When Rosenfeld returned to Copenhagen on the last day of February a storm had broken out there between Bohr and Landau and Peierls. Allegedly, Rosenfeld was briefed about the situation by Gamow who handed Rosenfeld a drawing of Landau gagged, pinioned to a chair and confronted by Bohr, with the comment that "he does not seem to agree [with the arguments in Landau and Peierls' paper] — and this is the kind of discussion which has been going on all the time".[81] The current crisis in the small group of theoretical physicists around Bohr had reached a climax; was the field concept not justified in a quantum frame?

A later reconstruction of Gamow's drawing of Landau gagged, pinioned to a chair and confronted by Bohr. Courtesy of The Niels Bohr Archive.

[80] *Ibid.*, pp. 88–89.
[81] Rosenfeld (1979m), p. 413.

The "Small War"

When Heisenberg constructed matrix mechanics in 1925, he had stressed the importance of observable quantities in the theory.[82] Initially, therefore, he was sympathetic to Landau and Peierls' critique of the many "unobservables" in relativistic quantum mechanics.[83] However, such signals were definitely not transmitted from Copenhagen. In Rosenfeld's words, "Landau and Peierls encroached, so to speak, on Bohr's domain, and he was immediately able to say, at least in the negative sense, that all this rough handling of measurements was certainly not allowed, that you could not pull any conclusions, either pro or contra, from such sloppy considerations".[84] Bohr acknowledged Landau and Peierls' discovery of the limits of space–time measurements, but he was of the opinion that in general "they overshot the mark and misunderstood what can be demanded of a physical theory".[85] Bohr distinguished between the presences of observables in a theory — this was a matter of *definition* — and the epistemological circumstance and conditions of actual *observation*, that is, measurability. In order to find out whether these quantities could be measured in practice, Bohr suggested analyzing idealized measurements of them.

Heisenberg soon came to agree with Bohr that Landau and Peierls' conclusion about the inconsistency of quantum electrodynamics was reached on the basis of a sloppy and naïve frame of argumentation and built on misunderstandings.[86] Bohr was quite upset because he found Landau and Peierls' "scepticism" about quantum electrodynamics "completely unfounded", as he wrote to Pauli later that spring.[87] The enormous stir which Peierls and Landau's paper aroused in Copenhagen was jokingly referred to as "our small war"

---

[82] Heisenberg (1925). Camilleri (2009b), p. 18.

[83] W. Heisenberg to L. D. Landau and R. Peierls, 26 Jan 1931, quoted in Lee (2007), Vol. 1, pp. 220–221. Also quoted in Darrigol (1991), p. 160.

[84] Kuhn and Heilbron (1963b), p. 15.

[85] N. Bohr to W. Pauli, 25 Jan 1933, in Hermann *et al.* (1985), Vol. 2, pp. 152–153.

[86] BSC, W. Heisenberg to N. Bohr, 23 and 30 Jan 1931.

[87] N. Bohr to W. Pauli, 21 Mar 1931, in Hermann *et al.* (1985), Vol. 2, pp. 68–69, on p. 68. See also Bohr to Pauli, 25 Jan 1933, *ibid.*, pp. 152–153, on p. 152.

by Oskar Klein in a card he wrote on behalf of Bohr to Landau and Peierls on 16 February 1931.

> Just a few lines to say that Bohr and all of us are happy, that you can be here soon. As for our small war, hopefully you did not take the bad jokes to be worse than they were intended. In particular, the Schiller quotation was only an ill-mannered impulse of mine, the exact meaning of which is unknown even to me.[88] It was meant to be a sort of substitute for the "destructive criticism" of your work that you demanded from Bohr and also as a small reminder of previous Landau language in Copenhagen. When you turn up here, we will treat you with far greater reverence than in our letters and coloured postcards, and we look forward to discussing the matter with you.[89]

Indeed, Bohr was eager to get an opportunity to discuss the issue with Landau and Peierls before they published their paper. The annual Easter conference was therefore rescheduled to the end of February in order that both of them could be present.[90] In spite of Bohr's famously strenuous efforts, as immortalized by Gamow's drawing, no consensus was reached at the meeting.

Pauli was not present at the meeting but visited Copenhagen later in the spring. He disapproved strongly of Bohr's criticism of his younger protégés' paper. Intimately connected with the problems that his and Heisenberg's theory had encountered, he was of the firm conviction that Landau and Peierls' inferences for quantum electrodynamics were completely justified. Incidentally, the Landau–Peierls paper was not the only matter that Bohr and Pauli quarrelled about at the time. With respect to the explanation of radioactive beta-decay processes, Bohr was once again prepared to give up the

[88]Unfortunately, the Schiller quote could not be found.

[89]BSC 23, O. Klein to L. D. Landau and R. Peierls, 16 Feb 1931. Originally in German.

[90]Bohr unpublished manuscript 252, "Copenhagen Conference" 1931, NBA. R. Peierls to G. Peierls, 3 Feb 1931 in Lee (2007), Vol. 1, pp. 227–230, on p. 228.

principle of conservation of energy.[91] Pauli considered Bohr's solution to this problem to be "on completely wrong track" and suggested instead the existence of a neutral particle (later named the neutrino) not yet detected which would ensure energy conservation in these processes.[92] In January 1931 Pauli wrote to Klein, "[i]t is of course clear to me that the [neutrino] hypothesis does not suit Bohr and the Bohrians. It gives me special pleasure just for that reason to discuss it".[93] Hence, Pauli seems to have had unusually strong objections to Bohr's physics at the time. Heisenberg made an attempt at mediating between Pauli and Bohr, writing to Pauli:

> Your critique of the Copenhagen physics is justified in as much as of course no one knows anything for sure. Still, Bohr's critique of the Landau–Peierls' work is quite interesting. Bohr is in agreement with L-P's uncertainty relations. He considers the derivation of those messy in a few places, but that is not essential. The main critique is much more directed at the inferences which L-P draw from the uncertainty relations. According to Bohr, the uncertainty relations mean in no way that the relativistic wave mechanics is too narrow and must give way to a more general formalism. Rather Bohr says: also in the non-relativistic wave mechanics only a small number of all operators are measurable... So all in all Bohr is of the opinion that current wave mechanics (Dirac + quantum electrodynamics) constitute a *satisfactory* schema... In details, Bohr makes some quite nice remarks.[94]

[91]The first time had been in the Bohr–Kramer–Slater paper from 1924, Bohr *et al.* (1984), p. 109[793]. In 1929, Bohr interpreted "the fatal transition from positive to negative energy [in Dirac's hole theory...] as a limitation in the applicability of the energy concept". N. Bohr to P. A. M. Dirac, 5 Dec 1929. Quoted in Moyer (1981b), pp. 1057–1058.

[92]W. Pauli to O. Klein, 18 Feb 1929, in Hermann *et al.* (1979), Vol. 1, p. 488 (emphasis in original). Enz (2002), pp. 213–214. Pais (1991), pp. 366–370.

[93]W. Pauli to O. Klein, 8 Jan 1931, in Hermann *et al.* (1985), Vol. 2, p. 51. English quotation in Pais (1991), pp. 368–369.

[94]W. Heisenberg to W. Pauli, 12 Mar 1931, in Hermann *et al.* (1985), Vol. 2, pp. 66–67. Emphasis in original. Originally in German.

During the year 1931, Pauli suffered from severe personal difficulties related to his recent divorce and his mother's suicide some years earlier.[95] While this did not prevent him from working, it may have amplified his choleric mood and made him even touchier than usual. Bohr tried to appeal to Pauli's rational sense: "We do understand each other, and you know it is not the intention to underestimate the difficulties".[96] Bohr may have hoped that he could persuade Landau and Peierls not to publish, or that they would at least correct some passages in their paper where references to their discussions with Bohr revealed to Bohr that they had misunderstood him. However, despite Bohr's objections to their paper, Landau and Peierls decided to publish anyway.[97] The paper was submitted from Zurich on 3 March 1931. Bohr may then have hoped that he could at least count on Pauli's understanding. However, Pauli would not listen to his mentor; he gave Landau and Peierls green light to publish their paper, provided that they corrected an error in their derivation of the uncertainty relations for the field strengths.[98] Bohr could not understand what he later called Pauli's "fanaticism about it all", and as a result Bohr stopped responding to Pauli's letters for one and a half years until in January 1933 he felt entirely confident about the conclusion of his examinations of the measurability of the electromagnetic field made in collaboration with Rosenfeld.[99]

It was soon after the Copenhagen meeting in 1931 that Bohr started working with Rosenfeld on his considerations of field measurements. With Rosenfeld's expertise on quantum field theory, this was a perfect match. Rosenfeld recollected, "My first task was to lecture Bohr on the fundamentals of field quantization; the mathematical structure of the commutation relations and the underlying physical assumptions of the theory were subjected to unrelenting scrutiny. After a very short time, needless to say, the roles were inverted and he was pointing out to me essential features to which

[95] Enz (2002), pp. 211–240.
[96] N. Bohr to W. Pauli, 27 Apr 1931, in Hermann *et al.* (1985), Vol. 2, p. 76.
[97] Peierls (1985b), pp. 228–229.
[98] W. Pauli to R. Peierls, 3 Jul 1931, in Hermann *et al.* (1985), Vol. 2, p. 91.
[99] N. Bohr to W. Pauli, 25 Jan 1933, in Hermann *et al.* (1985), Vol. 2, pp. 152–153.

nobody had as yet paid sufficient attention".[100] Bohr continued developing his ideas about a thought experiment for investigating the consistency of quantum electrodynamics and they began working out their joint paper on the measurability of quantized fields.[101] During 1931, Rosenfeld therefore became much more familiar with Bohr's way of thinking. He accompanied Bohr to Italy and on their return trip Bohr stayed in Belgium for a few days in order for them to continue their joint work. Rosenfeld later recalled this autumn as the time when he waved farewell to "Laplacia" forever, a metaphor for giving up determinism in physics, and accepting Bohr's idea that we are not only spectators but actors in the science of physics.[102] With specific reference to Bohr's Faraday lecture, Rosenfeld added a postscript to the proofs of his own review paper on quantum field theory to appear in *Annales de l'Institut Henri Poincaré* in November 1931.[103] The postscript even contained a sentence about Bohr's "well-known account of 'complementarity' which exists between the spatio–temporal description and the energy–momentum balance-sheet".[104] This was the first time Rosenfeld referred to Bohr's complementarity in writing. However, he did not embark on it any further.

Dirac, "the Ugly Duckling"

Like Landau and Peierls, Dirac was clearly of the opinion that a new approach to relativistic quantum mechanics was needed in order to overcome the paradoxes in quantum electrodynamics. He presented his bid at the 1932 Copenhagen Conference. However, Dirac did not get the applause he may have hoped for. Heisenberg found that it was just the old theory in a new form and Pauli was of the same opinion. In a letter to Lise Meitner, Pauli rejected it in his usual

[100]Rosenfeld (1979m), p. 414.
[101]BSC, N. Bohr to W. Heisenberg, 13 Mar 1931.
[102]Rosenfeld (1979a). As for Bohr's idea that we are both spectators and actors, see for example Bohr (1999b). Favrholdt (1999a), pp. xxxviii–xli.
[103]Rosenfeld (1932a), pp. 86–89.
[104]Rosenfeld (1932a), p. 88.

tone: It "cannot be taken seriously; neither does it contain anything new, nor is it justified to speak of a 'theory'".[105]

Up to that time, Dirac had not entered into elaborations on measurability of observables in his quantum algebra, neither in his papers nor in his textbook, and he was criticized for that by Pauli.[106] Dirac's paper published in March 1932 can be said to contain attempts at such considerations, most likely prompted by his discussions with Bohr during these years, but his considerations still did not satisfy Pauli.[107] More importantly, Dirac's paper contained a rather sharp critique of the Pauli–Heisenberg quantum field theory. Dirac stressed that, contrary to the Jordan–Pauli–Heisenberg approach, particles and fields should have different ontological status. He treated the electromagnetic field as a plane wave and the electron as a particle. In addition, Dirac returned to the operationalistic position of Heisenberg's 1925 work: "that one should confine one's attention to observable quantities, and set up an algebraic scheme in which only these observable quantities appear".[108] He introduced what was later called the *interaction picture*, which was an alternative to both the Schrödinger picture and the Heisenberg picture, and derived a wave equation for one dimension which described two electrons and their mutual interaction and their interaction with an electromagnetic field. The wave satisfying this equation was a function of the electromagnetic field and was itself an operator. The effect of the interaction was hence contained in the state function on which the operators acted. The matrix elements of this state operator between field states could be interpreted as probability amplitudes for transitions from an incoming field state to an outgoing one. Dirac attributed great importance to the fact that these scattering probabilities were observables in his theory. In the Heisenberg–Pauli theory, it was possible to calculate the evolution of the field during a scattering process, but it was

[105]W. Pauli to L. Meitner, 29 May 1932, in Hermann *et al.* (1985), p. 114
[106]Enz (2002), p. 252. Pauli (1964).
[107]BSC, N. Bohr to W. Heisenberg, 4 Jun 1930. N. Bohr to P. A. M. Dirac, 29 Aug 1930. Dirac to Bohr, 30 Nov 1930. Bohr to Heisenberg, 8 Dec 1930. Bohr to Heisenberg, 7 May 1931. Pauli to Dirac, 11 Sep 1932, in Hermann *et al.* (1985), Vol. 2, pp. 115–116.
[108]Dirac (1932), p. 456.

not represented by observables. Echoing Landau and Peierls' critique the year before, Dirac therefore stated that "the Heisenberg–Pauli theory thus involves many quantities that are unconnected with the results of observations and that must be removed from considerations if one is to obtain a clear insight into the underlying physical relations".[109] However, he did not address the problem of the infinite self-energy of the electron, which was conceived as the gravest difficulty with the Heisenberg–Pauli theory, nor did his theory solve this problem.[110]

Rosenfeld may have been concerned that Dirac's new theory would interfere with the results of his and Bohr's on-going investigations of the measurability of the electromagnetic field. However, after Dirac's presentation of his theory, Rosenfeld quickly succeeded in proving that Dirac's new theory was mathematically equivalent to the Pauli–Heisenberg theory. Rosenfeld circulated his proof among the physicists who were present at the 1932 Copenhagen meeting to hear their opinion.[111] According to Rosenfeld, "with a roguish smile" Bohr's comment to his equivalence proof was that: "Here you see what this quantum electrodynamics is really useful for, namely to refute every attempt at changing it".[112] Thus, Bohr seemed to have seen Rosenfeld's equivalence proof as a sign of the strength of the Pauli–Heisenberg field theory. The ensuing correspondence particularly among Pauli, Heisenberg, and Rosenfeld gives the impression of a rather sarcastic attitude towards Dirac's work, the style of his work, as well as a somewhat taunting attitude towards his introvert personality. Although Pauli had been the first to recognize the problems with his and Heisenberg's quantum field theory, this did not mean that he thought Dirac had come up with the right solution to the problems. Pauli's response to Rosenfeld was shaped in a sequence of aphorisms. The sarcastic tone is characteristic of Pauli.

[109] Dirac (1932), p. 457. Darrigol (1986), pp. 249–250. Schweber (1994), p. 49.
[110] Bohr and Rosenfeld (1979a), p. 360. W. Pauli to L. Rosenfeld, 2 May 1932, in von Meyenn (1993), Vol. 3, pp. 748–749.
[111] Rosenfeld (1932c). Rosenfeld (1963), pp. 71–72.
[112] Rosenfeld (1963), pp. 71–72. Originally in Danish.

(1) A Dirac quantum electrodynamics does not exist.
(2) There is only one quantum electrodynamics.
(3) It is self-identical.

(3a) This property to be self-identical is not characteristic of quantum electrodynamics].

(4) The difficulty with the self-energy cannot be removed by arranging a formal hiding game with it.
(5) Dirac's hope of a quantum electrodynamics exists temporarily.
(6) The existence [of this hope] is essentially conditioned by [his] insufficient knowledge of the works hitherto about quantum electrodynamics (since 1926).
(7) Your (doubtless correct) elaboration over quantum electrodynamics has no objective interest since it is not permitted to presume such a general spread of dilettantism in this area, as is the case with Dirac.
(8) Indeed, do send your elaboration to Dirac and draw it up as written private lessons about quantum electrodynamics which you inform Dirac about.
(9) Since Dirac also in recent years has shown a certain talent as regards the "hole world" and the magnetic monopole. There is nevertheless a chance different from zero that he can acquire knowledge of quantum electrodynamics, if he takes an interest in this area.
(10) I would seriously consider the question whether you ought to demand a fee from Dirac, as is after all the custom with private lessons, in case you can help him to this knowledge.

...

P.S. To your elaboration Dirac will probably answer you that it does not at all concern his theory. In any case, that would be a correct answer since his theory does not at all exist, only his hope of such. And the content of your comments is certainly not identical to this hope.[113]

[113]W. Pauli to L. Rosenfeld, 2 May 1932, von Meyenn (1993), Vol. 3, pp. 748–749. Originally in German.

When Rosenfeld showed this letter to Bohr, he interpreted it as a clear sign that Pauli had regained his high spirits and was recovering from his depression![114] Pauli was equally blunt in a letter to Dirac of September 1932.[115] Heisenberg conjectured about Dirac's likely reaction to Rosenfeld's equivalence proof in an equally cutting tone:

> You have of course now furnished the very proof which I tried in vain and "alles bleibt beim Alten". But what will Dirac say now? He will say the gauge invariance should be treated differently; then everything will come out nicely... Pauli phrases it less politely "he finds the whole thing a hair-rising insolence from Dirac's side and a rude dilettantism". I also do not understand Dirac at all; he appears to have learnt nothing from the whole failed development of quantum electrodynamics. But in any case it is gratifying that you have now slaughtered the Dirac theory.[116]

Max Delbrück, on the other hand, defended Dirac and questioned the strength of Rosenfeld's proof.[117] Rosenfeld sent his proof to Dirac and added that he "would be very pleased if you would tell me what you think about it in as <u>many</u> words as you can manage", thus, referring in a friendly tone to Dirac's well-known scantiness and taciturnity in his relations with people.[118] Dirac responded:

> Thank you very much for the paper you sent me. I found it very interesting. The connection which you give between my new theory and the Heisenberg–Pauli theory is, of course,

---

[114]Oskar Klein Papers, NBA, N. Bohr to O. Klein, 18 May 1932.

[115]W. Pauli to P. A. M. Dirac, 11 Sep 1932, in Hermann *et al.* (1985), Vol. 2, p. 115. See also RP, W. Pauli to L. Rosenfeld, 2 May 1932, Printed in von Meyenn (1993), Vol. 3, pp. 748–749, and W. Pauli to L. Meitner, 29 May 1932, in Hermann *et al.* (1985), Vol. 2, p. 114.

[116]RP, W. Heisenberg to L. Rosenfeld 9 May 1932. Originally in German.

[117]BSC 18, M. Delbrück to L. Rosenfeld, 3 May 1932.

[118]Florida State University Library, Dirac Collection, L. Rosenfeld to P. A. M. Dirac, 30 Apr 1932 (emphasis in original). For Dirac's taciturnity, see for example Crowther (1970), p. 107. Kojevnikov (1996), p. 7: Dirac to his wife: "people say that Dirac is a great silencer, that it costs tremendous effort to get a word from him, and that he talks only to children under ten". See also Farmelo (2009).

> quite general and holds for any kind of field (not merely the Maxwell kind) in any number of dimensions. This is a very satisfactory state of affairs.
>
> It does not seem certain to me that the singularities will cause an equal amount of trouble in both theories... Could it not be so, that a mathematical process which is convergent for the one theory is divergent for the other? (Nur um zu lernen).
>
> I have studied the Heisenberg–Pauli theory again and find it difficult to understand why their formalism is invariant under a Lorentz transformation. I can follow all their arguments except the last sentence in Section 5 on page 180 of *Zeits. f. Phys.* Vol. 59. I should be very glad if you could explain the sentence a little more fully...
>
> The paper, I believe, contains no mentioning of singularities and was merely intended to give a theory that is more closely connected with the results of observation than the preceding ones. I think you ought to publish your work.[119]

Allegedly, Ehrenfest was of the opinion that Dirac was treated as the ugly duckling on this occasion in Copenhagen.[120]

Later that fall Dirac published a second paper with the Russian physicists Boris Podolsky and Vladimir A. Fock in which he maintained that his new quantum electrodynamics based on an interacting particle picture was an improved departure from that of Heisenberg–Pauli on *physical* grounds. He considered Rosenfeld's equivalence proof "obscure" and gave a new simplified proof of the equivalence between the two formalisms.[121]

Pauli eventually considered the Dirac–Fock–Podolsky paper "a great improvement" of Dirac's theory because of the mathematical elegance with which a manifestly covariant formulation of quantum electrodynamics was achieved. However, Pauli's enthusiasm for

[119] RP, "Correspondance particulière", P. A. M. Dirac to L. Rosenfeld, 6 May 1932.
[120] RP, "Supplement: History of Quantum Theory", L. Rosenfeld, Ehrenfest as I saw him (An autobiographical chapter), May 1971, manuscript.
[121] Dirac *et al.* (1932), p. 468.

Dirac's new theory came too late for the theory to be included in Pauli's influential handbook article in 1933.[122] Indeed, if Dirac had been treated like the ugly duckling in 1932, his new refined quantum electrodynamics was soon recognized as a beautiful swan by his previous opponents. In addition, by 1933 it was realized that the discovery of the positively charged electron by Anderson, confirmed by Blackett and Occhialini, vindicated Dirac's "hole" theory. In 1935, Rosenfeld summarized the development of quantum electrodynamics up to that point in the following way in his first paper in Danish, which appeared in a Festschrift dedicated to Bohr on his 50th birthday:

> As Heisenberg and Pauli first showed, the quantum theory of radiation can... be immediately developed into an invariant and logically closed "quantum electrodynamics", which brings together the quantized electromagnetic field in the most general sense and the matter field associated with the point model as well as their interaction. The proof of the invariance of the field equations and of the quantum conditions in their original form demands rather complicated considerations, but later Dirac found a very beautiful way of presenting the theory, in which its invariance appears immediately.[123]

The Bohr–Rosenfeld paper

Bohr did not pay much attention to Dirac's new theory from 1932.[124] In the Bohr–Rosenfeld paper, Dirac's new approach was briefly

[122] W. Pauli to P. A. M. Dirac, 13 May 1933, in Hermann *et al.* (1985), Vol. 2, pp. 160–161. Pauli to Dirac, 10 Apr 1933, in Hermann *et al.* (1985), Vol. 2, p. 157. Kragh (2005), pp. 137–139. Darrigol (1986), p. 250.

[123] Rosenfeld (1935), p. 114. Originally in Danish.

[124] BSC 18, N. Bohr to P. A. M. Dirac, 14 Nov 1932. N. Bohr to W. Pauli, 25 Jan 1933, in Hermann *et al.* (1985), Vol. 2, pp. 152–153. BSC 19, N. Bohr to R. H. Fowler, 28 Oct 1932. At the Solvay meeting in October 1933, Bohr only discussed measurements in relation to the Pauli–Heisenberg field theory and Dirac's relativistic theory of the electron. Bohr MSS, "Solvay Conference" Oct 1930 (should be 1933) Morning October 28, microfilm 12, AHQP.

mentioned, but only by stating that it had not solved the problem of the infinite self-energy of the elementary particles. In addition, Bohr and Rosenfeld did not need an interaction theory of the electron and the electromagnetic field since for their investigations of the free electromagnetic field it sufficed to use extended charged test bodies that could be described classically. Therefore, they restricted themselves to the pure field theory.[125]

It was during the period from the fall of 1932 through 1933 that Bohr and Rosenfeld worked most intensely on their joint paper. The long paper was submitted to the Royal Danish Academy of Sciences and Letters on 2 December 1932. However, subsequently it went through about fourteen proofs before it was finally published at the end of the year 1933. In the process, Rosenfeld discussed the topic with his close friend Solomon. Solomon derived an uncertainty relation for the non-commuting field components that differed from Landau and Peierls' derivation. He suggested on that background in a paper that was published in July 1933 that it should be possible to find consistency between the predictions of the Heisenberg–Pauli field theory and idealized measurements; this was then confirmed by Bohr and Rosenfeld.[126] In November 1933, Rosenfeld gave Chandrasekhar a report of what took place at the Solvay Congress that year — the first Solvay congress Rosenfeld attended officially — ending the letter: "As to the work on field measurements I am sorry to say that it is still involved in a terribly slow semi-converging process".[127] The topic of the 1933 Solvay Congress was the structure and properties of atomic nuclei reflecting that the new research area of nuclear physics was becoming the main agenda of theoretical physics. During the discussion, however, Bohr found opportunity to criticize Landau and Peierls' ideas on measurement and reflected at length on the concept of measurement as a tool for investigating the consistency of a

[125]Bohr and Rosenfeld (1979a), p. 360.

[126]Solomon (1933), pp. 383–386. I want to thank Jean-Philippe Martinez for drawing my attention to this paper quite recently. Solomon referred to conversations with Rosenfeld about Landau and Peierls' paper, p. 386 footnote 2, and Solomon's paper was referred to in Bohr and Rosenfeld's paper, Bohr and Rosenfeld (1979a), p. 400.

[127]RP, copy of letter from L. Rosenfeld to S. Chandrasekhar, 19 Nov 1933. Original in CP.

theory: The "whole problem is a very subtle one; it lies in the word measurement itself. It is important to see the danger one can get into by using such a word... The whole thing is a problem of finding a consistent way to speak about the actual character of the theory... One tries to see that there is a complete consistency between the positive predictions and the possibility of testing these predictions".[128] Bohr repeated his view that "we can hardly, from considerations of measurements, find arguments against the theory of the type of Dirac, and of the field theory. The importance of the discussion on measurement is more that we get to know a number of things about the character of the restrictions in the use of classical pictures".[129] Hence, Bohr's message to Landau and Peierls was that the kind of logical investigation undertaken was independent of the prediction of negative energies and the infinite self-energy of the electron in Dirac's theory; Landau and Peierls imagined that the measurability problem was depending on the difficulties with the infinite self-energy of the electron. However, according to Bohr, the rational investigation of the theory could not authorize changes in the theory itself and Bohr and Rosenfeld's analysis did not, nor was it meant to, solve the problems with the infinities. In fact, contrary to Landau and Peierls, Bohr had great faith in the quantum field theory and was not inclined to question its consistency as long as the formalism obeyed the necessary relativistic invariance and rested on arguments of correspondence with classical electromagnetic theory; nor did he consider measurability examinations a possible means to disprove the formalism, nor did he think it was possible to find a formalism free of singularities.[130] According to Bohr, his imaginary measuring

[128]Bohr MSS, "Solvay Conference" Oct 1930 (should be 1933) Morning October 28, microfilm 12, AHQP.

[129]*Ibid.*

[130]Heisenberg to Pauli, 12 Mar 1931, in Hermann *et al.* (1985), Vol. 2, pp. 66–57. Bohr MSS, "Solvay Conference" Oct 1930 (should be 1933) Morning October 28, microfilm 12, AHQP. See also RP, "Supplement: History of Quantum Theory", L. Rosenfeld, Ehrenfest as I saw him (An autobiographical chapter), May 1971, manuscript. N. Bohr to W. Pauli, 25 Jan 1933, in Hermann *et al.* (1985), Vol. 2, pp. 152–153. Bohr to Pauli, 14 Feb 1934, in Hermann *et al.* (1985), Vol. 2, pp. 285–286. N. Bohr and L. Rosenfeld (1979a), p. 359. Rosenfeld (1979m), p. 430.

processes could be used *only* for providing a logical justification of the theory's statements, not for shooting down, or arguing for modifications, of the formalism, for example in terms of what was later denoted "hidden parameters".[131] As Bohr wrote in hindsight to Pauli in January 1933, "it is impossible to disprove a relativistically invariant formalism by simple relativity arguments".[132] That being said, had Bohr and Rosenfeld's examination of the measurability of field quantities *not* resulted in logical consistency with the predictions of quantum electrodynamics, one would suspect that it would have had some consequences for the theory.

Bohr and Rosenfeld's idealized experiment involved exceptionally complicated measuring devices with neutralizing bodies, springs compensating for the acceleration of the test charges and lever mechanisms too hard to visualize. They showed that there is nothing wrong with the logical consistency of the quantization of fields that had been done by Dirac, Pauli, and Heisenberg; the uncertainty rules for the electromagnetic field found when quantizing Maxwell's equations correspond well with the "real" ones. The zero-field fluctuations made it clear that the causal connection in the classical theory between the field and its sources was lost in quantum theory. However, the existence of field fluctuations did not prevent the measurability of field quantities as showed by Bohr and Rosenfeld. The implications of Bohr and Rosenfeld's paper for the interpretation of the free electromagnetic field in a quantum frame were that fields should not, as in classical electrodynamics, be thought of as probed by point charges such as electrons, but rather by extended test bodies. And in continuation hereof, field quantities should no longer be represented by point functions but by functions of averaged space–time regions.[133]

Because Bohr did not correspond with Pauli from the late spring of 1931 until January 1933, Pauli was excluded from following the

[131] See also Bokulich and Bokulich (2005).

[132] N. Bohr to W. Pauli, 25 Jan 1933, in Hermann *et al.* (1985), Vol. 2, pp. 152–153. See also Bohr to Pauli, 14 Feb 1934, in Hermann *et al.* (1985), Vol. 2, pp. 285–286. Pais (1991), p. 361.

[133] Bohr and Rosenfeld (1979a). Bohr (1996c), p. 198. Rosenfeld (1979m).

development of the Bohr–Rosenfeld paper closely.[134] Rosenfeld and Pauli did not discuss the topic either. In his *Handbuch der Physik* article, Pauli introduced Landau and Peierls' ideas about repeatability and predictability of measurements which led him to his ideas of measurements of the first and second kind. Pauli also referred to von Neumann's book, *Mathematische Grundlagen der Quantenmechanik* (1932) for an axiomatic elucidation of quantum measurements.[135] By contrast, Bohr never recognized von Neumann's axiomatic approach which nevertheless came to serve as a starting point for all the later discussions of *the* so-called measurement problem. Bohr never showed any interest in these discussions.[136] A letter from Pauli to Peierls of May 1933 indicates that after Bohr had resumed contact with Pauli, Pauli began to look at the Landau–Peierls paper with new eyes, realizing for example that "for the field measurement the negative energy states play no role [because the exchange relations] do not contain the charge, mass, or dimensions of the probe body".[137] However, only in early 1934, probably after he had studied the Bohr–Rosenfeld paper and after it had been discussed at the Solvay Congress in October 1933, did Pauli arrive at a "complete agreement with all of Bohr's assertions about *measurements*".[138] From then on he became a source of great encouragement to Bohr and Rosenfeld in their continued work on measurability.[139] Pauli had returned to the fold.

It ought to be noted that unlike so many other controversies in the history of science, the controversies that took place in the small community of theoretical physicists around Bohr in this period were not decisive for the mutual respect and friendship between

[134]This is evident from a letter from W. Pauli to W. Heisenberg, 18 Jan 1933, in Hermann *et al.* (1985), Vol. 2, pp. 148–151. See also Kalckar (1996), p. 11.

[135]Pauli (1980), p. 71. See also Jammer (1974), p. 487.

[136]RP, "Epistemology 1959–1964", Rosenfeld, Report on Louis de Broglie's *La théorie de la mesure en mécanique ondulatoire*, requested by Pergamon Press Limited, enclosed in correspondence between Miss S. Stratford-Lawrence and Rosenfeld, 26 Jan and 3 Feb 1959. See also Rosenfeld (1979x), p. 537 and Bokulich and Bokulich (2005), pp. 355–357.

[137]W. Pauli to R. Peierls, 22 May 1933, in Hermann *et al.* (1985), Vol. 2, pp. 163–165, on p. 165. English translation quoted from Miller (1994), p. 42.

[138]W. Pauli to W. Heisenberg, 23 Feb 1934, in Hermann *et al.* (1985), Vol. 2, p. 300. (Emphasis in original.)

[139]Bohr and Rosenfeld (1979b), p. 412. Rosenfeld (1979m), p. 424.

them in the long run. Bohr immediately developed warm and close friendships with both Landau and Peierls, for example, which were only strengthened with time.[140] Niels and his son Aage Bohr stayed with the Peierls when in Los Alamos during the war, and Peierls later served as the editor of volume 9 of the *Niels Bohr Collected Works* about nuclear physics. Bohr stayed with the Landaus when visiting Russia in 1961. Bohr excused Landau and Peierls in a letter to Gamow in 1933: "I hope it will be a comfort to Landau and Peierls that the stupidities they have committed in this respect are no worse than those which we all, including Heisenberg and Pauli, have been guilty of in this controversial subject".[141] In proposing Landau for the Nobel Prize in 1962, Bohr referred to Landau and Peierls' work in the following way: "By illuminating the known paradoxes of the radiation phenomena in a new way, Landau and Peierls' work gave rise to a resumed examination and clarification of the problems with measuring electromagnetic field quantities".[142] However, Landau and Peierls apparently never reconciled themselves with Bohr's "clarification" of their paper. When their diverging ideas were discussed at the Solvay meeting in 1933, Peierls, the only one of the two to be present, seems to have accepted that their requirements of repeatability and predictability had been asking too much of a measurement in Bohr's sense of the word.[143] However, Peierls did not acknowledge Bohr and Rosenfeld's sharp distinction between *idealized* thought experiments used to examine the measurement process logically in a classical setting in order to get a better *understanding* of the statements of quantum theory, and *real* experiments used to test the quantitative predictions of the theory in the laboratory.[144] In comparing Bohr and Rosenfeld's analysis with how an experimental

[140]See for example BSC, N. Bohr to L. D. Landau, 6 Dec 1933. Bohr to Landau, 25 Apr 1936.

[141]N. Bohr to G. Gamow, 21 Jan 1933, in *BCW*, Vol. 9, p. 571.

[142]BSC, N. Bohr, A. Bohr, B. Mottelson, C. Møller, and L. Rosenfeld to Kungliga Vetenskapsakademiens Nobelkommitté, 30 Jan 1962, carbon copy.

[143]RP, "Supplement: History of Quantum Theory", L. Rosenfeld, Ehrenfest as I saw him (An autobiographical chapter), May 1971, manuscript. Bohr MSS, "Solvay Conference" Oct 1930 (should be 1933) Morning October 28, microfilm 12, AHQP.

[144]Bohr and Rosenfeld (1979b), p. 411.

physicist would actually go about the problem, he later insisted that the operations involved in Bohr and Rosenfeld's measurement "look quite unlike any kind of measurement that an experimentalist would design".[145] He continued, "Bohr and Rosenfeld's arguments are based on the fundamental laws of the quantum theory of the electromagnetic field, without regard to the sort of small test bodies or particles the physicist actually has at his disposal".[146] Thus, Peierls did not accept Bohr and Rosenfeld's interpretation of the field as being probed by extended test charges that could be described classically, because test charges, Peierls argued, are particles, typically, that should be described by quantum theory. In the late 1940s, Rosenfeld had a collaborator in Manchester, Ernesto Corinaldesi, working on the calculations behind the considerations about the measurability of electromagnetic fields and charge–current components. When Rosenfeld and Corinaldesi discussed Corinaldesi's work with Peierls, Corinaldesi seems almost to have been taken hostage as the old disagreements resurfaced between Rosenfeld and Peierls. Peierls still maintained that field fluctuations were responsible for the lack of the measurability of field quantities such as suggested by Landau and himself in 1931.[147]

Landau's objections went along a somewhat different path although intimately connected with Peierls'. Rosenfeld later told the following anecdote about his and Bohr's confrontation with Landau when visiting Leningrad in May 1934 (about this Russian tour see the next chapter):

> [i]t was agreed that we [Bohr and Rosenfeld] should give a joint talk about our work for the physicists in Leningrad. The task was divided so that I should begin with the formal foundation, write the equations of the commutation relations, etc.,

[145] R. Peierls (1963), p. 36. Peierls (1985a), pp. 66–67. Peierls (1980), p. viii. Peierls (1985b), pp. 228–229. Other physicists shared this view, see for example Wigner (1963), p. 14.

[146] Peierls (1963), p. 36.

[147] RP, "Correspondance particulière: Corinaldesi" and "Manuscripts", Folder containing letters related to the second Bohr–Rosenfeld paper. E. Corinaldesi to L. Rosenfeld, 13 Jan 1952. Rosenfeld to E. Corinaldesi, 18 Jan 1952. E. Corinaldesi (1953).

> and Bohr should then "comment", as he expressed it, upon the epistemological aspects. But we had not anticipated the circumstance that Landau was present. Everything went well until I arrived at the commutations and remarked that since they were expressed by singular functions, we had to average the field quantities over finite space–time areas in order to arrive at physically meaningful finite quantities. Obviously I had not stressed this strongly enough, in Bohr's opinion. He jumped up and turned to Landau with the words: "Do you understand, Landau, this is the crucial point". Then Landau answered equally excitedly: "No, I completely disagree with you in this: since this area can be as small as one wishes, one may just as well talk about a point, it cannot make any difference". Then of course Bohr could not help answering: "No, that makes the whole difference: if one follows the method of Peierls and yourself, one obtains a radiation which can only be made to disappear, if one assumes a finite area". An animated discussion unfolded without anybody in the audience being able to understand what it was about. After this quarrel Bohr turned to me with a very lovable smile and said: "I didn't mean to interrupt". I was allowed to go on a little further, but it was not long before another crucial point arose and then the same game started all over again. Landau had definitely to be convinced about this second point before one could get any further, and even today I do not know what the Russian audience thought about our conclusions.[148]

In the middle of the 1950s, Landau followed up on his own ideas in a series of papers with his collaborators, A. A. Abrikosov, I. M. Khalatnikov, and I. Ya. Pomeranchuk. In these studies the authors maintained that the point interaction, for example between a point charge used to probe a field, and the field, was possible as the limit of "smeared" interaction when the smearing radius tends to zero.

[148] Rosenfeld (1963), pp. 72–73.

This made it possible to deal directly with finite expressions of the quantities in question.[149]

Among the majority of physicists, however, Bohr and Rosenfeld's paper was seen as having proved the consistency of the uncertainty relations of the electromagnetic field quantities with the possibility of measuring these quantities in an idealized measurement.[150] The paper had therefore secured the foundations of quantum electrodynamics, resulting in a revival of the physicists' faith in the theory after it had been shattered by Landau and Peierls' paper. This understanding of the Bohr–Rosenfeld paper may reflect that the theoretical physicists craved for a way out of the crisis they found themselves in. Bohr and Rosenfeld had assured that the elaboration of quantum electrodynamics could be continued without anybody having to worry about its foundations. This understanding of the Bohr–Rosenfeld paper may also explain why the second quantum generation to a large extent did not occupy themselves with epistemology; Bohr and Rosenfeld had said what needed to be said about that topic.[151]

David Kaiser has suggested that Bohr and Rosenfeld may have come out as winners of the "small war" with Landau and Peierls over the interpretation of measurement in quantum electrodynamics, but in the long run Bohr and Rosenfeld lost the game about what theoretical physics should be about. Instead, the new generation of physicists set that agenda, focusing on mathematical techniques and calculations devoid of deeper epistemological reflection. Quantum theory was to a large extent considered merely as a set of rules which prescribe the outcomes of experiments. Because of the political development in Europe the 1930s, this further development took place predominantly in an American context.[152] Rosenfeld deplored the development in America. Late in life he would complain

---

[149]Lifshitz (1989), pp. 16–17. Landau *et al.* (1956). See also Landau (1960), Landau (1955).
[150]See, for example, Heitler (1936), p. 81.
[151]Pais (1991), p. 362.
[152]Kaiser (2007a). Kaiser (2007b). Schweber (1986). See also Kaiser (2005). Kragh (1999), pp. 249–256.

"about the incredibly low level of epistemological thinking among American physicists".[153]

In Russia Landau too distanced himself from fundamental problems from the 1930s onwards. This is evident from textbooks he wrote with Evgenii Lifshitz for example. The reason for this may be for Landau to avoid getting into philosophical disputes with official Party ideologists. Landau advised his students to talk only in a phenomenological way about physics.[154] Bohr's Russian biographer Daniil Danin commemorated a conversation with Landau about this subject. According to Danin, Landau would say "in his swift style: 'Stop talking about waves and particles! There are no waves and no particles. It is a cheat of the working class!' "[155] As regards the critique of Bohr's complementarity interpretation which arose anew in the Soviet Union after the Second World War based on an ideological motivation and Rosenfeld's reaction to it, see Chapter 6.

The EPR paper

The Bohr–Rosenfeld paper did not attract much attention outside the narrow circle of theoretical physicists. A year and a half later, however, Bohr faced a new challenge, the publication of the so-called EPR-paper asking "Can Quantum Mechanical Description of Physical Reality be Considered Complete?" Contrary to the small war with Landau and Peierls about the interpretation of quantum electrodynamics, the spectacular debate between the two giants Bohr and Einstein about the interpretation of non-relativistic quantum mechanics attracted from the start huge attention among physicists, philosophers of science, and from the public.[156]

Einstein seems to have finally acknowledged the consistency of the uncertainty relations predicted by quantum theory when paired with analysis of idealized measurements of the quantities involved. During the Solvay conference in 1927, Bohr and Einstein

[153]RP, "Supplement B", L. Rosenfeld to E. Breitenberger, 29 Nov 1972.
[154]Hall (2005), pp. 269–271.
[155]RP, "Supplement D", D. Danin to L. Rosenfeld, 22 Dec 1971 and 22 Dec 1972.
[156]For the philosophers' interest, see Chapter 3. *New York Times* (1935).

had discussed the consistency of Heisenberg's uncertainty relations for momentum and position, and in 1930 they had debated the consistency of the uncertainty relations for the time parameter and energy. In addition, Bohr and Rosenfeld had proved Landau and Peierls wrong in contending that the available measuring procedures for field measurements are more restricted than is assumed by the uncertainty relations for the field quantities in the quantum electrodynamical formalism. Einstein now challenged the completeness of quantum mechanics, by suggesting that the available measuring procedures seem to be *less* restricted than assumed in the quantum mechanical formalism.[157]

In the spring of 1933, prior to his emigration to the US, Einstein had attended a lecture by Rosenfeld on his work with Bohr on quantum electrodynamics at Brussels. According to Rosenfeld, Einstein followed the arguments with great attention but had no comments specifically regarding this topic. Instead, Einstein approached Rosenfeld with an epistemological paradox similar to the one later presented in the EPR paper:

> "What would you say of the following situation?" he asked me. "Suppose two particles are set in motion towards each other with the same, very large, momentum, and that they interact with each other for a very short time when they pass at known positions. Consider now an observer who gets hold of one of the particles, far away from the region of interaction, and measures its momentum; then, from the conditions of the experiment, he will obviously be able to deduce the momentum of the other particle. If, however, he chooses to measure the position of the first particle, he will be able to tell where the other particle is. This is a perfectly correct and straightforward deduction from the principles of quantum mechanics; but is it not very paradoxical? How can the final state of the second particle be influenced

[157]Petersen (1968), pp. 167–168.

> by a measurement performed on the first, after all physical interaction has ceased between them?"[158]

Rosenfeld's recollections are of course subject to the limits of memory. However, they seem to be supported by Don Howard's study on the substance of Einstein's worries about quantum mechanics prior to the EPR-paper in 1935; it was the paradoxical non-separability of identical particles that led Einstein to arguing that quantum mechanics is incomplete.[159] The EPR paper appeared in the *Physical Review* in May 1935.[160] It was received by Pauli with the following comments to Heisenberg: "*Einstein* has once again made a public statement about quantum mechanics... As is well-known, that is a disaster whenever it happens".[161] The paper and the formulation of Bohr's response interrupted for a while Bohr and Rosenfeld's continued work on the measurability of charge and current distributions, the sources of the electromagnetic field. Rosenfeld later famously recollected the incident when Bohr received the news of the paper:

> This onslaught came down upon us as a bolt from the blue. Its effect on Bohr was remarkable. We were then in the midst of groping attempts at exploring the implications of the fluctuations of charge and current distributions, which presented us with riddles of a kind we had not met in electrodynamics. A new worry could not come at a less propitious time. Yet, as soon as Bohr had heard my report of Einstein's argument, everything else was abandoned: we had to clear up such a misunderstanding at once. We should reply by taking up the same example and showing the right way to speak about it. In great excitement, Bohr immediately started dictating to me the outline of such a reply. Very soon, however, he became hesitant: "No, this won't do, we must try all over again... we must make it quite clear...". So it went on for

[158] Rosenfeld (1967), pp. 127–128.
[159] Howard (1990), pp. 98–105.
[160] Einstein *et al.* (1935).
[161] W. Pauli to W. Heisenberg, 15 Jun 1935, in Hermann *et al.* (1985), Vol. 2, p. 402.

> a while, with growing wonder at the unexpected subtlety of the argument. Now and then, he would turn to me: "What *can* they mean? Do *you* understand it?" There would follow some inconclusive exegesis. Clearly, we were farther from the mark than we first thought. Eventually, he broke off with the familiar remark that he "must sleep on it". The next morning he at once took up the dictation again, and I was struck by a change in the tone of the sentences: there was no trace in them of the previous day's sharp expressions of dissent. As I pointed out to him that he seemed to take a milder view of the case, he smiled: "That's a sign", he said, "that we are beginning to understand the problem". And indeed, the real work now began in earnest: day after day, week after week, the whole argument was patiently scrutinized with the help of simpler and more transparent examples. Einstein's problem was reshaped and its solution reformulated with such precision and clarity that the weakness in the critics' reasoning became evident, and their whole argumentation, for all its false brilliance, fell to pieces. "They do it smartly", Bohr commented, "but what counts is to do it right".[162]

In July, Rosenfeld reported to Chandrasekhar "that Bohr is rather exhausted after that performance".[163] In early September, Bohr's reply was still not submitted. Rosenfeld informed Chandrasekhar about the "series of most regrettable accidents" causing this delay, *viz.*, first Bohr's excitement about the EPR paper and later illness. "We are now busy with the correction of the proofs of his answer to Einstein, which is also to appear in Physical Review; as usual, this correction amounts to re-writing half of the article".[164] Bohr's reply finally came out in October. For Bohr, the reply was an opportunity to promote anew his complementarity view of quantum phenomena.

---

[162]Rosenfeld (1967), pp. 127–129.

[163]RP, copy of letter from L. Rosenfeld to S. Chandrasekhar, 5 Jul 1935. Original in CP.

[164]RP, copy of letter from L. Rosenfeld to S. Chandrasekhar, 8 Sep 1935. Original in CP.

He did not specifically address "the interesting example" of non-separability of quantum systems which have interacted in the past.[165] Instead Bohr gave a critical assessment of the criterion of reality put forward by EPR: "*If, without in any way disturbing a system, we can predict with certainty (i.e. with probability equal to unity) the value of a physical quantity, then there exists an element of physical reality corresponding to this physical quantity*".[166] Bohr responded to this criterion by a repetition of his characteristic considerations of measurability in various examples of measuring arrangements. Thus, he repeated his views on the correlations between measurement apparatus and the quantum system and the correlations between canonically conjugate variables. In formulating this answer to the EPR paper, Bohr took the first step towards a refinement of his terminology by introducing the concept of a phenomenon which included both the quantum object studied and the method of its observation.[167]

During the 1930s, Bohr was very active in communicating the epistemological lesson of quantum theory to a broader audience. The next chapter focuses among other things on Rosenfeld's role in the popularization of complementarity in the 1930s.

[165]Bohr (1996a), p. 292[696].
[166]Einstein *et al.* (1935), p. 777. (Italics in original).
[167]Petersen (1968), p. 168.

Chapter 3
Physics, Philosophy, and Politics in the 1930s

The aim of this chapter is to illuminate Rosenfeld's maturing as an intellectual leftist connecting science with socialism during the 1930s. Meanwhile, there is only sparse testimony by Rosenfeld concerning the political development in that period and his views on the Soviet Union, the United States, fascism, and Nazism. Therefore the social and political contexts as Rosenfeld may have experienced them come to the fore, whereas Rosenfeld remains in the background. Since Rosenfeld collaborated very closely with Bohr in the 1930s, it is also relevant to treat some of Bohr's activities in more detail. For example, Bohr and Rosenfeld visited the Soviet Union together in 1934, but there are only few sources to document Rosenfeld's experiences during this trip, whereas Bohr's activities are documented in more detail.

In the first chapter it was described how Rosenfeld's interest in socialism was aroused in Paris in 1926–1927 and how his political interests were also pursued while he was in Göttingen and Zurich. Rosenfeld went further than most of his physics colleagues in pursuing his growing sympathy for socialism in that he, like his close friend Solomon, also became interested in Marxist thought, that is, dialectical and historical materialism. Contrary to Solomon, however, Rosenfeld never joined the Communist Party and that enabled him, in principle, to act politically as an independent intellectual of Marxist persuasion. In practice, however, this feature in Rosenfeld only became visible after the Second World War (Chapter 5). During the 1930s, Rosenfeld gradually became more aware of what socialism was about and, more importantly, how it related to his science. The political role, so to speak, he took upon himself in the cause of science and socialism came to expression in writings on history of science, in which he reflected on the mutual dependence of science and society, in a vigorous engagement in popularizing complementarity, and in fighting idealistic views on science. By the second half of the 1930s, he began to ask himself how complementarity and dialectical materialism could be merged. Along the way, he became inextricably involved in an extraordinarily wide intellectual network, a kind of intellectual Popular Front of the 1930s, in which, as described by historian Patrick Petitjean,

> it was difficult to dissociate academic, political, institutional, ideological, and even private, levels. Politics were never far away in professional relations, and vice versa. The networks were shaped by a continuity of commitments, from science to politics. The professional level was always present, and was often the starting point for an exchange. But these networks were never a mere extension of professional links. Some commitments were closely linked to the scientist's capacity: the struggle for the organization, planning and funding of science, the support of progressive uses of science for the welfare of humankind, the defense of science against Nazi and obscurantist attacks, the refutation of the abuses of science, scientific trade unionism, the Marxist history of science, the public understanding of science, etc. Other commitments were more in line with common features of intellectuals: pacifism, antifascism, relief for refugees, and fascination with the USSR.[1]

Science, moreover, was conceived as genuinely international. Many of these scientists and scholars ended up being persecuted for racial or political reasons.

One of Rosenfeld's acquaintances from the early 1930s was the Danish polymath Piet Hein who studied physics at Bohr's institute. Later he became a famous engineer, designer, and poet in both

---

[1]Petitjean (2008b), p. 254. See also Petitjean (2008a).

Denmark and in America.[2] According to Rosenfeld, Piet Hein began his studies at the institute in 1927: "It was his inexhaustible intellectual curiosity that brought him to the lion's den. He was promptly spotted by Gamow who saw in him a congenial companion, and I remember that when Ehrenfest came to visit Bohr at the time, he also was fascinated by this shy but remarkably clever young student".[3] Piet Hein had conceived an ingenious box with two dice illustrating the idea of complementarity, which Bohr was particularly fond of. It was described in a paper in the Copenhagen Student Magazine *Studenterbladet* in which Piet Hein introduced Bohr's epistemology of atomic physics, thus contributing to popularizing complementarity in the Danish intellectual milieu.[4]

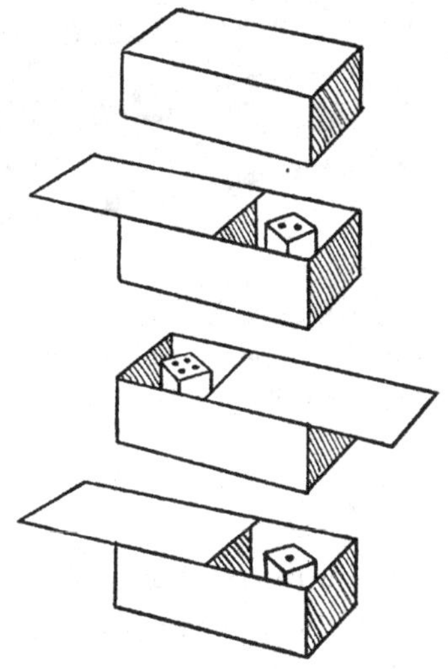

Piet Hein's drawing of the box with two dice meant to illustrate the complimentarity principle.

Piet Hein's drawing of the box with two dice.[5] Rosenfeld described the box: It "has two compartments, each of which contains

[2]Wamberg (1980).

[3]RP, "Supplement 1971–1974", L. Rosenfeld to D. Danin, 16 Jan 1973. RP, "Supplement: History of Quantum Theory", L. Rosenfeld, Ehrenfest as I saw him (An autobiographical chapter), May 1971, manuscript.

[4]Hein (1932).

[5]*Ibid.*, 1(8), p. 5.

a die. A sliding lid can at will uncover one of the compartments, while covering the other, thus allowing a reading of either die. However, when the lid is pushed from one position to the other, it presses a spring which shakes the box: thus, any attempt to read one of the dice destroys previous knowledge of the reading of the other".[6]

Participants in the Copenhagen Conference 1932.
Courtesy of The Niels Bohr Archive.

Trotsky in Copenhagen

As an active member of the university student council Piet Hein was directly involved with Leo Trotsky's spectacular visit to Copenhagen in late November 1932. Trotsky gave a public speech before an audience of about two thousand people, which took place on 27 November in a sports building very close to the Institute for Theoretical Physics.[7] Rosenfeld was in Copenhagen at the time, but it is unclear whether he attended Trotsky's speech.[8] What is clear, however, is that in 1946 Rosenfeld referred to an interview with Trotsky about

[6]RP, "Supplement: History of Quantum Theory", L. Rosenfeld, Ehrenfest as I saw him (An autobiographical chapter), May 1971, manuscript.
[7]Larsen (1986), pp. 146–148.
[8]The guest book of the institute, NBA.

students' relations with the revolutionary movement conducted by members of the student council during Trotsky's stay. The interview was printed in the Danish student organization's periodical *Studenterbladet* (the Students' Magazine).[9] Therefore there is reason to believe that Trotsky's visit made a lasting impression on Rosenfeld, and that his fascination with Trotsky as a historical figure and his affinity with Trotsky's political position stem from this event (see Chapter 5).

After Lenin's death in 1924, a bitter and ruthless fight broke out between the two crown princes of the Soviet Union, Stalin and Trotsky. When Stalin eventually won the power, the intellectual Left in the rest of the world was divided for and against him. The Jewish war hero and leader of the Red Army, Trotsky, opposed Stalin's rise and policies, for which he was expelled from the Communist Party and exiled in 1929. With remarkable success, communists were now portraying Trotsky as a richly facetted picture of the devil, the odious ideological center for all opposition towards the Soviet Union: public enemy number one. Stalin eliminated Trotsky from the official history of the Russian Revolution and Trotsky's name became a term of abuse. His ideas were marginalized and his political significance made negligible. This development coincided with a sharp leftward and self-isolating turn in the Comintern (the *Communist International* founded in Moscow in 1919) in which non-communist socialist parties such as the Social Democrats were condemned as "socialist fascists".[10]

It was in fact the Danish Social Democratic students who invited Trotsky to Copenhagen in late November 1932 to speak about the First Five Year Plan and the Russian Revolution in connection with its 15th anniversary. As for the Danish political background in the early 1930s, Rosenfeld gave a fairly accurate account of it in 1936.

> The economic development of Denmark, determined as it is by its geographical position and the limited resources of its territory, has given the social structure of the Danish nation a

[9] Neergaard (1932).
[10] Hobsbawm (1995), p. 74. Thing (1993), Vol. 1, pp. 399–401.

> quite accentuated character of progressive middle-class democracy. This character is accurately reflected in the composition of the political parties — the Social Democratic Party and the Radical Left — which have undertaken the government of the country without interruption since 1929 and with ever growing success. The Social Democratic Party not only groups the industrial and agricultural proletariat, but also the great number of craft industries, small commerce, and the lower middle class; the Radical Left represents the small farmers and a good part of the intellectual middle class. It is clear that the aspirations of such a grouping of interests are not specifically "socialist" in the proper sense of this term, but must offer, rather, a perfect image of the democratic ideal of a society, which does not very clearly reflect the contrasts between the various classes of which it is composed.[11]

As the paragraph suggests, the name of the Danish party, the Radical Left, is misleading; it should not be understood as socialist. Its orientation is liberal in both a cultural and economic sense.

Trotsky's speech was his first in the West since 1914, and it became the last of his greater speeches. He had been in exile since 1929 and at the time lived on a small Turkish island, Principo. He was invited in his capacity as "the greatest of the living figures of the Russian Revolution", and as such as a "historical figure", but also in his capacity as an able historian.[12] According to Perry Anderson, Trotsky's *The History of the Russian Revolution* (1930) remained for a long time "unique in the literature of historical materialism".[13] Needless to say, Trotsky's visit was a sensation in Denmark. It mobilized a strong opposition among the Danish communists and was also criticized by a member of the Royal family, Prince Aage, because he saw Trotsky as responsible for the death of the Russian tsar family. The visit was also strongly criticized in Germany and in the Soviet Union.[14]

[11] Rosenfeld (1936), p. 135. Originally in French.
[12] *Social-Demokraten* (1932).
[13] Trotsky (1932b). Anderson (1980), p. 154.
[14] Larsen (1986), pp. 150–151.

Trotsky was given permission by the popular, long-standing, and first-ever Social Democratic Prime Minister, Thorvald Stauning, to stay in Denmark for one week on the condition that he did not talk about politics! In his speech Trotsky mentioned his idea about "the permanent revolution" which differed radically from Stalin's idea of "socialism in one country".[15] Trotsky predicted that the Russian Revolution was only "the first stage of the socialist world revolution", and he continued with a quote from one of his works from 1905, which he considered as relevant in 1932.

> The present productive forces have long outgrown their national limits. A socialist society is not feasible within national boundaries. Significant as the economic successes of an isolated workers' state may be, the programme of "Socialism in one country" is petty-bourgeois utopia. Only a European and then a world federation of socialist republics can be the real arena for a harmonious socialist society.[16]

Although Trotsky insisted in his speech that he was a communist, communists demonstrated outside the sports hall accusing him of treachery. In the Danish communist journal, *Arbejderbladet* (the Workers' Magazine), he was called a social fascist along with the Social Democrats, a "watchdog of the bourgeoisie".[17]

Piet Hein took part in a meeting between Trotsky and student representatives, where he drew a characteristic sketch of Trotsky for the interview in the *Studenterbladet.* The audience took place in a private apartment where Trotsky stayed during his visit in Copenhagen.[18] Trotsky would have liked to obtain permanent asylum in Denmark, but at the end of his stay he was escorted by the police to his ship in the town of Esbjerg; the Danish authorities would not risk that this phantom of the Russian Revolution stayed longer than planned.[19]

---

[15]Trotsky (1932a).
[16]*Ibid.*, p. 10.
[17]*Arbejderbladet,* 19 Nov 1932, Quoted from Nygaard (2005).
[18]Neergaard (1932).
[19]*Ibid.*, p. 8. Larsen (1986), pp. 145–156.

Leo Trotsky speaking in Copenhagen. Robert Capa (c) 2001 by Cornell Capa/Magnum Photos.

Drawing of Trotsky by Piet Hein.[20]

Later Trotsky went into exile in Mexico. However, during the purges in Russia in the second half of the 1930s a price was put on Trotsky's head. In 1940, he was assassinated in his home.

[20]Drawing accompanying the article Neergaard (1932), p. 4.

Yvonne Cambresier

Soon after Trotsky's Copenhagen visit, another important event took place in Rosenfeld's life. In December 1932, Rosenfeld got engaged to the Belgian astrophysicist Yvonne Cambresier.[21] When in Liège, Rosenfeld frequently assisted at seminars organized by his friend, the professor of astrophysics Pol Swings at the Cointe Observatory, at Liège University. Here Yvonne was writing her PhD thesis on molecular abundances in stellar atmospheres.[22] Rosenfeld announced the news about their engagement in the following understated way to his close friend, Chandrasekhar, who was then in Copenhagen: "As to myself the only noteworthy detail in my new situation is that I am engaged to a girl student whom I met here some months ago. I authorize you to spread these news among our friends".[23] Léon and Yvonne got married on 8 July 1933.

Yvonne and Léon Rosenfeld (1935). From Rosenfeld album, Courtesy of The Niels Bohr Archive.

[21] RP, "Copenhague", S. Chandrasekhar to L. Rosenfeld, 28 Dec 1932. Chandrasekhar congratulates Rosenfeld with his and Yvonne's engagement.
[22] Swings (1974). Serpe (1980), p. 391.
[23] RP, Copy of letter from L. Rosenfeld to S. Chandrasekhar, 20 Dec 1932. Original in CP.

Chandrasekhar visited the couple in Liège in the spring of 1933 and gave a series of lectures about astrophysics at the astronomical observatory. While there he became closely involved with a publication of a joint work by Léon and Yvonne about the effects of temperature and pressure variations on the molecular abundances in stellar atmospheres.[24] This was at the time when the Nazis took power in Germany. For political reasons Rosenfeld therefore did not want to publish in *Zeitschrift für Astrophysik* and asked Chandrasekhar to communicate their paper to the Royal Astronomical Society in Cambridge. In the end, it was Dirac who presented their papers in the society. For unclear reasons, however, Yvonne refused to have her name on their second joint paper, which made Rosenfeld burst out to Chandrasekhar that "you never understand a woman's soul!"[25] He must have understood something of a woman's soul, however, for Léon and Yvonne seem to have lived in a very happy and harmonious marriage full of affection, love, and respect. They shared many interests in the arts, history, and literature including a shared political outlook. World events were discussed and interpreted in their home, and the children were encouraged to develop their own independent views. Yvonne did not continue her scientific career; apart from occasional teaching in secondary schools, she devoted herself to her family's well-being. She nearly always accompanied Léon on his travels. Later, she assisted with the editorial work of the journal *Nuclear Physics* founded by Rosenfeld in 1956.[26] The couple had two children, a daughter Andrée (August 1934–October 2008) and a son Jean L. J. Rosenfeld (born 29 September 1936). The spoken language in their home was French, although Andrée and Jean spoke at first Dutch and later English together.[27] It has been said about Yvonne that she was an indispensable support for Rosenfeld's tremendous, inexhaustible working spirit.[28] However, she would sometimes complain that it left

[24]RP, Chandrasekhar correspondence. Cambresier and Rosenfeld (1933). Rosenfeld (1933). Swings (1964). Wali (1991), pp. 102–104.
[25]RP, copy of letter from L. Rosenfeld to S. Chandrasekhar, 4 Jun 1933. Original in CP.
[26]Email correspondence, J. L. J. Rosenfeld to A. S. Jacobsen, 1 Aug 2009. Frank (1974).
[27]Email correspondence, A. Rosenfeld to A. S. Jacobsen, 13 Oct 2004.
[28]Rozental (1975), p. 100, Møller (1974), p. 4.

her feeling unhappily alone. Hilde Levi related the following anecdote from the late 1930s:

> Bohr and Rosenfeld's collaboration... frequently took place in Carlsberg [where Bohr lived] and often lasted until late at night. So Mrs. Rosenfeld was left alone with their two children and she often was quite unhappy, she hardly saw him and he always came home so late. In her desperation, she finally went to the family doctor and complained, and asked him to write a letter saying that it wasn't good for Rosenfeld's health to work so late, and the doctor would recommend that Dr. Rosenfeld had shorter working time... Bohr was of course impressed and regretted that he had used Rosenfeld so much at the cost of his family life.[29]

Everybody who knew Yvonne — Rosenfeld's colleagues, friends, and former students — adored her and always mentions this when asked about Rosenfeld. She was involved in the Resistance in the Netherlands during the war (see Chapter 4). As a Belgian, she was a brilliant cook, which, according to Andrée, made life more bearable while they lived in Britain.[30]

The Promised Land

Like Rosenfeld, Houtermans and Peierls married colleagues. As mentioned in Chapter 1, Peierls had even found a Russian girl friend, the physicist Genia Kanegisser, and he therefore visited the Soviet Union frequently.[31] As mentioned in the first chapter, Rosenfeld and many of his young colleagues were curious about the social and political development in the Soviet Union during the First Five Year Plan. The physicists wanted to see with their own eyes this unique,

[29] Recollections of H. Levi (1993), NBA.
[30] Email correspondence, A. Rosenfeld to A. S. Jacobsen, 13 Oct 2004.
[31] See RP, "Daily life in Copenhagen (1927–1938): Copenhague 1", R. Peierls to L. Rosenfeld, 19 Nov 1931.

practical demonstration of the relationship between science and socialism.

The world's first socialist state was gradually turned into the corner stone in the communist ideology. Despite being poor and underdeveloped, the country successfully realized a dream about Utopia. It achieved this central status, also encouraged by Stalin, who by the end of the 1920s launched the idea that the quality of a communist was measured by his or her relation with the Soviet Union. From then on, the Soviet Union became sacrosanct among communists. Elements of religion and the authoritarian Russian political culture blended with communist ideology, including an unconditional sacrificing of oneself for the party.[32]

Even for non-communists the Soviet Union stood out as the only existing alternative to capitalism which had escaped the economic crisis. From 1929 to 1940, Soviet industrial production tripled and there was no unemployment. By contrast, the West experienced very high unemployment rates. Small industrialized Belgium depended heavily on exports which dropped dramatically in 1930. By 1936, exports had shrunk to 43 percent of their 1929 level. The unemployment rate was 40 percent in 1932. In Denmark, it was 32 percent the same year. Britain experienced "Hunger Marches" by unemployed workers.[33] The Great Depression may have been the reason that the majority of foreign visitors to the Soviet Union were uniquely enthusiastic in reporting only about a country progressing by leaps and bounds. They were impressed by the novel developments and the goal-oriented use of science and technology in the country's violent industrialization process which caught the eye and pointed towards a new society. They saw what they expected and wanted to see and rarely noticed the impoverishment, primitiveness, and inefficiency of the Soviet economy, or the ruthlessness and brutality of Stalin's collectivization and mass repression.[34] Peierls, however, appeared

[32]Hobsbawm (1995), pp. 71–72, 74–75, 144. Thing (1993), Vol. 2, pp. 927–928.
[33]Cook (2004), p. 116. Halleux *et al.* (2001), Vol. 2, p. 36. Hobsbawm (1995), pp. 92–93.
[34]Werskey (2007b). Thing (1993), Vol. 2, pp. 927–928. Kuznick (1987), pp. 106–110. Hobsbawm (1995), p. 96.

level-headed, when in the late fall of 1930, he reported his observations from his recent trip to Russia to Rosenfeld:

> There were many interesting things to be seen in Russia, so many that it would be futile to begin a written account. In any case, one gets a very gratifying impression of the atmosphere all in all, in particular also from the manner in which one is received there as a foreigner, [and] from the tone between people which after all is much more unprejudiced than here with us. Economically, on the other hand, the situation is still quite bad and the people who live there (not foreigners who in every respect have a preferential position) have to put up with much. An extreme amount of construction work is done; to what extent it suffices, in order to meet the program, one cannot of course assess as an outsider. Next spring I will be in Leningrad again for two months, and in the meantime [I] study Russian intensively, because then one will get much more out of it.[35]

Indeed, the Soviet Union continued to be a poor and underdeveloped country while at the same time representing a utopian dream for many Westerners. Peierls aired some news that may have excited Rosenfeld, *viz.*, the prospect of being invited to Leningrad: "In September, after his return from America, [Yakov Ilyich] Frenkel intends to organize a theoretical conference in Leningrad, as he probably wrote to you, and [he] intends to invite you to it. However, as with all Russian plans, one must still consider it with a considerable amount of uncertainty".[36] As foreseen by Peierls, the invitation never seems to have materialized.

That Rosenfeld was eager to visit the Soviet Union may also be inferred from the following. In the spring of 1932, Landau and Bronstein promoted Gamow for election to the Soviet Academy of Sciences and in that connection attempted to establish an Institute of

---

[35] RP, "Correspondence particulière", R. Peierls to L. Rosenfeld, 23 Nov 1930. Originally in German.
[36] *Ibid.*

Theoretical Physics under the auspices of the Soviet Academy under Gamow's leadership. Gamow informed Rosenfeld, "When it is ready (there are still not enough soft sofas) a number of "foreign scholars" will be invited from time to time. Then you would certainly come for some months, wouldn't you?"[37] Rosenfeld finally got an opportunity to visit the Soviet Union in the spring of 1934. At this time he was "strongly coupled" with Bohr, as he called it, and he accompanied Bohr and his wife Margrethe.[38] Unfortunately, there exists to my knowledge no direct testimony from Rosenfeld about his impressions of the Soviet Union during this visit. However, given his left-wing standpoint and the sporadic hints noted above and below, we may infer that he was very keen to visit the Soviet Union. If Bohr was favorable to what he saw there, we may surmise that Rosenfeld would have been genuinely enthusiastic.

Bohr was invited to the Soviet Union by Abram Joffe on behalf of the Academy of Sciences and the Russian universities. Bohr greatly looked forward to the visit; he wanted to improve scientific liaisons with the Soviet Union and, like his younger colleagues, learn more about the state of things in the Soviet Union of which Dirac had spoken so favorably to Bohr.[39] "[A]nd also my collaborator Rosenfeld, who is particularly interested in the Russian development and even understands Russian, would like to join us on our trip to Leningrad", Bohr wrote to Joffe.[40] The troika left for Leningrad on 28 April 1934 by train from Helsinki, arriving in Leningrad on 30 April in the afternoon. The following day, they watched the May Day parade from a prominent guest stand. The number of tanks and other military equipment and soldiers in the parade on the central square of Leningrad made a deep impression on them, as must have the

[37]G. Gamow in a mixture of German and Danish to L. Rosenfeld, 14 Mar 1932. Quoted from Gorelik (1995), pp. 59–60. The idea of an institute of theoretical physics under the Academy was not materialized until about forty years later. Kojevnikov (2004), pp. 90–91.

[38]J. G. Crowther Archive, University of Sussex Library. Special Collections, L. Rosenfeld to J. G. Crowther, 21 Jun 1934.

[39]See for example Crowther, *Fifty Years*, p. 136. BSC, N. Bohr to W. Heisenberg, 20 Apr 1934. P. A. M. Dirac to N. Bohr, 20 Aug 1933. About Dirac's visit in the Soviet Union see Farmelo (2009), pp. 153–156, 175–176.

[40]BSC, N. Bohr to A. Joffe, 10 Mar 1934. Originally in German.

about one and a half million workers who, according to the newspaper *Leningradskaja Pravda*, participated in the demonstration.[41]

Rosenfeld, Niels, and Margrethe Bohr arriving in Moscow 1934. Courtesy of The Niels Bohr Archive.

Bohr cannot be described as a leftist. Yet he was clearly fascinated and impressed by what this country seemed to accomplish and how science and technology were employed in the rapid industrialization process the country was going through. In an interview conducted by the Soviet newspaper *Izvestia* (edited at the time by Nikolai Bukharin) on 12 May 1934, Bohr described his experiences thus:

> In all countries one naturally tries to apply science, but precisely here in the Soviet Union the conditions might be particularly suited for a fruitful connection between theory and practice. Everyone feels that here he is in close touch with the entire construction process. A particularly pleasing impression here is made by the enthusiasm that everyone puts

[41] I am grateful to J. Gregersen for allowing me to refer to his manuscript "Rejsen til Sovjetunionen i 1934", pp. 26–27.

> into his work... from what I have seen here, the great understanding with which your statesmen have supported science has made the greatest impression on me... [T]he close connection between pure science and its possible applications is never forgotten... I saw to what great lengths one goes here to enable the workers to increase their knowledge, and how one strives to satisfy the workers' needs. Particularly important in this regard is the circumstance that all this extends to the entire population. What one in foreign countries can only try to achieve is carried out here on a colossal scale... I would like to emphasize the most important among my impressions: the Soviet Union gives everyone something to think about. It is difficult to say to what an extent everything taking place here can be transferred to other countries, but there is no doubt that in many respects something wonderful is being created here... I now hope to strengthen these ties and visit the Union again. This happens to everyone who comes here. I know this from many scientists and especially from the young physicists. Professor Dirac has visited the Soviet Union several times and speaks of it with great affection.[42]

Bohr's report is reminiscent of the majority of travel literature about the Soviet Union from this time by scientists.[43] Here they witnessed big science projects with huge government support, an organizational model that would only appear in the West with the Manhattan project.[44] Chandrasekhar visited the Soviet Union a few months later strongly encouraged by Rosenfeld. He was equally excited about what he saw in Russia with respect to how the workers were treated.[45] After his return, Chandrasekhar wrote to his brother,

> Russia appeals to me most tremendously. Russia gives the impression of a young man, full of ideals and who has

---

[42] *Izvestia* (1934).
[43] McGucken (1984), pp. 74–75. As regards the British physicist P. M. S. Blackett's visits to the Soviet Union, see Nye (2008), p. 238.
[44] Kojevnikov (2002a).
[45] S. Chandrasekhar to his father, 20 July 1934. Quoted in Wali (1991), p. 119.

> such indomitable moral courage and indefatigable physical strength to go forward in spite of setbacks and who with his hopes derives consolation from his ideals during times of adversity.[46]

While in Leningrad, Bohr and Rosenfeld gave a joint talk about their paper on the measurability of quantum fields before the physicists there. As mentioned in the previous chapter, Landau was present and vigorously opposed Bohr's arguments against his and Peierls' paper from 1931.[47] On 15 May Niels, Margrethe, and Léon arrived in Moscow for three days, where they met among others the leading Marxist physicist, historian, and philosopher of science at Moscow State University, Boris M. Hessen. When participating as part of the Russian delegation — including among others Bukharin, Joffe, and the biologist N. I. Vavilov — at the Second International Congress of the History of Science and Technology at the Science Museum in London in July 1931, Hessen had, together with Bukharin, opened the eyes of the group of left-wing scientists in Britain as regards using history of science to exemplify the connection between science and socialism.[48] Hessen had boldly and provocatively taken Isaac Newton, the most celebrated British scientist of all times, as his case, and demonstrated the social and economic roots of his famous book *Principia.*[49] Rosenfeld may have learned about Hessen's influence on British scientific intellectuals from the prolific British science writer and journalist, and committed communist, J. G. Crowther. Crowther visited Bohr's institute several times in the 1930s, and Rosenfeld may have met him for the first time at Bohr's annual Easter conference in Copenhagen in 1932. Rosenfeld and Crowther seem to have found a common ground for their views on science, society, and

---

[46]S. Chandrasekhar to his brother Balakrishnan, 31 Aug 1934. Quoted from Wali (1991), p. 116.
[47]Rosenfeld (1963), p. 70.
[48]There is a wealth of sources treating this congress and Hessen's contribution. Werskey (1988), pp. 138–149. Mayer (2004). Chilvers (2006), pp. 179–206. Graham (1985). Chilvers (2003). Schaffer (1984). Nye (2008).
[49]Hessen (1971).

history of science, and immediately became friends. During these years Crowther published several books on the history of science inspired by historical materialism. Rosenfeld seems to have found his own path to Marxist history of science during the 1930s, independently of the British scientific intellectuals.[50]

In Moscow Bohr and Rosenfeld also met Igor E. Tamm, who later accompanied the Bohrs and Rosenfeld to Kharkov in the Ukraine.[51] Tamm and Rosenfeld had known each other since they met in Göttingen in 1928 (see Chapter 1). They remained friends until Tamm died in 1971.[52] During the meeting on nuclear physics at the Physico-Technical Institute in Kharkov, from 20 to 22 May, Bohr and Rosenfeld met Bronstein with whom Rosenfeld shared his interest in quantum field theory and quantum gravity. Bronstein was also one of the few Russians who, besides Landau, were familiar with Bohr and Rosenfeld's work on the measurability of the electromagnetic field.[53] Vladimir A. Fock also participated in the conference. Fock later became Bohr's spokesman with respect to quantum philosophy in the Soviet Union, when it was otherwise criticized for being bourgeois and idealistic after the Second World War (see Chapter 6). He translated many of Bohr's papers into Russian including the Bohr–Rosenfeld paper.[54] Also Rosenfeld's close friend, Solomon attended the conference in Kharkov, and Crowther was present in his capacity as *The Manchester Guardian*'s science

---

[50]Jacobsen (2008). Of J. G. Crowther's books, see for example Crowther (1930). Also the socialists Bernal and Blackett engaged enthusiastically in historical studies during the 1930s. Nye (2008), pp. 240–246.
[51]Kojevnikov (1996), p. 13. *Izvestija*, 16 May 1934. Cited in J. Gregersen, manuscript "Bohrs Første Rejse til Rusland (maj 1934)", pp. 39, 55.
[52]Rosenfeld (1979r), dedicated "To Igor Tamm, as a token of old friendship". RP, Supplement, Copy of letter from L. Rosenfeld to V. J. Frenkel, 26 May 1971.
[53]Gorelik (2005). Gorelik and Frenkel (1994). Stachel (1999), pp. 528–532.
[54]Gorelik and Frenkel (1994), p. 60. Fock suggested translating Bohr's answer to Einstein in *Physical Review* in 1935, BSC 19, V. A. Fock to N. Bohr, 4 Apr 1936. RP, "Copenhague: Correspondance générale 1950–1961", V. A. Fock to L. Rosenfeld, 18 Apr 1960. The Russian translation of the Bohr–Rosenfeld paper, "K voprosu ob izmerimosti elektromagnitogo polia", only appeared in Bohr (1971), pp. 120–162. I am grateful to Alexei Kojevnikov for drawing my attention to this.

correspondent.[55] As well as serving as Bohr's interpreter at the conference, Rosenfeld presented his joint work with Yvonne on the dissociation of molecules in the atmospheres of the carbon stars.[56]

Landau, Bohr, Rosenfeld, and Bronstein discussing during the Kharkov Conference. Published in the newspaper *Khar'kovskii rabochii* (Kharkov Worker) on May 20, 1934.[57]

Participants in the Kharkov Conference 1934. D. Ivanenko, L. Tisza, Rosenfeld, I. Rumer, Bohr, Crowther, Landau, M. S. Plesset, Ya. I. Frenkel, I. Waller, E. J. Williams, W. Gordon, Fock, I. E. Tamm.
Courtesy of The Niels Bohr Archive.

[55] Crowther (1970), pp. 95–99. Werskey (1988), p. 138. J. G. Crowther Archive, University of Sussex Library. Special Collections, L. Rosenfeld to J. G. Crowther, 21 Jun 1934.
[56] Bronstein (1934). I am grateful to J. Gregersen for translating Bronstein's review of the conference from Russian into Danish.
[57] Gorelik (2005).

Like his younger colleagues, Bohr was interested in following developments both socially and scientifically in the Soviet Union.[58] However, he was seriously concerned about the isolation enforced by the Soviet state upon its scientists. In his interview and in his speeches he stressed the importance of international collaboration among scientists and that especially young scientists should be allowed to travel abroad. However, for the Copenhagen Conference in June 1936 he invited Fock, Landau, Frenkel, and Tamm, but none of them were allowed to leave the Soviet Union.[59] As mentioned in Chapter 1, Gamow decided to remain in the West when he was allowed to attend the 1933 Solvay Congress in Brussels.[60] In the very summer of 1934, Bohr's close friend and colleague Pyotr Kapitza was not permitted to return to the Cavendish Laboratory in Cambridge after his vacation in the Soviet Union.[61] Bohr and Kapitza met again when Bohr returned to Russia in 1937. At this time Bohr experienced the other side of the country when he traveled by train through Siberia. Niels Bohr's son Hans, who was with Niels and Margrethe on this trip, kept a diary, in which he described the many transports of deported prisoners.[62] At this time, Niels Bohr was more concerned than ever about scientists being denied the possibility of going abroad. By 1938, no Russian scientist was allowed to travel abroad, and foreign scientists could not visit the Soviet Union. Russian scientists were compelled to stay put and do scientific work that benefited the socialist construction of the Soviet Union.[63]

Fascism and Nazism versus the Popular Front

In Europe, the Depression sharpened the political antagonisms prevailing since the Russian Revolution and the First World War. In

[58]Kragh (2003), pp. 66–67. Crowther (1970), p. 136.
[59]Kragh (2003), p. 68. BSC 19, N. Bohr to V. A. Fock, 13 Mar 1936. BSC, Supplement, V. A. Fock to N. Bohr, 11 Jun 1936.
[60]Kojevnikov (2004), p. 108.
[61]Josephson (1991), pp. 282–283. Kojevnikov (2004), p. 106. Aaserud (2005), p. 9.
[62]Bohr, H. (2008).
[63]Kojevnikov (2004), pp. 108–109.

countries where old regimes had collapsed, including the old ruling classes and their power structure, influence, and hegemony, the fear of social revolution and loss of faith in nineteenth-century style parliamentary governments gave rise to fascist and Nazi tendencies everywhere in society. Nationalism and anti-Semitism became noticeable even in groups not as far to the right as the proclaimed fascists and Nazis.[64] Belgium was in a sad state after the First World War. Of the participants in the war, it had suffered proportionally the greatest physical destruction and loss of life, even though it had not willingly taken part in the struggle. After the war, reforms secured the vote for all Belgian men. In 1932, the monopoly of French language was broken when Flemish was made the administrative language in Flanders. As a result of reforms, the dominance of the Catholic bourgeoisie within the Catholic Party was destroyed. The Labour Party grew almost as large as the Catholic Party. During the inter-war years, political life was marked by unstable coalitions due to competition between the antagonistic communities of the Flemish and the French, the workers and the upper middle classes. There were eighteen different governments between 1918 and 1940, sixteen of which were coalitions. Due to their dependence on export and transit trade to Germany, Belgium experienced economic crisis, even before 1929, following the German currency collapse in 1923. When the Belgian franc was devalued, it seriously affected the middle classes. As the economic crisis deepened, it was accompanied by political unrest and the growth of radicalism on the right. At the elections in 1936, Flemish extremist parties and the Belgian Communist Party experienced increasing support. The latter got six percent of the vote. However, the greatest surprise was the Rexist Party, founded by the Walloon politician Léon Degrelle, which gained 11.49 percent of the vote and 21 seats in the lower house and 12 in the senate. The ideology of Rexism took its origin in a Catholic youth organization at the University of Leuven. However, the Catholic Church condemned the Rexists in 1930. During the early 1930s, the Catholic origins were pushed into the background and the Rexists soon allied themselves

[64]Hobsbawm (1995), pp. 124–130.

with the extreme Flemish nationalists and with Nazi Germany. They incorporated Nazi-style anti-Semitism into their program and did their best to disrupt the workings of parliamentary government by introducing violence and populism into Belgian public life. Degrelle showed a remarkable Hitlerian flair for appealing to sentiments of violent nationalism, to conservative Catholicism, to the prejudices of the army-officer class and big industrialists, to the middle classes who suffered from devaluation, and to the grievances of the unemployed. Degrelle met both Mussolini and Hitler in 1936, each of whom supported Rexism financially and ideologically. The Rexist party was, however, won over by democratic means. A coalition government formed in 1936 by the Catholic Party, the liberals, the socialists, and even the communists managed, like the French Popular Front (see below), to improve conditions for the workers and the unemployed. At the next election in 1939, Degrelle suffered a crushing defeat and lost all but four seats. During the German occupation most Rexists supported and assisted Nazi Germany. Degrelle joined the Waffen SS, the front-line troops, in the fight against the Soviet Union during Operation Barbarossa in the summer of 1941.[65]

When the Nazi Law for the Restoration of the Civil Service, better known as the *Beamtengesetz*, was promulgated in Germany on 7 April 1933, it dictated the dismissal of civil servants, including university faculty, who were Jewish or who had undesirable political connections, including leftists. This mobilized scientists outside Germany to help their German colleagues. Rosenfeld became involved with the British Academic Assistance Council (AAC), established on the initiative of the Hungarian physicist Leo Szilard, to secure German refugee physicists jobs abroad.[66] Among other things,

[65]Cook (2004), pp. 113–119. Arblaster (2006), pp. 215–219.

[66]RP, "*History of Science* (2), Vol. 7", copy of letter from L. Rosenfeld to F. Herneck, 12 Jan 1962. Lanouette (1992), pp. 120–121. Szilard contacted Rosenfeld, who communicated to him the link to the rector of Liège University, J. Duesberg and J. Willems, the very influential Director of the University Foundation. Szilard to Dr. D., May 7 1933, reprinted in Weart and Szilard (1978), pp. 32–33. Rosenfeld also suggested Szilard to

Rosenfeld approached Einstein in Brussels in the spring of 1933, when Einstein attended Rosenfeld's lecture (see last section in Chapter 2) and was otherwise on the verge of emigrating to the US, in order to seek Einstein's support for the AAC.[67]

Many physicists who fled from Germany because of race or politics found shelter at Bohr's institute in Copenhagen for shorter or longer periods until they found a more permanent position abroad, often through Bohr's intervention. Bohr was involved with the Danish Committee for the Support of Refugee Intellectual Workers, established in October 1933 to help refugees obtain permission to enter Denmark and to seek financial support for them. Among the physicists helped by Bohr were Otto Robert Frisch, Hilde Levi, Victor Weisskopf, Max Delbrück, James Franck, Walter Gordon, Guido Beck, and David Herrmann Martin Strauss.[68] Needless to say, the political development deeply affected life at Bohr's institute. Rosenfeld later recalled that when the Copenhagen conference was convened in September 1933, the "exodus of many of our German colleagues and friends fleeing persecution, and the plight of the few who had not left, cast a deep gloom on the reunion".[69] For Ehrenfest, who was particularly sensitive and self-tormenting, these events seemed to have been the last straw. After returning home from the Copenhagen Conference in 1933, he committed suicide after first shooting his retarded son.

take contact to key figures with important connections to both University and financial circles as well as to the central figures responsible for similar initiatives taken in Paris. The American Philosophical Society, copy of letter from L. Rosenfeld to L. Szilard, 24 Apr 1933. Copy of letter from L. Rosenfeld to L. Szilard, May 2, 1933. L. Szilard to W. Beveridge, 4 May 1933, reprinted in Weart and Szilard (1978), p. 31.

[67]RP, "*History of Science* (2), Vol. 7", L. Rosenfeld to F. Herneck, 12 Jan 1962. RP, "Supplement 1971–1974", copy of letter from L. Rosenfeld to B. H. Muller, 1 Sep 1972, with enclosed notes from an interview with Rosenfeld at the Varenna Summer School 12 Aug 1971.

[68]Aaserud (1990), pp. 105–164. Aaserud (2005), p. 11. Hoffmann (1988), p. 50.

[69]RP, "Supplement: History of Quantum Theory", L. Rosenfeld, Ehrenfest as I saw him (An autobiographical chapter), May 1971, manuscript.

Copenhagen Conference 1933. Courtesy of The Niels Bohr Archive.

Part of the German professoriate remaining during the Nazi regime attempted to foster the ideal of being apolitical and conducting pure science, which they claimed could be seen as entirely value-free. In the end, however, this position amounted to support for the regime.[70] Some German scientists such as Rosenfeld's former supervisor, Pascual Jordan, openly expressed their Nazi sympathies. Jordan joined the NSDAP on 1 May 1933 and in December that year he joined the Sturmabteilung (SA).[71]

As a response to the growing threat from fascism, the Comintern changed tactics away from its ultra-left politics. At the 7th World Congress in the summer 1935, it began endorsing broader alliances, United Fronts, of workers and their allies.[72] Thus, the Communist Party now welcomed middle class sympathizers whom it had hitherto

[70]Beyerchen (1977), pp. 4, 10–11, 58, 199, 206–210. Heilbron (1986), pp. 141, 150. Cassidy (1992), pp. 323–331.
[71]Beyler (1994).
[72]Hobsbawm (1995), p. 71.

branded as "social fascists", such as radicalized students, teachers, academics, and artists. Those sympathizers in turn tended to see the Communist Party as the only organized effort against fascism and poverty. This development laid the foundation for political alliances with liberals and social democrats in joint defense of science and democracy, the so-called Popular Front, which was established in France and Spain initially with great political success. In May 1935, France and Russia signed a defense treaty. In France, the Popular Front won a majority of parliamentary seats in the elections of May 1936, resulting in the French Government being led for the first time by a socialist, the intellectual Léon Blum. Through this government, the group of left-wing scientists in Paris, whom Rosenfeld knew so well from his earlier stays in that city, including the Curies, Pierre Biquard, Jean Perrin, and Paul Langevin, became involved in politics.[73] In Spain, the Popular Front came to power in February 1936. These victories filled the socialist movements with hope and euphoria and the belief that the ultimate triumph of socialism was near. The Left could no longer be ignored, they thought.[74] Rosenfeld was clearly pleased with this development, including the broad political coalition formed in Belgium in 1936. At the end of 1936, he found that Chandrasekhar took "too gloomy a view" on "the situation of the world":

> Not that I have any hope that war can be avoided, but I think that just war is [not] the only means now of crushing and, I hope, definitely exterminating fascism. The last developments, both diplomatic and military, must be, even by the most pessimistic, characterized as a victory of USSR over Nazi Germany. Of course, nothing decisive has been reached as yet, but a prelude, so to speak, has been played and its outcome gives some comfort regarding the future outlook.[75]

[73]Pinault (2000), pp. 81–89.
[74]Hobsbawm (1995), p. 148. Werskey (1988), p. 136. Kojevnikov (2008), p. 128.
[75]RP, copy of letter from L. Rosenfeld to S. Chandrasekhar, 30 Dec 1936. Original in CP.

In Britain, the Social Relations of Science (SRS) Movement arose in response to the Depression and the rise of fascism and Nazism with its persecution of scientists and what in this connection was conceived as abuse of science. The movement was driven by a desire for social improvement and democratization of society and it was believed that this could be accomplished if science was given the means to thrive and deliver solutions to social needs. It drew inspiration from the enthusiastic cultivation and application of science in the new Soviet state and appears to have been started among a group of left-wing scientists who attended the 1931 congress of the history of science in London, *The Visible College*, as Garry Werskey has called them.[76] At this congress the Russian delegation had introduced a novel philosophy and historiography of science and reported about the planned economy and the close relationship between science and technology in the Soviet Union. The SRS movement was spearheaded by the charismatic crystallographer John Desmond Bernal, whose book *The Social Function of Science* (1939) became the manifesto of the movement.[77] The book summarized the problems with the current organization of science and put forward a strategy for changing it so that research would support social welfare (more about this in Chapter 5). Prominent spokesmen of the SRS movement included, besides Bernal, Crowther, the physicist P. M. S. Blackett, and the biologists J. B. S. Haldane and Joseph Needham. The movement expressed itself largely through Britain's major scientific organizations, such as the Royal Society of London, the British Association for the Advancement of Science, the Association of Scientific Workers, and government organizations such as the Scientific Advisory Committee and Britain's War Cabinet, among others. The movement's principal concern was to strengthen the bonds between science and government and integrate science into society in order to help the nation in crisis. It highlighted the social responsibility of scientists as citizens and workers. The SRS movement's scientific worldview was disseminated to other European countries including the Netherlands, where it was

[76]Werskey (1988).
[77]Bernal (1939).

taken up and transformed in the Breakthrough Movement during the 1930s and early 1940s. In Denmark, the SRS movement's ideas for reform of the sciences and higher education and the advancement of the scientist as a worker, were taken up during the Second World War by progressive scientists who were also active in resistance work.[78]

In the 1930s, scientists' sense of social responsibility tied in with a strong prevailing scientism. Scientists possessed extreme self-confidence with respect to offering guidance in matters of the development of society. They had great confidence in scientific method and principles as a means to provide the answers to societal problems. In fact, scientific laws, methods and the quest for reason and verification enjoyed very high prestige in all areas of intellectual life including political ideology.[79] This was evident on both the radical Left and Right. Both communists and fascists claimed that their ideologies were based on scientific laws, which supposedly determined the development of human society, whether it was scientific Marxism or eugenics and racial science. Bohr attempted to undermine the Nazis racist ideology by arguing that race prejudices and national conflicts might be overcome in a way similar to that by which physicists, at the advent of quantum theory, had overcome their nineteenth-century prejudices in favor of causality; namely by recognizing the relationship between different human cultures as complementary. Being part of one culture, for example English culture, might disqualify a man from understanding another culture such as the Chinese, but that had nothing to do with biological racial character, Bohr stressed, and such prejudice could be overcome.[80] In the Soviet Union, science and technology appeared to be prerequisites for the revolution to survive in this underdeveloped country. Elsewhere, for example in the Netherlands, a value-free scientific politics was seen as a tool to avoid and transcend political and religious ideology. Since, however,

[78]McGucken (1984). Kuznick (1987). Landström (1996). Edgerton (1996). Werskey (1988). Knudsen (2010), pp. 216–266. Molenaar (1994). Somsen (2008).
[79]Kuznick (1987), pp. 7–8, 38, 46–54. Kojevnikov (2008), pp. 115–116, 123.
[80]Bohr (1999d). See also Crowther (1970), p. 159 and Rosenfeld (1967), p. 135. Scientific refutation of racist theories was also a topic Bohr discussed with some of the Russian scientists in 1934. Kojevnikov (1996), p. 13.

politics is not value-free but requires an attitude for or against moral and normative issues, this approach was a dead end; science could not offer guidance as to what the political ends should be; it could only provide means to reach a goal.[81]

While the economic crisis, extremist movements and the threat of war darkened the sky over Europe, a reckless and satirical attitude to political events seemed to reign among the young physicists at Bohr's institute. The *Journal of Jocular Physics* was an internal medium for the communication of the informal and humoristic tone which spiced daily life there. The first volume appeared on Bohr's 50th birthday in 1935 and was followed by issues on his 60th and 70th anniversaries. In the 1935 issue, the role of scientific principles behind Mussolini's fascist policy was satirized in a short article by Oskar Klein. The title of Klein's article was "On political quantization".

> As is well-known, Bohr has been able to create almost complete harmony in the atomic world (including Pauli), by the introduction of the Quantum of Action into the classical physical laws, without which neither the Atom nor the World would be stable. In Politics, however, the process of Quantization has not been used, which is most strikingly shown by the classical formula[1] quite recently published by B. Mussolini, which has proved rather unsuited as a basis for Political Stability.
>
> 1) With, without or against Genève.[82]

"Genève" refers to the Geneva Protocol which was a treaty signed by the League of Nations, including Italy and Ethiopia, at Geneva on 17 June 1925, prohibiting chemical and biological warfare. However, Mussolini broke the Geneva Protocol by using mustard gas against the Ethiopians, including civilians and Red Cross camps, in the Abyssinian War from October 1935 to May 1936.

This issue of *Journal of Jocular Physics* reflected some of the younger physicists' responses to the blending of ideologies such

[81] Somsen (2008), pp. 234–236. See also Landström (1996), p. 55.
[82] Klein (1935a). Emphasis in original.

as Nazism, fascism, and dialectical materialism with science. As well as Klein's contribution, Gamow volunteered a rambling piece, "Dialectics of Atomic Nuclei", interweaving dialectical materialism, Soviet style, with the detention of Russian physicists, his theory of alpha-decay, "escaping particles" (probably a metaphor for his own experience), Bohr's "idealistic tendencies", and ideas by the "Nazi-physicist Heisenberg".[83] Rosenfeld, who was the editor of the issue, contributed a satirical article in German about the dispute between Niels Bohr's brother, the mathematician Harald Bohr and the German spokesman of so-called Deutsche Mathematik (as opposed to Jewish mathematics), Ludwig Bieberbach. According to Bieberbach, the more intuitive reality-oriented style of doing mathematics, as represented by Felix Klein, was superior to modern, formalist mathematics.[84] Rosenfeld's paper, "Zur Rassentheoretischen Mathematik", ended with "Heil Hitler. L. Rosenfeld, unordentliches Mitglied der NSDAP".[85] Despite its grossly satirical style, this part of the 1935 issue of *Journal of Jocular Physics* was considered controversial because of its sensitive statements which might compromise for instance German physicists who visited on a regular basis, such as Jordan and Heisenberg. It was therefore only shown to a selection of people.[86]

---

[83] RP, "Correspondance particulière", G. Gamow to L. Rosenfeld, 10 Sep 1935. Gamow (1935). RP, "Correspondance particulière", L. Rosenfeld to O. Klein, 26 Sep 1935. About Gamow's experiences, see for example Kojevnikov (2004), p. 106. Crowther (1970), pp. 147–151.

[84] Mehrtens (1987), pp. 196, 221, 225–226. Ramskov (1995), pp. 337–343.

[85] *Journal of Jocular Physics*, 1935, NBA.

[86] "Interesting papers of high standard have been contributed by G. Gamow, O. Klein and L. Rosenfeld. Since, however, the possibility of misinterpretation in a political and, therefore, not purely jocular sense could not be entirely excluded, we regret that they could not be published in the frame of this volume". *Journal of Jocular Physics*, Niels Bohr Celebration Number, 7 Oct 1935 (Institute of Theoretical Physics, Copenhagen), NBA. Reprinted in *Faust and Journal of Jocular Physics*, Vol. I, II, and III, on the Occasion of Niels Bohr's Centenary, 7 Oct 1985, NBA. Much later S. Rozental added the following note to this issue of Jocular Physics: "This supplement of *Journal of Jocular Physics*, Vol. 1 (1935) was considered secret and was only given to a few chosen people. [This was] in order to avoid political complications at a time where the general situation was somewhat delicate; besides there was also guests from Germany which could run into difficulties". (Originally in Danish.)

Battling for the Ideological Rights to Complementarity

With respect to the development of Bohr's thought, Rosenfeld has seen the 1930s as the period of "Consolidation and extension of the conception of complementarity".[87] As mentioned in the previous chapter, Bohr's discussion with Einstein and his controversy with Landau and Peierls were important for Bohr in refining his epistemological terminology. However, Bohr was determined to popularize the "epistemological lesson" of modern physics beyond the narrow theoretical physics community. In reaching a wider audience for his quantum epistemology Bohr benefited from his younger foot soldiers, or as Rosenfeld called them, Bohr's "sectators". Bohr's attempt at popularizing complementarity caused a lot of debate about its meaning and its implications for other areas of scientific, intellectual, and political thought. In connection with the broader reception of Bohr's ideas, a fight arose for what we might call the ideological rights to complementarity. Bohr and his disciples were busy trying to maintain control of the situation, which, however, was difficult since among the same disciples there were protagonists of the antagonistic ideologies. Rosenfeld was involved in this enterprise, as we will see in the next section. However, let me begin with Bohr's own tribulations.

Incidentally, philosophers of science felt a strong affinity at the time with modern physics. In general, Bohr had a rather skeptical attitude towards professional philosophers; he lamented in particular their insensitivity to the rapid developments in modern physics. At the end of his life he complained that philosophers did not understand quantum physics and complementarity.[88] When in Russia in 1934, Bohr was asked what he thought of the relationship between modern physics and Marxist philosophy. A heated debate about this issue had been going on for years in Russia between physicists and philosophers.[89] Bohr's contributions to and views on quantum theory

[87] Rosenfeld (1967).
[88] Kuhn *et al.* (1962), p. 3.
[89] Josephson (1991), pp. 247–275. Joravsky (1961), pp. 275–295.

of course epitomized more than anything the radical novelties of modern physics. In the interview for *Izvestia*, Bohr answered,

> When one raises the question of which philosophical consequences arise from modern physics, one may not thereby understand the question to mean which old philosophical schools comply with modern physics. Every new generation of philosophers learns from the new discoveries of other sciences of its time. Although some consequences of modern physics have something in common with the viewpoints of many great philosophers, yet it seems to me, that if men such as Spinoza or Marx were alive today they would probably, together with the rest of us, enjoy learning new things from modern physics of relevance for general philosophy.[90]

This view corresponds exactly with Rosenfeld's view on the matter as we find it in his writings and letters from the late 1940s (see Chapters 5 and 6). What Bohr knew about Marxist literature was what he had heard from Rosenfeld. However, as Rosenfeld later reported to the East German historian and philosopher of physics Friedrich Herneck, "I cannot say that it roused his enthusiasm".[91] According to Rosenfeld, the strong attraction to the theories of Marx and Engels by leftists puzzled Bohr; contrary to natural science and its epistemology, he thought, these theories had not made any progress since their original works appeared, yet they claimed to be scientific.

Nor was Bohr particularly well-versed in the philosophical tradition. He had a long-term interest in problems of a philosophical kind going back to his youth when he attended the Danish philosopher Harald Høffding's lectures. Høffding also introduced his students to German idealism. Høffding advocated the identity thesis, according to which mind and matter are but two attributes or aspects of one substance, a view that goes back to Benedict Spinoza, but was also

---

[90] *Izvestia* (1934), *BCW*, Vol. 11, p. 200. See also RP, "*History of Science* (2), Vol. 7", L. Rosenfeld to F. Herneck, 20 Oct 1964, in which Rosenfeld refers to this incident.
[91] RP, "*History of Science* (2), Vol. 7", L. Rosenfeld to F. Herneck, 20 Oct 1964.

put forward, for example, in F. W. J. Schelling's *Naturphilosophie*.[92] Bohr was also familiar with German idealism through his reading of the Danish romantic philosopher and poet Poul Martin Møller's *Tale of a Danish Student* (The Adventures of a Danish Student).[93] According to David Favrholdt, Bohr linked his reflections upon free will to this novel, which allegedly every Danish child received as a Confirmation present in those days.[94]

Despite his skepticism of philosophers, Bohr may have seen the interest of some of them in modern physics as an opportunity for communicating his new quantum epistemology to a broader audience. This seems to be the background of Bohr's involvement with the logical positivists in the mid-1930s.[95] The Vienna Circle sought to reform traditional philosophy by looking to modern science. Apart from attempting to secure the scientific status of philosophy and promoting their critique of traditional philosophy, the Vienna Circle sought, through Otto Neurath's Unity of Science movement, to reach a wider public in order to advance their alternative scientific worldview. This involved communicating science and epistemology to lay people in order to stem the influence of reactionary obscurantism and antiscientific views. As such, it was a modern Enlightenment movement building on twentieth-century science, logic, social thought, and politics. Most of the members of the circle were socialists of some variant.[96]

As mentioned in the previous chapter, Bohr gave a popular exposition of his quantum epistemology in a booklet, *Atomic Theory and the Descriptions of Nature*, published in Danish in the Year Book of Copenhagen University for 1929, of which Rosenfeld provided a French translation in 1932. Right away, however, this book was cited in support of idealistic viewpoints and thus fell directly into the categories fought about amongst the polarized world ideologies. There were several statements which could be interpreted

[92]Favrholdt (1999a), p. xliii.
[93]Møller (1925).
[94]Favrholdt (1999a), pp. xxxi, xlv.
[95]Faye and Folse (1998), p. 8. BSC, P. Frank to N. Bohr, 9 Jan 1936.
[96]Reisch (2005), p. 3. Richardson (2008), pp. 90–91.

as idealistic in Bohr's book. It was particularly the following words that gave the book a bad press among materialists: "[W]e have been forced step by step to... reckon with a free choice on the part of nature between various possibilities".[97] Later in the book Bohr suggested that the quantum of action had implications for the free will of the mind.[98] Following the indeterminist interpretation of quantum mechanics such ideas were quickly picked up in what was often referred to as "Eddington-Jeans" idealism and mysticism, *viz.*, the idealism exposed in popular science books written by the famous astrophysicists Sir Arthur Eddington and James Jeans in the late 1920s and the 1930s. Eddington's 1927 Gifford Lectures were published under the title *The Nature of the Physical World.*[99] This book became one of the most influential popular books on science in Britain. It was translated into eight languages. It advanced an explicitly anti-materialist philosophy, and Eddington discussed the advantageous implications of the collapse of determinism in physics for the free will of the human mind as well as of the physical world. Jeans' Rede Lecture was published as *The Mysterious Universe* in 1930. In his book, Jeans took recent advances in quantum physics to support the existence of God in the shape of a "pure mathematician". Both books sold tens of thousands of copies during the 1930s and 1940s. Jeans and Eddington were seen as defending religion and traditional values, and their books had a tremendous impact on theologians and public opinion, not only in Britain.[100] For left-wing materialists, Eddington and Jeans represented all the evils of bourgeois science with their idealistic philosophy, their religious beliefs, and their ignorance of the need for applying scientific ideas to society.[101]

[97]Bohr (1934), p. 4. BGC, N. Bohr to O. Neurath, 24 Oct 1934.
[98]Bohr (1934), pp. 100–101. Favrholdt (1999a), p. xxv. Many years later, the philosopher of science A. Grünbaum pointed out to Rosenfeld the following sentence, as he saw them, in Bohr's book on p. 116: "the theory of relativity reminds us of the subjective character of all physical phenomena". According to Grünbaum such statements had done much harm. RP, "Epistemology 1955–1958", A. Grünbaum to L. Rosenfeld, 20 Apr 1957.
[99]Eddington (1930).
[100]Stanley (2007), pp. 195–213.
[101]*Ibid.*, pp. 224–225.

Like Jeans and Eddington, also Pascual Jordan, with whom Bohr had carried on an extensive correspondence on philosophical matters in 1931, combined idealistic and vitalistic views, which he claimed were based on Bohr's quantum epistemology, with Nazi ideology. Jordan detested the materialistic world view and he claimed it was founded on a belief in complete causal determinism in nature. Nevertheless, Jordan gained at the time a reputation as the most outspoken positivist among German physicists, and he used positivism to argue against materialism. Yet, Jordan was strongly criticized for his metaphysics by the Vienna Circle. Like Jeans and Eddington, Jordan argued that the discovery of quantum indeterminism offered the means to secure the scientific legitimacy of free will.[102]

When Bohr became aware of the ideological minefield wherein his statements had landed, he maintained that his viewpoints had been misunderstood and abused, and he was more than eager to clarify them. Therefore, he willingly approached the logical positivists. He was anxious to have his ideas and concept formation discussed by competent people, and he was alert to their criticism. Two Danish professors at the University of Copenhagen were involved with the movement, the philosopher Jørgen Jørgensen and Bohr's cousin and close friend, the psychologist Edgar Rubin. In 1934, Jørgensen invited Otto Neurath to Copenhagen where he gave some lecturers which Bohr also attended. Shortly after he had met Neurath in Copenhagen in 1934, Bohr sent him a copy of his booklet in German and asked if its introduction was "as bad as its reputation".[103] In May 1936, Bohr wrote to Philipp Frank: "I am anxious, first of all, to clarify the misunderstandings which appear too often in discussions over the meaning of atomic physics".[104] In his reply, Frank assured Bohr that they could reach an agreement about the issues in question. It was Frank's opinion that the responsibility for the fact that the emerging national-socialistic journals of science in Germany

[102]See for example Zilsel (1935). Neurath (1935). Aaserud (1990), pp. 82–93. Beyler (1994), pp. 7–12, 154. See also Heilbron (1985), pp. 213–219, 221–222. As for how Bohr and the logical positivists disagreed with Jordan, see also Favrholdt (1999b), pp. 17–19.
[103]O. Neurath to R. Carnap, 14 Nov 1934, quoted in Faye and Folse (1998), p. 8.
[104]BSC, N. Bohr to P. Frank, 27 May 1936.

were able to abuse scientific statements in their favor rested partly with the physicists who had formulated their statements so that they could be abused.[105] Although both Frank and Neurath addressed the political dimensions of the Unity of Science movement in their correspondence with Bohr and suggested a connection between these values and Bohr's enormous significance in physics, Bohr remained silent and avoided this perspective altogether.

In order to clarify the misconstrual of his quantum epistemology as supporting idealism, vitalism, and mysticism, Bohr decided to host the Second International Unity of Science Congress at the Carlsberg Mansion in June 1936. The conference was arranged by Neurath and Jørgensen. In order to attract as many physicists as possible to attend the conference Bohr moved the annual "Easter" conference to take place immediately before the Unity of Science conference. The theme of the Unity of Science Congress was the problem of causality, particularly pertaining to physics and biology.[106] There were more than one hundred participants at the conference, particularly many Americans it was reported in the positivists' journal *Erkenntnis* subsequently.[107] Rosenfeld seems not to have been present at the conference, but Jordan, Delbrück, Piet Hein, and Crowther were present, as were the philosophers Karl Popper, Neurath, Carl Gustav Hempel, Frank, Rubin, Alf Ross, and many others. During the conference the news of the assassination of the founder of the Vienna Circle Moritz Schlick by a Nazi student in Vienna was received with horror.[108]

One of the speakers at the conference was Martin Strauss, who sought to combine the foundations of quantum mechanics with logical positivism, and who was a close friend of Rosenfeld.[109] Strauss was imprisoned twice in Germany in 1933 and 1935 because of his left-wing sympathies. He then stayed at Bohr's institute from September 1935 to December 1936.[110] Subsequently, Strauss went to Prague

[105]BSC microfilm 19, P. Frank to N. Bohr [not dated]. Frank (1936a).
[106]*Erkenntnis* (1936).
[107]Carnap *et al.* (1936), p. 276.
[108]Carnap *et al.* (1936). Crowther (1970), pp. 166–167. Hoffmann (1988).
[109]Strauss (1972).
[110]BSC, Strauss Curriculum vitae.

to work with Frank at the Institute of Theoretical Physics there. Bohr deemed it worthwhile to support Strauss, because he thought Strauss could teach the philosophers the physical background proper of their philosophy. In September 1938, Rosenfeld could therefore inform Strauss from Copenhagen: "We all think that you could play a meritorious role with Carnap by teaching him a little about the seriousness and real significance of the epistemological problem of quantum theory! I am therefore pleased that I can now officially inform you that Bohr will be pleased to support a proposal from you for the purpose mentioned in your letter to the Rockefeller Foundation".[111]

Participants in the Second International Unity of Science Congress at the Carlsberg Mansion in June 1936. The picture shows Jørgensen, Frank, N. Bohr, George Hevesy, H. Bohr, Popper, Hempel, Neurath, Hein, Delbrück, and Jordan, among others.

---

[111]RP, "Correspondance particulière", L. Rosenfeld to M. Strauss, 19 Sep 1938. Originally in German.

Bohr opened his address "Causality and Complementarity" at the 1936 congress by stressing that since "the opinion has been expressed from various sides that [the epistemological] attitude would appear to involve a mysticism incompatible with the true spirit of science, I am very glad to use the present opportunity of addressing this assembly of scientists working in quite different fields... to come back to this question, and above all to try to clear up the misunderstandings which have arisen".[112] Bohr clarified the characteristic logical structure of quantum mechanics in terms of complementarity. He clearly distanced himself from the kind of argument for free will put forward by Jordan, and rejected the "widespread opinion that the recent development in the field of atomic physics could directly help us in deciding such questions as 'mechanism and vitalism' and 'free will or causal necessity'. The renunciation of causal description in atomic physics did not amount to an argument for spiritualism he stated.[113] A lively debate took place at the conference involving both speakers and other participants, among them notably the German mathematician, physicist, and philosopher Grete Hermann.[114] Hermann had studied under the mathematician Emmy Noether and the neo-Kantian philosopher Leonard Nelson in Göttingen, where she received her PhD in 1926. She took an interest in the interpretation of quantum theory, particularly the causality problem and the significance of modern physics for the theory of knowledge. As a philosopher, Hermann combined neo-Kantianism with the views of the early nineteenth century philosopher of mathematics Jakob Friedrich Fries. Following Kant, she attempted to maintain causality as a transcendental requirement for knowledge and experience.[115]

At the end of the conference, Frank concluded that there was no antagonism between Bohr's interpretation of quantum mechanics and logical empiricism. Bohr had clarified the characteristic logical structure of quantum mechanics in terms of complementarity in his talk. Furthermore, Strauss had justified the theory of complementarity

---

[112] Bohr (1999c), p. 39.
[113] *Ibid.*, pp. 39[289], 45[295], 47[297].
[114] Carnap *et al.* (1936), p. 275.
[115] Herzenberg (2008). Heisenberg (1971). Hermann (1935).

by means of a certain logical syntax. As yet another indication of the correspondence between logical positivism and modern physics, Frank argued that both Einstein and Heisenberg had been inspired by Ernst Mach when they formulated their theories of relativity and quantum mechanics.[116] As a result of this conference, Bohr may have succeeded in distancing himself from idealistic connotations of complementarity to a certain degree. It has been suggested that Bohr's contact with the logical positivists in the early 1930s influenced his way of expression in his works from then on, so that he would avoid statements that could be interpreted as metaphysical such as references to a microphysical world in-itself behind the measured phenomena.[117] At least Bohr and the logical positivists seem to have found a common ground in some of their philosophical views which, however, they had reached in totally different ways. As stated by Jan Faye and Henry Folse, Bohr was "sympathetic with the positivists' concern to limit the meaningful use of language on matters beyond human experience. For the positivists, this limitation was drawn on the basis of a verifiability criterion of meaning, whereas in Bohr's case, it was a consequence of the physical conditions necessary for the well-defined applicability of the unavoidable classical descriptive concepts".[118] For Bohr, physics was about what can be said about nature, it was a question about which concepts could be used in an unambiguous way, not what *is* in an ontological sense. All the same, later in the century Bohr's interpretation of quantum mechanics was considered positivistic by leftist scholars, as we shall see.

Bohr's French Connection

As mentioned in the beginning of this chapter, Rosenfeld became part of an extraordinarily wide network of physicists, historians, and philosophers of science across Europe and particularly in Paris, within which he spread the gospel of Bohr's complementarity whenever the opportunity arose. The English and German

[116] Frank (1936b). Strauss (1936). Bohr (1999c).
[117] Faye and Folse (1998), pp. 8–12. Faye (2010).
[118] Faye and Folse (1998), p. 12.

translations of Bohr's booklet from 1929 included an "Addendum", in which Bohr specified that his suggestion that the epistemological lesson of quantum physics may lead to a deeper understanding of problems in biology too, should not be misunderstood as to say that biology could be reduced to chemistry and physics. Rosenfeld's French translation of Bohr's booklet from 1929 included an extra passage written by Bohr added at the end of the original "Addendum", stating the following: "Notwithstanding the intrinsic interest the biological and psychological questions have, even for those who like me [Bohr] are strangers to these fields, my primary aim in dealing with them in these articles has been to throw light on the physical and epistemological problems met with in the atomic theory".[119] This can be interpreted as an attempt at beating a retreat from Bohr's extension of complementarity into the realms of biology and psychology and his statements about free will. Contrary to Bohr, at the time Rosenfeld plainly denied that complementarity was relevant in domains such as biology and psychology.[120]

Not surprisingly, then, Rosenfeld was eager to save Bohr's reputation from Jordan's anti-materialistic views on Bohr's extension of complementarity to biology and psychology. In November 1937, Rosenfeld warned Hélène Metzger, the French philosopher of science and historian of early modern chemistry and natural philosophy,[121] "that Bohr's sectators (if one can call people so who endeavour not to be sectarian) cannot recognize as orthodox the considerations of Jordan (for example in his book *Anschauliche Quantenmechanik*) which claims to attach complementarity to neo-positivism. (Besides, poor Jordan seems, so Ph. Frank told me, to be disavowed also by the positivists!)"[122] Odd as it seems, however, Rosenfeld was never to comment more explicitly on Jordan's viewpoints, and the two seem never to have interacted after the 1930s. In print, Rosenfeld only

[119]N. Bohr (1932), p. 21. English translation quoted from *BCW*, Vol. 10, p. xxvii.
[120]RP, "*History of Science* 1: Histoire des sciences, etc. 1930–1940", L. Rosenfeld to H. Metzger, 3 Nov 1937.
[121]About H. Metzger, see for example Chimisso and Freudenthal (2003). Chimisso (2008), pp. 109–123.
[122]RP, "*History of Science* 1: Histoire des sciences, etc. 1930–1940", L. Rosenfeld to H. Metzger, 3 Nov 1937. Originally in French. Jordan (1936).

commented on what he considered Jordan's sympathetic sides.[123] The reason may be that Rosenfeld deeply admired Jordan as a physicist; he had been Rosenfeld's teacher while in Göttingen. Or, as indicated in the above quotation, perhaps Rosenfeld pitied Jordan as a tragic figure, whose extremely reactionary viewpoints were not worth challenging. Or he thought it the best strategy to ignore Jordan's reactionary views in order to attempt to prevent them from attracting a lot of attention.

Rosenfeld seems to have considered what he later called Heisenberg and Carl Friedrich von Weizsäcker's "irrelevant idealistic stuff" much more dangerous than the views of Jordan.[124] By the mid-1930s, influenced by discussions in Leipzig with his student and close friend Weizsäcker and the Kantian scholar Hermann, Heisenberg began considering Kant's philosophy relevant for understanding quantum mechanics.[125] Hermann joined Heisenberg's seminar as a visitor at the Physics Institute in Leipzig in the spring of 1934 where she worked with Weizsäcker and B. L. van der Waerden. Rosenfeld strongly disapproved of Heisenberg's support of the Kantian study group in Leipzig. As he complained to Strauss, "I also cannot agree with Heisenberg's attitude, and Bohr is also somewhat sad about that, because, he says, it is psychologically (and of course in the first instance substantively!) important that one does not enter into any compromises with philosophers".[126] This was the same view Bohr had expressed in the interview with *Izvestia*, quoted above; philosophers should adapt to the new discoveries in science, not vice versa. However, Rosenfeld did not enter into an open dispute with Heisenberg at the time. After the Second World War, Rosenfeld every now and then criticized Heisenberg's idealism "with unsparing frankness" through reviews of Heisenberg's books.[127] At the same

---

[123] See for example, Rosenfeld (1951). See also Kuhn and Heilbron (1963a), pp. 14, 18–20.
[124] RP, Supplement, L. Rosenfeld to H. P. Stapp, 4 May 1971.
[125] See for example, Heisenberg (1934), pp. 700–701. Heisenberg (1971). Camilleri (2009b), pp. 5–6, 133–151.
[126] RP, "Correspondance particulière", L. Rosenfeld to M. Strauss, 16 Nov 1935. (Emphasis in original.) Originally in German.
[127] Rosenfeld (1960). As usually was the case with reviews of works by his close colleagues, Rosenfeld sent the review to Heisenberg before its publication. Heisenberg responded to

time Rosenfeld attempted to overcome political barriers in order to keep his friendship and professional relation with Heisenberg, while he seems not to have interacted with Jordan.[128]

As regards Hermann, Rosenfeld was rather critical of her Kantian interpretation of quantum mechanics put forward in her treatise *Die naturphilosophischen Grundlagen der Quantenmechanik* (1935).[129] She had among other things discovered a flaw in von Neumann's proof of the impossibility of so-called hidden variables, which, however, seems to have been largely ignored at the time.[130] Weizsäcker reviewed Hermann's treatise favorably,[131] but she did not stand a chance with Rosenfeld, who wrote to Strauss:

> Any discussion with Miss Hermann seems to me idle not only in and of itself, but even directly damaging for the lady herself; that is to say this could strengthen her even more in her belief in the importance of her work... As you see, I estimate Miss Hermann's work more as an individual-psychological document than a scientific contribution; one could perhaps give it the title "The sorrows of young Grete" with the subtitle "How she has overcome them". From this it appears only that Friesian philosophy (whatever it is) is not as malicious a scourge as Neo-positivism.[132]

some of Rosenfeld's criticism in RP, "Correspondance particulière", W. Heisenberg to L. Rosenfeld, 16 Apr 1958. See also Rosenfeld (1952). Rosenfeld (1979t), pp. 480–481. Rosenfeld (1979b), and Jacobsen (2007), pp. 14, 27–28.

[128]Heisenberg stayed with the Rosenfelds when he visited Manchester in 1947. Rozental (1971), p. 13. Heisenberg also visited the Rosenfelds in Manchester in February 1956. RP, "Supplement H 1969–1971", W. Heisenberg to L. Rosenfeld, 17 Feb 1956.

[129]Hermann (1935).

[130]*Ibid.*, Hermann was also an active socialist. When Hitler came to power, she took part in the underground movement against the Nazis, but fled from Germany in 1936 to Denmark and later England. In Denmark she stayed at Østrupgaard, where her former colleague Minna Specht had opened a socialist school for children. Herzenberg (2008). Von Neumann seems to have been considered a representative of the "orthodoxy" in Copenhagen at the time, at least by Rosenfeld. RP, "Correspondance particulière", L. Rosenfeld to M. Strauss, 19 Sep 1938. About von Neumann's no-hidden-variables proof, see for example Breuer (2001).

[131]Mehra and Rechenberg (2001), pp. 712–713.

[132]RP, "Correspondance particulière", L. Rosenfeld to M. Strauss, 16 Nov 1935. Originally in German. Rosenfeld of course referred to Goethe's famous novel *The Sorrows of Young Werther*.

Thus, Rosenfeld considered her thesis a personal judgment rather than objective science. The quotation also suggests that the logical positivists did not escape criticism from Rosenfeld either. Nevertheless, Rosenfeld seems to have regarded Frank as an ally. After the Second World War, Rosenfeld asked Frank if he would consider taking up a position at the University of Utrecht.[133]

The astrophysicists Jeans, Eddington, and A. A. Milne were portrayed as the three archangels in the "Blegdamsvej Faust".[134] With their blending of religion and science, their impact on religious circles, and insensitivity to the social relations of science, they were considered reactionary by left-wing materialists, including Rosenfeld, to whom they constituted one of the greatest dangers to the emergent culture of science. Rosenfeld clearly felt committed to fight such idealistic views on science and turn the profession in a progressive direction.[135] The three astrophysicists were subject of discussion in Chandrasekhar and Rosenfeld's correspondence in the mid-1930s. Chandrasekhar was not motivated by Marxism, and Rosenfeld was not able to convert him to it, but they shared repulsion for viewpoints of the venerable "high priests".[136] In April 1935, Chandrasekhar expressed how bitterly disappointed he was about Milne's use of astrophysics to defend the existence of God.[137] Rosenfeld in return deplored that "[i]t is of course no use attacking him [Milne], because the people who ought to be warned against that evil kind of misuse of science, namely the laymen, finding more appeal in shallow speculations than in earnest thinking, will take the offence for themselves, I mean they will consider you to be the offender, not Milne or Eddington (not to speak of Jeans!)".[138] Speaking of Jeans, however,

[133]RP, "Utrecht (1940–1947): Political (1945–47)", P. Frank to L. Rosenfeld, 15 May 1946.
[134]*Faust and Journal of Jocular Physics Volumes I, II, and III.* Reprinted on the Occasion of Niels Bohr's Centenary, 7 Oct 1985, NBA.
[135]Werskey (1988), p. 152.
[136]Wali (1991), pp. 123–146. RP, copy of letter from L. Rosenfeld to S. Chandrasekhar, 14 Jan 1935. Original in CP.
[137]RP, "Copenhague (1): Chandrasekhar 1932–1936", Chandrasekhar to Rosenfeld, 26 Apr 1935.
[138]RP, copy of letter from L. Rosenfeld to S. Chandrasekhar, 7 May 1935. (Emphasis in original.) Original in CP.

three months later Rosenfeld informed Chandrasekhar how he had taken steps in a historical paper, "The first phase in the evolution of the quantum theory", "to reduce Jeans's part in that drama to its proper value, which is rather that of the villain".[139] Thus, Rosenfeld had taken up his own fight against the reactionary "charlatans". In the paper mentioned Rosenfeld introduced Jeans' reactionary position in the following way:

> This picture of the acquisition of a great advance in our knowledge of nature would lack a very human feature if, alongside of the enthusiastic efforts of pioneers like Planck and Einstein, the crisis generated by the problem of black-body radiation had not also provoked a reactionary attitude in other minds. Such an attitude consists in refusing to see that the solution to a crisis requires renunciation of previously accepted ideas; consequently, the reactionary tries, in order to deny the difficulties, to disguise their nature with the help of some distinctions which can only be purely verbal. In the example which concerns us, it is J. H. Jeans, the currently fashionable author of several popular works, whom we find as a representative of this tendency.[140]

Rosenfeld was referring to Jeans' persistent resistance until around 1913 to the transition from classical to modern physics, that is, the new quantum view of radiation. After that time he developed into one of quantum theory's leading spokesmen in Britain, which, however, seems to have pleased Rosenfeld even less because of the idealism Jeans used as a framework.[141]

The French translation of the collection of Bohr's essays from 1929 did not stand alone for long. Rosenfeld also translated Klein's popular Swedish books on relativity theory and quantum theory from 1933 and 1935 into French. These books were written in the form of dialogues and contained, according to Rosenfeld, "the necessary

[139] RP, copy of letter from L. Rosenfeld to S. Chandrasekhar, 8 Sep 1935. Original in CP.
[140] Rosenfeld (1979w), p. 217.
[141] Kuhn (1978), pp. 231–232.

antidote against Jeans and Eddington".[142] He asserted further that "[t]hey would come in very handy in popularizing modern physics in an authoritative way among a French audience. I also think especially of our students; in the beginning of their study of physics they need exactly what you have in these small books — here I think not only of the presentation of facts, but also of the scientific spirit itself, which one feels so beautifully and vigorously in them".[143]

Rosenfeld's one-volume translation of Klein's popular texts was published in 1938.[144] Even before the book came out, he introduced it to Hélène Metzger whom he met at the 4th International Congress for the History of Science in Prague in September 1937. At that congress Rosenfeld may have presented his paper, "Le dualisme entre ondes et corpuscules", which is a historical introduction to Bohr's complementarity and hence another contribution to promoting Bohr to a French lay audience. The paper was published in the journal of the International Academy for the History of Science, *Archeion*.[145] Rosenfeld clearly managed to arouse Metzger's interest in Bohr's philosophy. They continued discussing quantum theory, complementarity, and history and philosophy of science during the fall of 1937 and spring of 1938.[146]

Among the problems Rosenfeld was faced with clarifying when propagating Bohr's quantum epistemology was the theoretical status of Bohr's complementarity. Complementarity should be understood

[142]RP, "Correspondance particulière", L. Rosenfeld to O. Klein, 28 May 1936. See also Rosenfeld to Klein, 28 Apr 1936, in which Rosenfeld sympathized with Klein's "beautiful presentation" of Rayleigh's law, the law which is often denoted Rayleigh–Jeans' law. Rosenfeld found that Jeans did not deserve credit for this law.
[143]RP, "Correspondance particulière", L. Rosenfeld to O. Klein, 26 Sep 1935. Originally in Danish.
[144]Klein (1935b). Klein (1933). Klein (1938). RP, "Correspondance particulière: Klein".
[145]Rosenfeld (1937b). Rosenfeld published another paper in that journal: Rosenfeld (1938).
[146]RP, "*History of Science* 1: Histoire des sciences, etc. 1930–1940". Despite being Jewish, Metzger courageously refused to hide after the occupation of France, and even as a kind of statement took openly part in an informal group of Jewish scholars and public servants, the *Bureau d'études juives* (Office for Jewish Studies), who met on a weekly basis in order to learn more about the history of the Jewish tradition to which they belonged and which was their "crime". In February 1944, she was arrested by the Gestapo and deported to Auschwitz where she was killed. Freudenthal (2010).

as a "logical method, and not as 'hypothesis' or 'postulate' or 'principle'," according to Rosenfeld.[147] Moreover, he was met with puzzlement about the purpose and strength of Bohr's logical analyses of measurement situations. Rosenfeld told Metzger that "absence of formulas does not mean verbiage. In fact, "scientists" often reproach "philosophers" with their verbiage without realizing that a cluster of formulas is quite as reprehensible as a cluster of words as a means of compensating for one's lack of comprehension".[148] Rosenfeld was in favor of Bohr's simplified accounts of the physics involved when explaining complementarity to lay people like philosophers, a situation which Strauss also faced in his intimate collaboration with the logical positivists. Rosenfeld wrote to Strauss in 1935, "I believe more generally that it would be advantageous to present the complementarity argument first of all with complete simplicity and with the renunciation of axiomatic rigor, and then to bring in the axiomatic subtleties. For it seems to me that precisely in order to instruct philosophers, a particular emphasis on the simple physical background is urgently necessary; should one not attempt to teach the philosophers that first of all they have to understand the new physical situation, before they begin axiomatizing about it?"[149] Another problem was how to understand the relationship between physics and mathematical formulation in quantum theory. All the talk about interpretation mystified observers. Rosenfeld reassured Metzger that the mathematical formalism of quantum mechanics played exactly the same role as mathematics in Einstein's and Maxwell's theories: "[i]t is a technique... which does not dispense from understanding [the underlying physical ideas] (which quite a few mathematicians lose sight of)".[150]

---

[147] RP, "*History of Science* 1: Histoire des sciences, etc. 1930–1940", L. Rosenfeld to H. Metzger, 3 Nov 1937. Emphasis in original. Originally in French.
[148] RP, "*History of Science* 1: Histoire des sciences, etc. 1930–1940", L. Rosenfeld to H. Metzger, 1 Dec 1937. Originally in French.
[149] RP, "Correspondance particulière", L. Rosenfeld to M. Strauss, 16 Nov 1935. Originally in German.
[150] RP, "*History of Science* 1: Histoire des sciences, etc. 1930–1940", L. Rosenfeld to H. Metzger, 1 Dec 1937. Emphasis in original. Originally in French.

Whether or not Rosenfeld succeeded in convincing the French and others about Bohr's viewpoints, with time he certainly succeeded in becoming well-known as Bohr's spokesman in a French context.[151] Apart from fighting reactionary idealists and promoting complementarity, Rosenfeld also became concerned with establishing a dialectical materialist understanding of complementarity. In connection with their discussion of the reality criterion put forward in the EPR paper versus complementarity, Strauss suggested to Rosenfeld that the only philosophical position he could think of as forerunner of the complementarity view was dialectical realism.[152] The first time we encounter Rosenfeld's attempt at amalgamating complementarity and dialectical materialism is in May 1937. Rosenfeld jotted down some comments in his notebook under the title "La physique atomique et le déterminisme" (Atomic Physics and Determinism). Here he stated that "complementarity is a logical relationship between systems of concepts. It makes it possible to use systems of concepts in the limit of their validity while avoiding all contradictions. It is thus not an arbitrary renunciation of the 'comprehension' of the phenomena, but on the contrary an enrichment of thought... The idea of complementarity is of general applicability... It concerns a method. The only value of philosophy is the theory of knowledge... Science has a materialist content and a dialectical method".[153] The passage shows that Rosenfeld stressed the element of anti-reductionism in dialectical materialism and complementarity. He held an epistemological position and assigned to dialectical materialism a cognitive, logical, and methodological role. Dialectics was applicable as a scientific method in the same way as complementarity that was introduced by Bohr to account for wave–particle-dualism in quantum mechanics, for example. Contrary to Schrödinger, who wished to reduce quantum mechanics to wave

---

[151] See for example RP, "Correspondance particulière", A. Proca to L. Rosenfeld, 20 Feb 1950. Proca denotes Rosenfeld as Bohr's spokesman. George (1949), in "Manchester 7: Epistemology 1947–1958".

[152] RP, "Correspondance particulière", M. Strauss to L. Rosenfeld, 2 Nov 1935.

[153] RP, Rosenfeld notebooks, box IV notebook containing notes on "La physique atomique et le déterminisme" May 1937. Originally in French. See also RP, "*History of Science* 1: Histoire des sciences, etc. 1930–1940", L. Rosenfeld to H. Metzger, 3 Nov 1937.

mechanics, and other physicists such as David Bohm later in the century who wished to reduce quantum mechanics to a particle theory, Bohr held that both aspects of quantum phenomena were important for an exhaustive account, even though the two aspects were simultaneously mutually exclusive. Complementarity could account for both aspects without reducing the one to the other. Rosenfeld later called complementarity "the modern form of dialectics".[154]

Walter Heitler and Rosenfeld at the Copenhagen Conference in 1937. Courtesy of The Niels Bohr Archive.

Rosenfeld 1938. Courtesy of The Niels Bohr Archive.

[154] RP, "Epistemology 1955–1958", Draft of letter from L. Rosenfeld to J. Koefoed, 20 Feb 1957.

Dreary Times

In the fall of 1938, the Munich Agreement allowed Germany to annex the Sudetenland in southern Czechoslovakia.[155] At that time Rosenfeld lost his optimism about the success of the Popular Front and in preventing war. As he wrote to Strauss who was in Czechoslovakia at the time: "As for the general situation, I am too depressed at the moment to comment about it in any way".[156] Through his connection to Langevin and Edmond Bauer, Rosenfeld attempted to help Strauss move to Paris in early 1938 without success. In September, Rosenfeld deplored the desperate situation of his friend and that he could do nothing to help. Fortunately for Strauss, he got an opportunity to go to The Hague in the Netherlands, invited by Neurath, before the Nazis closed the Czechoslovakian border. During the war, Strauss, like Neurath, had the good fortune to escape from the Netherlands to Britain; Frank immigrated to the United States.[157]

In the late 1930s the great terror swept Soviet society, and physicists were by no means immune to the purges. Leningrad and its leading theoreticians were particularly in the danger zone even though the terror also struck in Moscow and Kharkov. In the process, the center of gravity of Soviet physics moved from Leningrad to Moscow.[158] Physicists in the West became immediately aware of the detention and imprisonment of their colleagues.[159] Rosenfeld and Peierls felt completely helpless when hearing about the fate of their friends and colleagues in Russia. In November 1938, Peierls expressed his concern to Rosenfeld about the destinies of Landau and Iurii Rumer in Moscow: "However, I fear that in these cases one can do nothing".[160] Rosenfeld agreed that "it's no use lamenting

[155] Hobsbawm (1995), p. 146.
[156] RP, "Correspondance particulière", L. Rosenfeld to M. Strauss, 19 Sep 1938. Originally in German.
[157] RP, "Correspondance particulière", L. Rosenfeld to M. Strauss, 10 Feb 1938. Rosenfeld to Strauss, 19 Sep 1938. BSC, M. Strauss to N. Bohr 1946. Reisch (2005), p. 53.
[158] Josephson (1991), p. 276.
[159] See for example W. Pauli to P. Epstein, 10 Dec 1937, in Hermann *et al.*(1985), Vol. 2, pp. 541–543. Pauli was in Moscow in September 1937 for a congress on nuclear physics.
[160] RP, "Correspondance particulière", R. Peierls to L. Rosenfeld, 5 Nov 1938. Originally in German. About Landau's and I. Rumer's "crime", see Hall (2008), p. 257.

over them".[161] Landau and Rumer were arrested on the same day in 1938. Rumer was accused of being a German spy. Landau was accused of all sorts of crimes including the approval of a leaflet opposing the regime. Kapitza made several efforts to free Landau; including writing to Stalin and Molotov that he needed Landau's help for his research on liquid helium. Kapitza succeeded in the end and Landau ended up spending "only" one year in prison while Rumer spent ten years.[162] Boris Hessen was arrested on 21 August 1936 and he was shot on 20 December 1936.[163] Towards the end of 1937, the German physicist Frederick (Fritz) Houtermans, who had immigrated with his family to the Soviet Union in December 1934 (after he had escaped from the Nazi regime to England in 1933 because of his leftist sympathies) and worked at Kharkov, was imprisoned under the suspicion of being a spy.[164] A campaign to free him was initiated among physicists in the West, including Houterman's wife Charlotte, Bohr (who helped Charlotte Houtermans to escape from Russia with the couple's small children), Blackett, Bernal, Irène and Frédéric Joliot-Curie, and Jean Perrin.[165] Rosenfeld was a close witness to the campaign and informed Pauli about the situation in January 1938.[166] Fock was arrested in 1937 and imprisoned for about one week, but was set free, helped by Kapitza's intervention on his behalf.[167] Bronstein was arrested on 6 August 1937. The alleged

[161]Bodleian Library, Sir Rudolf Peierls papers, L. Rosenfeld to R. Peierls, 9 Nov 1938. Originally in German.
[162]Josephson (1991), pp. 311–312. Kojevnikov (2004), pp. 91, 117–120. Aaserud (2005), p. 10.
[163]Chilvers (2003), p. 433. Kojevnikov (2004), p. 91. Josephson (1991), pp. 309–310.
[164]Khriplovich (1992).
[165]Werskey (1988), footnote p. 210. Sheehan (1993), pp. 234, 417. Khriplovich (1992), p. 33.
[166]W. Pauli to V. Weisskopf 13 Jan 1938, in Hermann *et al.* (1985), Vol. 2, pp. 547–548. See also pp. 551, 596. In April 1940, Houtermans was extradited to the Gestapo together with other German and Austrian refugees from Nazism. In Germany he was saved by the intervention of the physicist Max von Laue. Being kept under observation by the Gestapo, he was able to get a position in a private laboratory with the help of von Laue and Weizsäcker. Khriplovich (1992), p. 35.
[167]Josephson (1991), pp. 312–314.

crime was Trotskyism and he was executed in a Leningrad prison in February 1938.[168]

Fission — A Question of Priority

In the midst of the grim times and the danger of war breaking out, Rosenfeld visited the US with Bohr in early 1939, where they stayed at the Institute for Advanced Study in Princeton from 16 January to 20 April. Before leaving the US on 5 May, Rosenfeld made short visits to a number of places, *viz.*, the University of Rochester and Cornell University in Ithaca, both in the state of New York; Harvard University and M.I.T.; he attended the Washington meeting of the American Physical Society; and finally visited McGill University in Montreal, before returning to Europe.[169] His trip was funded by the Belgian American Educational Foundation.

Just before their departure for the US on 7 January 1939, Otto Robert Frisch, who worked at the time at Bohr's institute, had told Bohr about his and his aunt Lise Meitner's theoretical explanation of Fritz Strassmann and Otto Hahn's latest experiments in Berlin. Strassmann and Hahn had found that bombarding uranium with slow neutrons resulted in the formation of products behaving chemically like barium. Calculations based on Bohr's liquid drop model by Meitner and Frisch showed that it was in fact possible that a uranium nucleus could split into two pieces of about equal size. Bohr reacted with great enthusiasm to the news, but promised Frisch not to tell anyone until Frisch and Meitner had published a note about their theoretical explanation in *Nature*. However, six weeks went by before the note was published, because Frisch wanted first to confirm Hahn and Strassman's discovery and investigate the phenomenon further by setting up experiments in the basement of Bohr's institute.[170] In the meantime, Bohr and Rosenfeld discussed the new discovery on their way to the United States: "As we were boarding the ship,

[168] *Ibid.*, pp. 314–315. Gorelik and Frenkel (1994), pp. 144–145. Gorelik (2005).
[169] RP, "Liège (1922–1940): Conférences etc.: Princeton (1939)", Report for the Belgian American Educational Foundation about Rosenfeld's stay in America in 1939.
[170] See for example, Rhodes (1986), pp. 262–264.

Bohr told me he had just been handed a note by Frisch, containing his and Lise Meitner's conclusions; we should "try to understand it". We had bad weather through the whole crossing, and Bohr was rather miserable, all the time on the verge of seasickness. Nevertheless, we worked very steadfastly and before the American coast was in sight Bohr had got full grasp of the new process in its main implications".[171] According to Rosenfeld, it was his impression that the news had already been published or was just about to be published in *Nature*, and he was not aware of Bohr's promise to Frisch.[172] When Bohr and Rosenfeld arrived in New York, Bohr remained there while John Wheeler brought Rosenfeld directly to Princeton. At a meeting here in the "Journal Club" in the evening the next day, Rosenfeld was asked if he had any news from Europe. "Well, I had: I told them all about the problem we had struggled with during the journey. I did not know that Bohr had no intention of giving out the news so quickly, because he was anxious that Frisch's note should first come out in print".[173] And so Rosenfeld broke the news to the audience unaware of the commotion it would bring about. A storm broke loose: "The effect of my talk on the American physicists was more spectacular than the fission phenomenon itself. They rushed about spreading the news in all directions".[174] A veritable race began among the experimentalists to observe the splitting of the atomic nucleus. Bohr became extremely concerned when he heard what had happened and sent several telegrams to the institute in Copenhagen trying desperately to convey the urgency of the situation; it was important that Meitner and Frisch's results be published immediately in order to secure priority for the discovery.[175] Even though there is no evidence that Bohr ever blamed Rosenfeld, the latter got quite nervous and desperate about the whole situation; not only had he let Frisch and

[171] Rosenfeld (1979k), p. 342. Rosenfeld (1963).
[172] Kuhn and Heilbron (1963c), p. 17.
[173] Rosenfeld (1979k), p. 343.
[174] *Ibid.*
[175] For a very good and gripping description of the train of events as well as Bohr's distress, see Stuewer (1985). See also Peierls (1986), pp. 52–76, as well as the correspondence between Bohr and E. Fermi, and between Bohr and R. Frisch in the same volume.

Meitner down, but he had also let Bohr down and was now constantly witnessing how upset Bohr was about giving "everyone concerned proper credit".[176] As a result Rosenfeld must have felt a heavy burden of responsibility and was extremely anxious that the people at Bohr's institute should grasp the urgency of the situation. At the same time, he probably witnessed at close hand how slowly the correspondence between Bohr and Frisch developed. When in a letter to Rosenfeld Møller threw in the casual remark, "I suppose you have already heard about Frisch's amusing experiments here concerning the splitting of the very heavy nuclei",[177] the tension Rosenfeld felt himself seems to have tipped over and he responded while he was still quite agitated. I quote Rosenfeld's letter here at length as well as Møller's response because the former conveys well the circumstances Rosenfeld found himself and Bohr in, and the latter provides a good impression of the atmosphere and the research situation at the Institute in Copenhagen:

> Now, my dear Møller, please do not think that I am crazier than usual if I request you to respond to this letter immediately and to make sure that the response is sent by a fast ship. Just read the rest of this letter and you will understand where this anxiousness about the speed of correspondence derives from. Because I will now attempt to give you a pale introduction to the hardships which Bohr and I have had to go through the last two weeks, solely because the importance and difficulty of maintaining the connections between us and the Institute do not appear to have been sufficiently appreciated in the Olympian clouds above Blegdamsvej. I was completely shocked when I read the last sentence of your letter: "I suppose you have heard about Frisch's amusing experiments..." I felt like an inhabitant in Chile in the midst of the ruins of his home receiving a letter with the comment: "you may have heard that there has been an earthquake in

[176] N. Bohr to E. Fermi, 1 Feb 1939, in *BCW*, Vol. 9, pp. 550–551, on p. 551.
[177] RP, "Copenhague: Møller (1935–1939)", Møller to Rosenfeld, 1 Feb 1939. Originally in Danish.

Chile". How is it possible that you people in Copenhagen show such an incredible lack of imagination?

So you say "amusing experiments". It does not occur to you to send us the news about it immediately. Why would we find it amusing to hear about it? And telegrams are so terribly expensive, and it would be so exhausting to compose a 25-word night letter. Let's instead wait a couple of weeks, that would make it even funnier. We went ashore without the least suspicion about the real state of the question. We only know about Hahn's experiments and about Frisch and Meitner's explanation, but not about the crucial triumph: the "amusing experiments". There are also physicists in this country; it is perhaps strange, but it is a fact nevertheless. They hear immediately about Hahn's experiments (how strange!) and at the same time through Bohr and me about Frisch–Meitner's explanation. But then comes the most improbable of all: when they hear about Frisch and Meitner's explanation, they also have the idea that it could be "fun" to make the same experiments! (Really the chain of events is a little more complicated, but I take the liberty to schematize in order to spare your failing imagination.) So we witnessed (very bad) experiments in Washington, without knowing, that already two weeks in advance Frisch had done everything much better (how "funny", isn't it?). The experiments took place in this way: first oscillograph-kicks that were somewhat more violent than the usual were observed; as soon as this observation was made for about 5 minutes, a connection was established with Science Service by means of a common phone.[178] Within a few hours all newspapers were full of amusing descriptions of the experiments, and even funnier considerations about the possibility of immediately blowing up the whole world.

[178]Science Service was formed in 1920 in order to meet the American public's hunger for news about science. Its purpose was expanding and upgrading the coverage of science news in the nation's press. Kuznick (1987), p. 13.

In Washington, Bohr attempted the best he could to secure at least the recognition Frisch and Meitner's explanation of Hahn's experiments deserved. The next day, back in Princeton, ... a letter arrived from Hans Bohr with a completely casual remark about how it was so "funny" to hear Frisch tell about his experiments! Now the case turned much more serious and more difficult for us (as you may understand despite your Olympian calm). Now it was up to us to secure Frisch the deserved recognition for the experiments: but some of the physicists in question (I will not mention who) proved disinclined to do that. They claimed that the experiments were done independently, and they went so far as to add that the very explanation by Frisch and Meitner was "obvious" and that it was enough to have heard about Hahn's chemical experiments, which were "public property". In order to fight this idea we had first of all to have precise information from Olympus. We therefore sent several telegrams and had partial answers. The last telegram unfortunately contained the sentence "KEEP SENDING TELEGRAPHIC INFORMATION", as an immediate result of which no further information was sent to us.

Now Bohr has done very "funny" theoretical considerations about the matter — which of course I will not tell you about — and submitted it to *Physical Review*, where they will be published already on 15 February. In this way he had the opportunity to give an account of the history of the discovery. But the others also have notes in the same issue, and I fear that an awkward situation may arise. Let us hope for the best, however.

Have I impressed upon you the graveness of this whole matter? If that is the case, which I dare hope, would you do us the favor to spread all over the institute the crucial "slogan" for the salvation of our souls':

KEEP SENDING TELEGRAPHIC INFORMATION[179]

---

[179]MP, L. Rosenfeld to C. Møller, date missing, 1939. Emphasis in original. Originally in Danish. Rosenfeld kept a copy of this letter in his drawer. See also the letter from

Møller responded rather matter-of-factly and in a completely unsentimental way:

> [C]an I just present the story as it looks from "Olympian calm". The reason that I, like others from the Institute merely mentioned briefly Frisch's "amusing" experiments was that I honestly thought that Frisch had sent a telegram to America because that is what he promised. Of course we were not ignorant of the fact that there were also physicists in America and that they might perhaps get the same idea as Frisch, but this is a risk which unfortunately will always be there in science, and I cannot believe that you would propose that we should have employed the same method, which you describe so well in your letter and which apparently was used by the Americans. After all, here at "Olympus" it has always been customary to send notes to Nature and not to "Ekstrabladet" [the Danish tabloid] and that is also really what happened in this case. Naturally I realize that it would have been best if Frisch had sent a telegram immediately. However, as things are, I can't really see it in any other way than that the Americans have made the same discovery independently.[180]

Møller's easy-going attitude seems to be part of the spirit of what, according to Aaserud, marked the working atmosphere at Blegdamsvej.[181]

It was a questionable pleasure but an instructive experience for Rosenfeld to be the cause of the leakage of the news of the discovery of fission, and the subsequent uproar in connection with the race to the experimental confirmation of it. Despite this unfortunate experience, in the end Rosenfeld seemed rather content with his first visit to the US.[182] He wrote to Møller that he "had collected many variegated

---

N. Bohr to R. Frisch, 3 Feb 1939, and Frisch to Bohr, 15 Mar 1939, in *BCW*, Vol. 9, pp. 563–566.

[180]RP, "Copenhague: Møller (1935–1939)", Møller to Rosenfeld, 26 Feb 1939. Originally in Danish.

[181]Aaserud (1990), pp. 246–247.

[182]RP, "Liège (1922–1940): Conférences etc.: Princeton (1939)", Report for the Belgian American Educational Foundation about Rosenfeld's stay in America in 1939.

impressions and lost a few prejudices about the Americans as well as about us".[183] As we shall see in Chapter 5, however, Rosenfeld distanced himself radically from the Americans during the Cold War. He gave several talks about "the field theory of nuclear forces" on his tour, visited laboratories with "interesting equipment";[184] at Cambridge, Massachusetts, he met the nestor in history of science and fellow national, George Sarton, and apparently he discussed the possibility of taking up a position at the Macdonald Physics Laboratory at McGill University in Montreal and in Québec, neither of which, however, materialized.[185] We shall see, however, that Rosenfeld was becoming very serious about leaving Belgium in order to nurture his career in physics and perhaps to raise his social standard, but most importantly to leave behind what he found an antiquated Belgian teaching system; hence he continued to look for opportunities to go abroad.

Meanwhile, the political situation in Europe became even more threatening. Rosenfeld wrote to Møller that although he and Yvonne had planned some vacation in May, Yvonne would not visit the US under the current circumstances and that he would therefore return to Europe in the first week of May. With reference to the German invasion of Czechoslovakia in March and Mussolini's occupation of Albania in early April 1939, Rosenfeld sarcastically remarked to Møller: "Let's hope that before this time [May] there will not be a 'peaceful' occupation of Denmark by German 'protectors'."[186]

The Popular Front strategy of the Left failed; Hitler was unstoppable, Nazism advanced and war seemed inevitable. The final blow to the Popular Front came with the Molotov–Ribbentrop pact of

[183]MP, L. Rosenfeld to C. Møller, 19 Apr 1939. Originally in Danish.
[184]RP, "Liège (1922–1940): Conférences etc.: Princeton (1939)", Report for the Belgian American Educational Foundation about Rosenfeld's stay in America in 1939.
[185]RP, "Liège (1922–1940): Conférences etc.: Princeton (1939)", A. N. Shaw to L. Rosenfeld, 4 Jul 1939. RP, "Correspondance particulière", G. Gamow to L. Rosenfeld, 18 Apr 1939. About Sarton, see Chapter 1.
[186]MP, L. Rosenfeld to C. Møller, 19 Apr, 1939. Originally in Danish. RP, copy of letter from L. Rosenfeld to S. Chandrasekhar, 10 Feb 1939. Original in CP. Rosenfeld tells Chandrasekhar about his and Yvonne's plans for a vacation and expresses the wish that they could meet while he is in the US.

23 August 1939 between Nazi Germany and Russia. With this pact Stalin hoped to keep the USSR out of the war. For Hitler the pact seemed simply opportune at the time. As would become clear he had not given up his idea of conquering the European part of Russia. Since the Comintern served as an instrument of Soviet foreign policy, it followed suit and changed its views overnight on Nazism and on the war that loomed large in the horizon. It was now regarded as an "imperialistic" war in which the imperialistic powers Britain and France wanted to enforce a new division of the world market. Not surprisingly, the communist parties experienced some problems with this radical political turn. Even so, they too followed suit.[187] It is unclear how Rosenfeld reacted to the pact, but given his consistent anti-Nazi and anti-fascist position as well as his earlier and later enthusiasm for the Popular Front he was most likely more than frustrated about this development. Eleven days later, the Second World War broke out.

[187]Hobsbawm (1995), pp. 147–151. Thing (1993), Vol. 1, p. 349.

Chapter 4
Surviving the War in Utrecht

This chapter treats Rosenfeld's time in Utrecht, the Netherlands, where he took up a position as professor of theoretical physics simultaneously with the German occupation of the Low Countries. Rosenfeld never wrote down his own account of the war time in Utrecht, but his recollections about Heisenberg's visit in Holland in 1943 were recorded by Rosenfeld's friend and colleague, Stefan Rozental, thirty years later. Through correspondence between Rosenfeld and colleagues in Copenhagen attention is also drawn to the conditions at the Institute for Theoretical Physics during the war. Even though the Institute for Theoretical Physics in Copenhagen was isolated from the world outside Germany and occupied Europe, work seems to have continued without notable disruptions during most of the war. This was even true for experimental work, because when the war began investments in new experimental equipment for research in nuclear physics had just been made.[1] Apart from experimental physics, which has drawn most attention because of its potential use in war efforts, scholars have mainly focused on Bohr's activities during these years. It was therefore quite surprising to learn about the extent of contact and collaboration that took place among theoretical physicists across occupied Europe despite the hardships of occupation. This chapter begins with Rosenfeld's choice to take up a position in Utrecht in the first place.

[1] Rozental (1967), p. 157. Aaserud (2005), p. 12.

The Second World War

On 1 September 1939, Germany attacked Poland, which resulted in a declaration of war from Great Britain and France two days later. The Second World War had begun. The Low Countries attempted for as long as possible to stay neutral but mobilized their armies.[2] At the end of January 1940, Rosenfeld complained to Møller about the way academics were treated as a consequence of the mobilization.

> Completely ruthlessly all professors and assistants liable for military service are ordered away and those left behind are assigned all the duties which those mobilized cannot fulfil themselves during their weekly leave. As payment for the extra work one receives enormous salary cuts and extra taxes. Personally I am so far one of the more privileged. Apart from the complaints one hears about the huge losses and victims which the mobilization costs the atmosphere here is, however, much more exhilarating than I had imagined. It can be characterized as determinedness, peacefulness, and confidence. The latter feeling is aroused I think (how justified I don't know) by the enormous military measures which can be observed everywhere. Along railway lines, highways, canals, beside bridges, and across the fields endless trenches spread all over the country — and barbed-wire roadblocks and vast tank roadblock. Everywhere small fortresses with machine guns, anti-aircraft batteries, etc., are hidden. The whole country looks like a fortress and swarms with military.[3]

Apart from this nuisance and the fear of war, Rosenfeld had on his mind his unsatisfied professional ambitions. He was more or less the only representative of modern physics in Belgium, and even though he had good opportunities for research and travel, he felt quite isolated.[4] In a letter to the influential director of the University Foundation, the National Fund for Scientific Research (FNRS), and the

[2]Arblaster (2006), p. 221. Cook (2004), p. 123.
[3]MP, L. Rosenfeld to C. Møller, 28 Jan 1940. Originally in Danish.
[4]BPC, Y. Rosenfeld to M. Bohr, 27 Dec 1940.

Francqui Foundation,[5] Jean Willems, written after he had in fact accepted a position in Utrecht, Rosenfeld explained that "it is correct that my situation at the University of Liège leaves nothing to be desired from the point of view of my personal possibilities of scientific work. But I have not been able to content myself with the egoistic satisfaction it is to continue my personal research in isolation".[6] What seems to have weighed him down was that he was frustrated as regards the Belgian system of education. During the preceding ten years he had witnessed the fruitful collaboration between experimentalists and theoreticians at Bohr's institute in Copenhagen and had become convinced that such conditions were worth striving for everywhere. He had just come back from the US where such collaboration was widely institutionalized, and this aspect is emphasized in Rosenfeld's report about his journey for the Belgian American Educational Foundation.[7] In this light Rosenfeld explained to Willems the specific difficulties he had faced in Liège: "I had a teaching mission to fulfil too and felt responsible for the formation of new recruits for science, a task which in the domain of physics demands close collaboration between the theoreticians and experimentalists. Now, despite my persistent efforts in the aim to coordinate the teaching in the different aspects of physics in a harmonious way, it has been impossible for me to find this collaboration in Liège".[8] It seems to have been what he elsewhere denoted "the vested interests of antiquated tradition" in the teaching system in Belgium which were the main reason for him to seek a job abroad.[9] Several years later he again reviewed what he had felt was the administrative mistake in the Belgian university system: "The teaching duties of a professor are much too

[5]About these institutions, see Halleux *et al.* (2001), Vol. 1, pp. 82–83.
[6]RP, "Utrecht: Nomination (1940)", Copy of letter from L. Rosenfeld to J. Willems, 14 Sep 1940. Originally in French.
[7]RP, "Liège (1922–1940): Conférences etc.: Princeton (1939)", Report about Rosenfeld's stay in the US for the Belgian American Educational Foundation. Schweber (1986), pp. 57–59.
[8]RP, "Utrecht: Nomination (1940)", Copy of letter from L. Rosenfeld to J. Willems, 14 Sep 1940. Originally in French.
[9]RP, "Liège (1922–1940): Conférences etc.: Princeton (1939)", Report for the Belgian American Educational Foundation about Rosenfeld's stay in America in 1939.

narrowly limited by the terms of his appointment, which minutely specify with which of the legally prescribed courses of lectures he is entrusted... As a result relations between professors tend to develop on a feudal level. Any suggestion of cooperation is usually looked upon as an encroachment on the jealously guarded domain within which autonomy is confined".[10] A consequence was that alternative initiatives were given up beforehand, and that teachers preferred to concentrate on their own research. "Hence Belgian universities offer the paradoxical spectacle of a high standard of scientific and scholarly activity flourishing on a thoroughly bad system of education".[11]

Rosenfeld's Belgian students seem, however, to have been fond of their teacher. A characterization of Rosenfeld and his examinations by one of his students in an article in the university periodical *L'Etudiant Libéral*, leaves no doubt that he was adored as a teacher.

> Mr. Rosenfeld is a small, plump man with a big bottom, whose skull is laid bare like a glossy steppe and [whose] chin is rosy and cleft like a small child's behind.
>
> A profound materialist, he possesses this remarkable originality to mix the mathematical development in his course with his personal philosophical remarks. Between two gradients he talks about the economic production of the slaves or the conception of Poisson [equations] in geometry, then he returns to Carnot and Maxwell.
>
> This is where things deteriorate, however. Because he does not only consider the analysis and the calculus of vectors as instruments; Mr. Rosenfeld considers them known in their remotest nooks. The result is that his course is much harder if not almost incomprehensible, and when the students set out for the examination it is almost like a lottery.
>
> The principal characteristic of this examination is that they can go to it issued with all the antiquarian books they can

[10] Rosenfeld (1949), p. 597.
[11] *Ibid.*

carry off from the different libraries in town or from elsewhere. Indeed, it is not a question of simply learning a series of ready-made formulas or arguments by heart but to understand the course and to know how to apply the notions in question in a concrete case.

This way of interrogating is certainly much more rational and its general application should perhaps be recommended, but the originality of it confuses and the results...

The results, well! Nobody knows them.

Behind his spectacles Mr. Rosenfeld remains impenetrable; he watches you with a slightly ironic air and one can find out neither what he thinks, nor how he gives marks. Some assert that he has never failed anyone, others that he prudently takes refuge behind his colleagues in order to fail [someone] in silence. There is a third hypothesis which probably reconciles the two preceding: Mr. Rosenfeld gives everyone an average mark which definitely finishes off those who are already afflicted, but which does not fell those who pull through with the other professors. Needless to add, it is much more likely that he fails those who, according to him, deserve it: it is much more simple and more normal than you think.[12]

Rosenfeld used history of science as a tool in a genetic epistemology which satisfied his own quest for understanding contemporary scientific concepts in the light of their origin, and which he therefore also found a good idea to use as a pedagogical tool in teaching physics to students. It was his hope that knowledge of the historical development of scientific concepts would make scientists more open-minded with respect to introducing new scientific concepts in the never-ending dialectical development of science. As the article suggests, that part may have gone over the students' heads. His teaching was quite challenging.[13] The underlying criticism of the Belgian teaching system in the article did not escape Rosenfeld's attention

---

[12]Salamandre (1940) in RP, "Liège 1922–1940". Originally in French. As regards Rosenfeld's use of history of science in his teaching, see also Serpe (1980), pp. 390–391.
[13]Hooyman (1979).

Monsieur Rosenfeld

Sketch of Rosenfeld by a Belgian student in 1940.

when he criticized the Belgian education system severely from the students' point of view; there was too much learning by heart, too little emphasis on the links between subjects, and no general supervision of the students.[14] The type of blunt humour contained in the article seems to have been highly estimated by Rosenfeld. He carefully saved the clipping for his records. When he later regretted that the relations between students and professors were "fundamentally unsound" in both Belgium and Holland, he noted, however, that "the Belgian student finds some relief in a sense of humour painfully absent from the Dutch environment".[15]

Uhlenbeck's Successor

In the neighbouring country, the Netherlands with its proud tradition in physics, a position in theoretical physics became vacant after

[14] Rosenfeld (1949).
[15] *Ibid.*, p. 597.

George E. Uhlenbeck left for the US in 1939. The Faculty of the University of Utrecht preferred a native Dutchman for the post and had Hendrik B. G. Casimir in mind, but he declined the offer. In June 1939, the Faculty prepared a list of other possible candidates, proposing in the following order: Walter Heitler, who, however, had left the Continent for Britain, Heisenberg then in Leipzig (and determined to stay in Germany and do what he could for physics and his country),[16] Schrödinger then in Ghent but on his way to Dublin, and finally Rosenfeld in Liège.[17] Of course Heisenberg and Schrödinger were famous for their contributions to quantum mechanics, and each had already received the Nobel Prize in 1932 and 1933 respectively. Besides, Heisenberg and Heitler were both students of Sommerfeld, a merit which was much coveted at European universities.[18] Apart from being Belgian and thus overlooked at first, Rosenfeld's highly abstract mathematical style in physics as described in Chapter 1, combined with the philosophical orientation he developed while in Copenhagen, did not make him the first choice for a practically minded experimental physicist in charge of a physical laboratory. However, in January 1940, Leonard Salomon Ornstein, the professor of experimental physics at Utrecht University, contacted Rosenfeld on behalf of the Faculty, asking for his CV and list of publications.[19] Rosenfeld was subsequently invited by Ornstein on behalf of the Physical Society to come to Utrecht in February to give a colloquium. On 15 February, Rosenfeld gave a talk there about his recent collaborative work with Møller on the meson theory of nuclear forces.[20] In the early days of April, Rosenfeld received a request from

[16]Mott and Peierls (1977).

[17]RP, "Utrecht: Nomination (1940)".

[18]The following quote of a letter from the Oxford physicist F. Lindemann to A. Einstein in 1933, when Oxford University looked for a new theoretical physicist, serves to testify that this may have been a common priority at several European universities when looking for candidates: "I have the impression that anyone trained by Sommerfeld is the sort of man who can work out a problem and get an answer, which is what we really need at Oxford, rather than the more abstract type who would spend his time disputing with the philosophers". Quoted in Moore (1992), pp. 269–270, and in Seth (2010), p. 3.

[19]RP, "Utrecht: Nomination (1940)", L. S. Ornstein to Rosenfeld, 2 Jan 1940.

[20]RP, "Utrecht: Nomination (1940)", L. S. Ornstein to Rosenfeld, 3 and 31 Jan 1940. Pais (1997), pp. 38–39.

the curators of Utrecht University asking if he would be willing to accept a nomination for the position in question, and since he had a favorable opinion of the Dutch milieu, of which he mentioned in particular Ornstein and Hendrik Kramers, Rosenfeld gladly accepted.[21] Ornstein was happy about Rosenfeld taking over the chair in theoretical physics and it was agreed that he should begin on 1 September. As soon as the nomination was announced on 7 May 1940, Rosenfeld was greeted warmly by his future Dutch colleagues and students, among them the student Abraham Pais.[22]

Given Rosenfeld's dissatisfaction with Liège, did he find what he was looking for in Utrecht? Indeed he did. "If [a would-be reformer of the Belgian universities] ventured across the Northern border, he would have the opportunity of observing a university system at work from which the glaring defects of the Belgian one are absent", he later wrote.[23] Rosenfeld emphasized the principles of freedom in Dutch university education, particularly the student's freedom to choose his subject and the autonomy of the professors in determining the extent and scope of their teaching. The former fostered the student's "intellectual curiosity, thoroughness, and independence of judgment".[24] On the other hand, it "requires from the students a good deal of steadfastness, initiative, and determination; it is very hard on students of weak character and mediocre intelligence".[25]

As for the Physical Laboratory at the University of Utrecht, Ornstein had been the leader of it for twenty years, and under his leadership it had become a first class research centre — well-equipped, highly organized and well-managed. With quite some success the laboratory had specialized in measuring intensities of spectral lines and other properties related to the structure of atoms and molecules. Even so Ornstein was skeptical about quantum theory. This was especially the case when the theory challenged causality

[21]RP, "Utrecht: Nomination (1940)", Curators, President and Secretary of the University of Utrecht to L. Rosenfeld, 30 Mar 1940.
[22]RP, "Utrecht: Nomination (1940)".
[23]Rosenfeld (1949), p. 597.
[24]*Ibid.*, p. 598.
[25]*Ibid.*, p. 599.

and determinism which Ornstein swore by all his life. It would have been interesting to know whether Rosenfeld and Ornstein discussed this issue. Ornstein had a strong personality. Besides, he was known for his sincere concern about the education and future career of his students and for his innovative ideas in didactics. He was concerned with cultivating the parts of technical physics that could be useful in society and was a pioneer with respect to cooperating with industrial research. He valued a practical education which gave his students good prospects for getting jobs in industry.[26] When Rosenfeld talked about the "fundamentally unsound" relationship between professors and students also in Holland, he ought perhaps to have mentioned the Laboratory in Utrecht as an exception. On the other hand, he never really came to experience Ornstein in action, as will become clear.

Although Rosenfeld was about to spend "some of the pleasantest years of my academic life" at the Utrecht faculty of science, as he wrote in 1949, for political reasons it turned out not to be the best time to take up a position in the Netherlands.[27] German troops invaded Belgium and the Netherlands in the early morning on 10 May 1940, just three days after Rosenfeld was appointed to his new job.[28] Belgium capitulated on 28 May, and a military administration was imposed on the Belgian territories. In Holland, on the other hand, the Germans installed a Reichskommissar appointed personally by Hitler, *viz.*, the infamous Nazi and Chancellor of Austria since the *Anschluss*, Arthur Seyss-Inquart, who imposed a ruling system which proved to be much worse than a military administration. Whereas the aim of the military was to keep things quiet so that troop movements would not be hampered, the institution of civil leadership in Holland was merely the first step toward the Nazi aim of incorporating the country into a greater German *Reich*. In addition, Dutch efficiency, in contrast to the "deliberate incompetence cultivated by so many

[26]Heijmans (1994), pp. 162–166. I am grateful to Leo Molenaar for calling my attention to this work.
[27]Rosenfeld (1949), p. 595. Yvonne also reported to Margrethe Bohr during the war that Rosenfeld was much more content in Utrecht than in Liège. BPC, Y. Rosenfeld to M. Bohr, 17 Dec 1941.
[28]Presser (1988), p. 7.

Belgian functionaries", resulted in the end in a smaller percentage of the Jewish population dying in Belgium than in Holland.[29]

Mail delivery between Holland and Belgium and between Belgium and Denmark (which had been invaded already on 9 April)[30] was interrupted for several months, but Rosenfeld's aunt in Prague, Ada Pierre, managed to write to Møller in Copenhagen on behalf of her nephew to ensure Møller that Rosenfeld and his family had come safely through the initial terror of war.[31] The mail situation was normalized again around August, and Pais then reported to Rosenfeld that he had acquired the assistantship at the institute in Utrecht; because of the capitulation of Holland, Pais' Jewish predecessor had committed suicide.[32]

As agreed, Rosenfeld arrived with his family in Utrecht in September 1940. He wrote Møller and Bohr each a letter on 1 October to get news from Denmark and to announce his decision "despite the current uncertainty to accept an offer to come to Utrecht as Uhlenbeck's successor. We are already provisionally settled in Utrecht".[33] Rosenfeld was anxious to have his and Yvonne's furniture sent to Utrecht from the house they had rented in Copenhagen the previous decade. The furniture was sent by means of Bohr's and Møller's interventions and arrived safely at the Utrecht address around Christmas the same year.[34] As it happened, Bohr also sent a letter to Rosenfeld on 1 October, but probably to Belgium as he was unaware of Rosenfeld's new position in Utrecht.[35] Perhaps also in doubt about how to tackle the new situation under German occupation Bohr wrote the letter in German. All letters between Rosenfeld and colleagues in Copenhagen and Sweden (and even Kramers) were otherwise written in Danish. When he received news from Rosenfeld, Bohr warmly

---

[29]Arblaster (2006), p. 225. Pais (1997), p. 52.
[30]BSC, Supplement 1910–1962, L. Rosenfeld to N. Bohr, 9 Apr 1940.
[31]MP, A. Pierre, Prague, to C. Møller, 12 Aug 1940.
[32]RP, "Utrecht: Nomination (1940)", A. Pais to L. Rosenfeld, 11 Aug 1940. Pais (1997), p. 39.
[33]MP, L. Rosenfeld to C. Møller, 1 Oct 1940. Originally in Danish. RP "Copenhague: Møller (1940–1943)". Møller to Rosenfeld, 9 Oct 1940.
[34]MP. RP, "Copenhague: Bohr (1940–1948)" and "Copenhague: Møller (1940–1943)".
[35]RP, "Copenhague: Bohr (1940–1948)", N. Bohr to L. Rosenfeld, 1 Oct 1940.

congratulated him on the new position and expressed his relief that Rosenfeld and his family were safe. Bohr hoped that it would be possible for Rosenfeld to make a trip to Copenhagen in the near future, and he was also anxious to resume their collaboration on the measurability in quantum electrodynamics. Although Bohr returned to this issue in more or less every single letter throughout the next decade, Bohr and Rosenfeld only resumed their joint work on measurement problems in quantum electrodynamics in 1949. Bohr ended the above-mentioned letter by emphasizing "how happy I will be to hear from you often, about how work goes and how all of you are and in return [I] promise to strive to improve in the same manner".[36] The contact between Rosenfeld and the physicists in other occupied countries probably strengthened him psychologically; making him feel less isolated at a time when "patience", as he told Møller more than once, became "a virtue". It also served as motivation for continuing work.[37] However, the absence of direct dialogue could make the illusion of contact difficult to maintain. As Møller once excused himself to Rosenfeld: "Forgive me for the hopeless handwriting by which I jot down my letters — but the only way I can keep the illusion of contact is to write with a speed comparable to the speed with which I speak".[38]

Life in Holland under Nazi Occupation

In October 1940, the German authorities introduced race laws in Holland. The Nazis used race and not religion as their chief criterion for who was to be identified as a Jew. According to a circular from the German authorities distributed to the heads of all Dutch departments of education in October, 1940, Jews were defined to be everybody with one parent or grandparent known to have been a member of the "Jewish community" at one time or other. For a person of mixed descent to be termed Jewish he or she should have at least

[36]RP, "Copenhague: Bohr (1940–1948)", N. Bohr to L. Rosenfeld, 23 Oct 1940.
[37]MP, L. Rosenfeld to C. Møller, 23 Mar 1943, 8 Feb 1944.
[38]RP, "Copenhague: Møller (Jan–Apr 1940)", C. Møller to L. Rosenfeld, 17 Feb 1940. Originally in Danish.

three grandparents who were Jewish or if he or she was a practicing Jew or married to one, it sufficed with two Jewish grandparents.[39] By the end of the same month all employees at the universities had to complete declarations of descent. During the weeks of November, the measures that were expected to follow from this registration were awaited with great anxiety.[40] Rosenfeld's father was Jewish and Rosenfeld was therefore in a vulnerable position. In a letter to Møller of 3 November, Rosenfeld remarked that he hoped for the best! "You will understand that we are not without fears for the future and that the uncertainty of the moment spoils a little the joy of the establishment of our new home. But we do not lose courage".[41] Later in November, the Germans issued a decree banning Jews from all civil service positions, including all academic posts. In Leyden, the students called a strike which resulted in the Germans closing the university there. The moves by the Germans against Leyden University during the war were consistently and courageously opposed by the faculty of the university. For instance, when in March 1942 two faculty members were dismissed, the majority of the remaining faculty (56 in all) handed in their resignations, among them the physicist Kramers.[42] At the University of Utrecht the rector, the chemist H. R. Kruyt, took a compromising line with the German authorities.[43] Here three Jewish professors were dismissed in November 1940, among them Ornstein, the professor of experimental physics, who had been instrumental in Rosenfeld's employment. He became a broken man and committed suicide on 20 May 1941. Four assistants were dismissed, among them Pais.[44] Luckily for Rosenfeld he was not dismissed; hence he was considered Aryan by the Germans. This probably was related to his mother being Aryan, that he was married to a non-Jew, and lack of knowledge of his father's ancestors in Russia. According to Jean L. J. Rosenfeld, "my mother told

[39]Beyerchen (1977), pp. 13–14.
[40]Presser (1988), pp. 17–25. van Walsum (1995).
[41]MP, L. Rosenfeld to C. Møller, 3 Nov 1940. Originally in Danish.
[42]Dresden (1987), pp. 498–499.
[43]Rozental (1971), pp. 7–8.
[44]Pais (1997), p. 40.

me on several occasions much later that it was because there were separate German bureaucracies in different occupied territories and that communication between them was inefficient. Any information that may have been available to the Germans in Belgium (where the civil records concerning Father's parents would have been) may not have been transmitted to their counterparts in Holland... since our grandfather had died long before the war and our grandmother was non-Jewish, there would be no particular reason for the authorities in Belgium to initiate an investigation into the family records and send the information to Holland (they would not necessarily even be aware that Father was there), and if there was no evident connection between Father and Jewish communities or organisations to attract the attention of the Gestapo in Holland (other than the family name itself), they may not have sent a request to the Belgian authorities either, or if they did, bureaucratic inefficiency prevented any information from being transmitted".[45]

Due to delays in mail delivery, Møller did not immediately receive news from Rosenfeld until January 1941 and became quite worried.[46] When contact was resumed, Rosenfeld told Møller that "[w]e had in November some frightening weeks but now it is over and as long as the current state continues we must count ourselves happy".[47] However, as the Dutch learned the hard way in those years filled with terror, the "current state" never continued; the situation always grew worse by "[t]he cautious way in which the Nazis manoeuvred with us in Holland".[48]

In spite of the outcome of the descent declaration evaluation, Rosenfeld may have felt he was in an extremely vulnerable situation and could not afford to attract any kind of attention. According to Rosenfeld's daughter Andrée, they did not have their family name on

[45]Email correspondence J. L. J. Rosenfeld to A. S. Jacobsen 12 Jan 2005. Rozental (1971), p. 7.
[46]RP, "Copenhague: Møller (1940–1943)", C. Møller to L. Rosenfeld, 4 Jan, 22 Jan 1941. MP, "Rosenfeld 1935–1941", Rosenfeld to Møller, 7 Dec 1940.
[47]MP, L. Rosenfeld to C. Møller, 14 Jan 1941. Originally in Danish.
[48]L. Rosenfeld to R. Peierls, 4 Jun 1948, in Lee (2009), Vol. 2, p. 149. Rozental (1971), pp. 7–8.

the door.[49] Even so, during the war years the Rosenfelds made courageous efforts to help others in dangerous positions such as Pais; Julius Podolanski, a Polish-born German Jewish physicist who had assisted Kramers, and who hid in a windmill during the war; Wladyslaw Opechowski, a Polish physicist and assistant first of Kramers and later of Adriaan Daniel Fokker and supervised by Rosenfeld; and Jozeph Kazimir Lubański, a Polish physicist who stranded in Holland on his way to Copenhagen at the outbreak of the war, and was extremely unhappy in Holland during the occupation. Since Jews were soon forbidden admittance at the University, Rosenfeld for a while gave seminars secretly at his home, thus enabling Jewish physicists and students to take part. Meanwhile Yvonne was courier for the local resistance movement.[50]

Rosenfeld continued for several months to commute back and forth between Liège and Utrecht to take care of administrative and teaching duties.[51] From around the turn of the year 1940, mail was subjected to censorship which meant long delays in correspondence between Belgium and Utrecht as well as between Copenhagen and Utrecht, to Rosenfeld's great regret. There appeared to be problems with sending printed material, and in April 1941 Rosenfeld was informed that mathematical formulas were no longer considered admissible in letters. However, there seems to have been no problem with enclosed manuscripts containing formulas, except for the slow delivery.[52] Despite these nuisances Rosenfeld and Møller continued their collaboration steadfastly with the assistance of their respective assistants, Pais in Utrecht, Jean Serpe in Liège, Lamek Hulthén in Lund, Sweden (it was possible until the summer 1943 to travel between Denmark and "neutral" Sweden if in connection

[49]A. Rosenfeld in email correspondence with A. S. Jacobsen, 12 Jan 2005.
[50]MP, L. Rosenfeld to C. Møller, 20 Feb 1942. BPC, Y. Rosenfeld to M. Bohr, 4 Sep 1942. Rosenfeld (1948b). Walker (1989), p. 111. Dresden (1987), pp. 498–499. Rozental (1971), pp. 1, 6–8, 13. L. Rosenfeld (1955). Frank (1974), p. ix. A. Rosenfeld in email correspondence with A. S. Jacobsen, 15 Oct 2004. Podolanski became Rosenfeld's assistant after the war and followed him to Manchester. RP, "Correspondance particulière: Opechowski".
[51]Serpe (1980), p. 396.
[52]MP, L. Rosenfeld to C. Møller, 14 Jan, 12 Apr 1941.

with work), and in Copenhagen Rozental and Møller's assistant Ib Nørlund (Bohr's wife's nephew).[53] Via Hulthén in Sweden the group also had occasional contact with Gregor Wentzel in Zurich.

A further German decree in Holland soon meant that the final date for Jews to be permitted advancement for their doctor's degree was July 1941, and this had serious implications for the pace with which Pais had to finish his dissertation. In addition it required most of Rosenfeld's attention in this period.[54] The topic of Pais' dissertation was to formulate Møller and Rosenfeld's version of meson theory in terms of a specific five-dimensional description, the so-called de-Sitter space. In this formulation the universe is supposed to possess a fifth dimension confined to a very small extension.[55] After Pais' successful graduation, Rosenfeld sought help from Bohr and Oskar Klein (the son of a prominent rabbi in Sweden) to secure Pais' future both as a physicist and in general. Thus upon sending copies of Pais' dissertation to both Klein in Stockholm and colleagues in Copenhagen Rosenfeld asked whether Pais could come to Copenhagen or Sweden to continue his studies under Klein's or Bohr's supervision.[56] Of course the situation was difficult, as Rosenfeld wrote to Klein:

> Perhaps this thought will not seem reasonable at first during the current circumstances and, by the way, I do not even know if he could obtain a permission to travel at all; but if you think about it a little more carefully, you will see that there can be good reasons for him to try at least.
>
> What I would like to ask you today is first of all whether it would be possible for you to receive him in your circle; he is very clever, even if not yet completely mature; I think that he would benefit from working under your supervision. It is

[53]MP, L. Rosenfeld to C. Møller, 14 Jan 1941, 20 Feb 1942. RP, "Copenhague: Møller (1940–1943)", Møller to Rosenfeld 26 Nov 1940, 4 Jan, 22 Jan, 14 Oct, 8 Dec 1941. RP, "Correspondance particuliére: Hulthén and Serpe". Nørlund (1991), p. 113.
[54]MP, L. Rosenfeld to C. Møller, 12 Apr 1941. Pais (1997), p. 41.
[55]Pais (1997), p. 39.
[56]RP, "Copenhague: Møller (1940–1943)", C. Møller to L. Rosenfeld, 25 Aug 1941. Rosenfeld also sent a copy of the dissertation to A. Proca in Paris. RP, "Correspondance particulière: Proca", A. Proca to L. Rosenfeld, 15 Apr 1942.

> in addition my duty to inform you that he is Jewish and I therefore have to ask you if this circumstance will mean any insuperable difficulty as regards you or the authorities.[57]

Klein cannot have been in a position to help because nothing came out of this inquiry. A year later, after the Jewish deportations from Holland had started, Klein anxiously asked Rosenfeld how he, Pais, and Podolanski were doing, probably quite worried about their destiny in Holland.[58] Unfortunately, Bohr's reply was also negative: "I fear that the current circumstances are too difficult for the realization of such a plan".[59] Rosenfeld had failed to find refuge for Pais in Denmark or Sweden, but throughout the war Rosenfeld continued to promote his bright student to his Danish colleagues and even to Heisenberg in Berlin (see below). Rosenfeld submitted Pais' dissertation to the Royal Danish Academy and also helped Pais to publish several papers during the war based on his thesis.[60] In this way, Rosenfeld's colleagues were made aware of Pais' existence, his talents, and of course the fact that he was in an extremely dangerous position. Rosenfeld also tried to arrange for Lubański to go to Copenhagen in March 1942 but Møller and Bohr considered this impossible too under the current circumstances. During 1942 and 1943, Bohr and Kramers attempted to get a permission from the Germans for Kramers to visit Copenhagen but they did not succeed in that either.[61]

Although conditions in Denmark were never as bad as in Holland (or any other occupied country) because of the cooperation between the Danish and German governments until 1943, the situation got worse here too. After Germany declared war on Russia in June 1941, Nørlund, who was a prominent member of the Danish Communist

---

[57]RP, L. Rosenfeld to O. Klein, 1 Aug 1941. Originally in Danish.
[58]RP, O. Klein to L. Rosenfeld, 22 Jun 1942.
[59]RP, "Copenhague: Bohr (1940–1948)", N. Bohr to L. Rosenfeld, 26 Aug 1941. Originally in Danish.
[60]Pais (1997), pp. 41–43. MP, L. Rosenfeld to C. Møller, 7 Jan 1942, 6 Aug 1943. RP, "Copenhague: Møller (1943–1955)", Møller to Rosenfeld, 2 and 9 Apr 1943.
[61]RP, "Copenhague: Møller (1940–1943)", C. Møller to L. Rosenfeld, 25 Mar 1942. BSC 22: Bohr–Kramers correspondence.

Party, knew instantly he had to hide. Møller informed Rosenfeld in January 1942 that Nørlund was "now unfortunately absent from the Institute".[62] In September 1941, Heisenberg and Weizsäcker arrived in Copenhagen, officially in order to participate in an astrophysics conference at the German Cultural Institute, which was one of several such institutions with the purpose of encouraging cooperation between Germany and the occupied countries. During this visit Heisenberg had his later much discussed private meeting with Bohr in which Heisenberg hinted at an on-going German atomic bomb project. Heisenberg also met Møller, Rozental, and other colleagues at the institute, but nothing about this visit is mentioned in the letters from Bohr and Møller to Rosenfeld.[63]

Yvonne frequently corresponded with Bohr's wife, Margrethe, about life on a day to day basis, the difficulties, weight losses, how the children managed, Niels Bohr's eye accident, etc.[64] In September 1941, Rosenfeld asked Møller for bicycle tyre, which was now an article in short supply in Holland, but Møller was unable to solve Rosenfeld's bicycle problem due to an export ban on rubber from Denmark.[65] The problem was later "solved" when the Germans collected all bikes in Holland for their own war use. Rosenfeld's son Jean remembered that the Germans found and confiscated "our bicycles that had been hidden under a tarpaulin in the attic (at the time, one was supposed to donate such metal items to the German war effort)".[66]

[62]RP, "Copenhague: Møller (1940–1943)", C. Møller to L. Rosenfeld, 26 Jan 1942, 8 Feb 1943. Nørlund was imprisoned by Danish police one of the first days of 1942. He escaped from the prison, went into hiding for a while but then took active part in the Danish resistance. In January 1945, he was imprisoned by the Gestapo until the war ended. Nørlund (1991), pp. 115–154.

[63]Aaserud (2005), p. 13. Walker (1995), pp. 124, 144–151. Pais (1991), p. 483. Rechenberg (2005).

[64]RP, "Copenhague: Bohr (1940–1948)", M. Bohr to Y. Rosenfeld, 10 Oct 1942. N. Bohr to L. Rosenfeld, 24 Dec 1942. BPC, Y. Rosenfeld to M. Bohr, 17 Dec 1941, 4 Sep 1942, 15 Dec 1943.

[65]MP, L. Rosenfeld to C. Møller, 8 Sep 1941. RP, "Copenhague: Møller (1940–1943)", Møller to Rosenfeld, 14 Oct 1941. In this letter Møller suggested they finally became "Dus"! See Chapter 2.

[66]Email correspondence J. L. J. Rosenfeld to A. S. Jacobsen, 12 Jan 2005.

In January 1942, the Rosenfeld family ran out of coal for heating and therefore assembled in a single room in the home which they heated by means of wood burning.[67] "Luckily we have so many other sad things to think about that we forget about the cold for a while", Rosenfeld wrote to Møller.[68] A few times during those five years Rosenfeld (and other Dutch physicists) received Red Cross packages containing food from the Bohrs and other colleagues in Copenhagen. This help was mediated by the Danish former physicist Sophus T. Holst Weber, now an official in the Danish Consulate in The Hague, with whom Rosenfeld was acquainted through Bohr.[69]

Rosenfeld busied himself with work and published relatively extensively during the war years.[70] Working hard was a way to take one's mind off the difficult circumstances they were living amidst. Yvonne had much greater difficulty adjusting to the circumstances it would seem from her letters to Margrethe. Rosenfeld was not the only one who buried himself in work. Bohr wrote to the Japanese physicist Yoshio Nishina in September 1940: "Such interests [physics] are indeed the only way sometimes to forget the great anxieties under which all people in Europe are living at present".[71] According to Hilde Levi — who had been George Hevesy's assistant at the Institute for Theoretical Physics, but worked at the Carlsberg Laboratory during the war — "the only way to get through all these troubles was to work. Work is the only help out of a constant pressure, fear

[67]MP, L. Rosenfeld to C. Møller, 7 Jan 1942.
[68]MP, L. Rosenfeld to C. Møller, 20 Feb 1942. Originally in Danish.
[69]Holst Weber graduated at the University of Copenhagen in 1910, the same year as Niels Bohr. He defended his doctoral dissertation in 1916. He worked a couple of years as an assistant in physics at the University of Copenhagen and at the Polytechnic, and later for Philips in Eindhoven. From 1917, he worked as a managing director in various Dutch companies. Hansen (1984). RP, "Copenhague: Copenhague (1942–1947)", H. Weber, Denmarks consulary representative of the Netherlands, to L. Rosenfeld, 13 Oct 1942, 9 May 1944. RP, "Møller (1943–1955)", C. Møller to L. Rosenfeld, 1 Jun 1944. MP, L. Rosenfeld to C. Møller, 29 Jul 1942. L. Rosenfeld to C. Møller, 27 Apr 1944. See also BSC 22: Kramers–Bohr correspondence. Weber also helped the Rosenfelds to get permission to travel to Denmark in 1945 to attend Bohr's 60th birthday. MP, L. Rosenfeld to C. Møller, 7 Sep 1945. BPC, Y. Rosenfeld to M. Bohr, 4 Sep 1942.
[70]Rosenfeld (1979c), p. 913.
[71]N. Bohr to Y. Nishina, 14 Sep 1940, quoted in *BCW*, Vol. 8, p. 233.

and worry".[72] Pais too occupied his mind with work during these years.[73] In contrast to Kramers, who had taken an active part in the resistance at the University of Leyden and therefore was considered rather problematic in the eyes of the German authorities, Rosenfeld was, according to himself later, "all right" and had "tolerable" and "rather normal" working conditions, at least until the spring or summer of 1944 (see below).[74] In February 1942 Rosenfeld finally gave his inaugural lecture entitled "Ontwikkeling van de causaliteitsidee" (The Evolution of the Idea of Causality).[75] The topic of the lecture reflects how Rosenfeld besides his concern with painstaking and detailed mathematical analyses in meson theory got more and more preoccupied with Bohr's quantum epistemology in which causality was given up in favor of complementarity. Indeed, according to Pais, "already in 1946 I knew more about complementarity than most of my generation, because my teacher Rosenfeld had often talked to me about that subtle subject... I must confess that in my Utrecht years complementarity interested me as much as communism — very little".[76] In early March 1943, Møller and Rosenfeld submitted their paper "Electromagnetic Properties of Nuclear Systems in Meson Theory" to the Royal Danish Academy.[77] Bohr intervened to speed up the publishing process and on this occasion told Rosenfeld that "I admire tremendously that despite the times you can accomplish such extensive and valuable investigations and maintain such an animated interest within your whole circle".[78]

As a result of the deportations of Jews from Holland, which had begun in the summer 1942, Pais went into hiding on 19 March

---

[72]Recollections of H. Levi (1993), NBA, p. 44.
[73]Pais (1997), pp. 109–110.
[74]Rozental (1971). BPC, Y. Rosenfeld to M. Bohr, 17 Dec 1941.
[75]RP, "Utrecht: Oratie (1942)". MP, L. Rosenfeld to C. Møller, 20 Feb 1942. Rosenfeld (1979v).
[76]Pais (1997), pp. 160–161.
[77]Rosenfeld and Møller (1943).
[78]RP, "Copenhague: Bohr (1940–1948)", N. Bohr to L. Rosenfeld, 13 Mar 1943. Originally in Danish. RP, "Copenhague: Møller (1943–1955)", C. Møller to L. Rosenfeld, 9 Apr 1943. See also RP, "Correspondance particulière: Hulthén", L. Hulthén to L. Rosenfeld, 5 Jul 1945, in which Hulthén expressed the same.

1943. Rosenfeld mentioned it in a letter to Møller: "My wife very often misses the animated quarrels with you. Here, she says, there is nobody, who she finds sympathetic enough to fight. An exception was of course Pais, but he is gone now".[79] In April 1943, Dutch students were asked to sign a declaration of loyalty to the German Reich or else they would be enlisted for forced labour in Germany. Protesting students who refused to sign resulted in "wholesale arrests and deportations of students and also to the persecution of a small number of professors who had been prominent in supporting the students".[80] In most of the universities the leaders called on the students to sign the loyalty declaration. In mid-December, Yvonne reported to Margrethe that there were hardly any students left since they had all gone to Germany.[81] In August, Rosenfeld informed Møller that Boris Kahn, a former assistant of Uhlenbeck who had refused to hide despite being Jewish, was in Poland, meaning he had been deported, and that his wife and two children, one of them newly born, were in Amsterdam "so far".[82]

Later that autumn, the political situation in Denmark changed dramatically with the collapse of the collaboration between Germany and the Danish Government, and as a result the Nazis immediately planned deporting all Jews in Denmark. Bohr was warned and managed to escape to Sweden on 29 September. Møller informed Rosenfeld about Bohr's escape in November: "Niels Henrik and his family too are well, as perhaps you know".[83] Indeed, Rosenfeld was already informed of "Niels' escape through Heisenberg who visited the Dutch physicists in October (see next section).[84] From Sweden Bohr and his

[79]MP, L. Rosenfeld to C. Møller, 22 Apr 1943. Originally in Danish. Pais (1997), p. 94. However, according to Yvonne, Pais and Podolanski had disappeared already in September 1942. BPC, Y. Rosenfeld to M. Bohr, 4 Sep 1942.
[80]L. Rosenfeld to R. Peierls, 4 Jun 1948, in Lee (2009), p. 149. Warmbrunn (1963), pp. 151–153.
[81]BPC, Y. Rosenfeld to M. Bohr, 15 Dec 1943.
[82]MP, L. Rosenfeld to C. Møller, 6 Aug 1943. Pais (1997), p. 84. As for Bohr's further activity, see Aaserud (2005), pp. 14–49. Aaserud (1999).
[83]RP, "Copenhague: Møller (1943–1955)", C. Møller to L. Rosenfeld, 18 Nov 1943. Originally in Danish.
[84]MP, L. Rosenfeld to C. Møller, 1 Dec 1943. BPC, Y. Rosenfeld to M. Bohr, 15 Dec 1943.

son Aage went on to London. Here they were briefed about the development of the atomic bomb by James Chadwick already in the airport.[85] In the summer of 1942, the Manhattan project was launched under the scientific direction of Robert Oppenheimer and the military direction of Leslie Groves. Niels and Aage were enrolled in the bomb project on the British side and at the end of November, they left for the US.[86]

Heisenberg's Visit

The German authorities in Holland had suggested that Heisenberg visit the Dutch physicists officially to re-establish contact and collaboration with German colleagues. On the part of the Dutch physicists the aim of Heisenberg's visit was to improve their working conditions during the occupation. Heisenberg accepted the invitation on the condition that Kramers personally approve of the visit, which he did.[87] The visit took place from 18 to 26 October 1943.[88] On this occasion, Heisenberg had opportunity to present his recent work on the S-matrix to the Dutch physicists Casimir, Ralph de Laer Kronig, Kramers, Fokker, and Rosenfeld during three private meetings where they all gathered to have dinner and discuss physics and on-going projects.[89] The first of these meetings took place in Rosenfeld's home, where Heisenberg stayed while in Utrecht, the second in Leyden at Kramers' home, where Heisenberg also stayed, and the third in Eindhoven at Casimir's home.[90] Rosenfeld later remembered that it was very cold at the time, and in order to host Heisenberg and the first meeting Rosenfeld was allowed by the Town Hall to receive extra electricity during Heisenberg's visit. The discussions about Heisenberg's

[85] Aaserud (2005), p. 15.

[86] *Ibid.*, pp. 14–49.

[87] Heisenberg Nachlass, H. Kramers to W. Heisenberg, 29 Jul 1943. Heisenberg to Kramers, 20 Aug 1943. Copy of letter from H. Kramers to L. Rosenfeld, 11 Oct 1943. Walker (1989), p. 111. Rechenberg (2005), pp. 181–183. Rozental (1971), p. 1.

[88] Heisenberg Nachlass, W. Heisenberg to Hernn Reichminister für Wissenschaft, Erziehung und Volksbildung, 10 November 1943, "Bericht über eine Reise nach Holland vom 18.-26.10.43".

[89] MP, L. Rosenfeld to C. Møller, 8 Feb 1944.

[90] Rozental (1971). Dresden (1987), pp. 454–457.

work had been initiated by mail in the summer of 1943 and continued after Heisenberg had left the country.[91] Heisenberg's main concern was clearly the experimental work that took place in Holland at the time, and he visited among other places the research laboratories of Philips (where Casimir worked) and the Kamerlingh-Onnes Laboratory.[92] As regards to the political situation, Heisenberg appeared rather "outspoken" and "uncalculating"; "he was just speaking his mind about the situation", as Rosenfeld later characterized it. This impression is supported by the report Heisenberg sent subsequently to the German authorities and also through testimonies by other physicists in the occupied countries who met Heisenberg. Heisenberg was not a Nazi, but he was supportive of the German war efforts in particular on the East Front and in the early war years gave the impression that Germany would win the war. While in Holland in 1943, however, he regretted that Germany could no longer win the war. Heisenberg seemed to have suffered from a grave lack of sensitivity when he as the envoy of the oppressors openly expressed his patriotism during his visits in the occupied countries, and this behavior deeply hurt the feelings of his colleagues there.[93]

As for private matters, Heisenberg helped Rosenfeld get permission to visit his mother, who was ill, in Belgium. Rosenfeld had a lot of trouble to get permission to go and see her. Heisenberg approached the Reich's commissioner about it on his return trip and he later informed Rosenfeld about the successful meeting and remarked that hopefully Rosenfeld would experience no more difficulties.[94] Rosenfeld was indeed able to visit his mother and was very grateful for Heisenberg's intervention. Rosenfeld had also tried to get permission for his mother to move to Utrecht, but that was not allowed. This may have been for the best since during his trip Rosenfeld observed

---

[91]MP, L. Rosenfeld to C. Møller, 8 Jun 1943. Rechenberg (2005), pp. 182, 183. Rozental (1971).
[92]Casimir (1983), p. 191.
[93]Rozental (1971). See also Dresden (1987), p. 499. Casimir (1983), pp. 207–209. NBA, Unsent letters from N. Bohr to W. Heisenberg in late 1950s and early 1960s. See http://www.nba.nbi.dk/papers/docs/cover.html (accessed 10 May 2011). Carson (2010), p. 23.
[94]RP, W. Heisenberg to L. Rosenfeld, 28 Oct 1943. Walker (1989), pp. 110–113.

that "contrary to the situation a year ago people now live better in Belgium than here in Holland".[95] And the situation in Holland was going to get even worse.

Since Rosenfeld had heard about the allied bombings of Berlin, he expressed his concern for Heisenberg and his family as well as for the "beautiful Dahlem Institute" in his next letter to Heisenberg.[96] After Heisenberg's visit and Bohr's escape, Rosenfeld appeared to be more concerned about the experimental work taking place in Copenhagen.[97] In fact, things took a bad turn at Bohr's institute during the winter 1943–1944 when it was occupied by the German military police from 6 December 1943 till 3 February 1944.[98] Møller informed Rosenfeld that "various things have happened which to a great extent have made work more difficult for us — meanwhile we try to arrange ourselves at the Polytechnic where they have shown great flexibility with respect to premises etc. As a theoretician I guess I am not the worst off since my fountain pen is still intact".[99]

In March 1944, Rosenfeld sent a manuscript by Pais about the theory of the electron to Heisenberg and asked him to comment on it. In this way Heisenberg's attention was indirectly called to Pais' existence and talent as a physicist.[100] With this move Rosenfeld seems to have insisted on maintaining a collegial connection across oppressor and victim roles in the midst of war terror, and deportations. Heisenberg expressed his delight to hear from Rosenfeld "one more time" and on this occasion mentioned his recent visit to Bohr's institute in Copenhagen where the work, according to him, now continued "undisturbed".[101] Partly because of his intervention (Møller suggested to the university authorities to contact Heisenberg about

[95]MP, L. Rosenfeld to C. Møller, 1 Dec 1943. Originally in Danish.
[96]Heisenberg Nachlass, L. Rosenfeld to W. Heisenberg, 10 Dec 1943.
[97]MP, L. Rosenfeld to C. Møller, 1 Dec 1943.
[98]NBA, folder concerning the occupation of the Niels Bohr Institute 1943–1944.
[99]RP, "Copenhague: Møller (1943–1955)", C. Møller to L. Rosenfeld, 4 Jan 1944. Originally in Danish. See also Rozental (1967), pp. 149–190.
[100]Probably it was the manuscript later published as Pais (1948).
[101]RP, W. Heisenberg to L. Rosenfeld, 10 Mar 1944. Heisenberg to Rosenfeld, 5 May 1944. Heisenberg Nachlass, Rosenfeld to Heisenberg, 14 Apr 1943; see also copy of letter from Heisenberg to H. Kramers, 10 Jan 1944. According to Dresden (1987), p. 455, Heisenberg also mentioned his visit to Copenhagen in a letter to Kramers.

the episode), the institute was freed from its occupiers at the beginning of February.[102] In April, Møller saw light at the end of the tunnel: "I suppose we have now started the fateful year and if we all live a reunion should probably not be far away. Our work at the institute is again normalized".[103] In the meantime, during this spring Bohr sought secret information through his wife Margrethe in Sweden and Møller in Copenhagen about the activities of the German physicists. Margrethe often met the Austrian physicist Lise Meitner, who had been in Stockholm since 1938 when as an Austrian Jew she had to escape from Germany because of *Anschluss* of Austria. Meitner had some contact still with Otto Hahn, and his reports about the activities of Heisenberg and his collaborators at an institute in southern Germany worried Meitner and Margrethe. In May 1944, Bohr wrote secretly to Møller about his concern about the German physicists: "Over here, it is of particular interest to know whether there is any information at all in Copenhagen about German physicists, which could possibly give us a hint about where they are working and what they are doing... This is of course much to ask, but we would be grateful for even the smallest piece of information, however insignificant it might seem".[104] In June 1944, Heisenberg again visited the institute in Copenhagen and officially discussed the S-matrix theory with Møller.[105] The information Møller acquired through his contact with Heisenberg about his and the other German physicists' activities was that they "do not seem to have worked with the question during the last year", which he reported back to Bohr on 29 September 1944.[106] This judgement was later seriously questioned by the British Secret Service.[107]

---

[102] Rozental (1967), pp. 171–172. NBA, folder concerning the occupation of the Niels Bohr Institute 1943–1944. "Report about the events during the occupation of the university's institute of theoretical physics from 6 December 1943 until 3 February 1944". Rechenberg (2005), p. 183.
[103] RP, "Copenhague: Møller (1943–1955)", C. Møller to L. Rosenfeld, 11 Apr 1944. Originally in Danish.
[104] Aaserud (2005), pp. 28–29, 243–244.
[105] RP, "Copenhague: Møller (1943–1955)", C. Møller to L. Rosenfeld, 1 Jun 1944. Rechenberg (2005), p. 183.
[106] Aaserud (2005), p. 29.
[107] *Ibid.*

Back in the Netherlands, Rosenfeld had been able to see his mother again in Belgium in April.[108] A few months later "life" was no longer "funny" in Holland, but "everything is quiet — so far!" he reported to Møller.[109] Rosenfeld received one more letter from Møller dated 5 July 1944, but after that date their correspondence, and indeed Rosenfeld's correspondence with anybody else, stopped for almost a year.[110]

Hiding

Rosenfeld was driven into hiding probably sometime in the late spring of 1944.[111] According to his son Jean, he was hiding in the attic and only emerged again when the allied forces entered Utrecht 10 May 1945.

> ...in the spring of 1944...as far as I was concerned he [Rosenfeld] disappeared. Mother told us that if asked, we were to say that he'd gone to Friesland, (a province in the Northern part of Holland) to get some potatoes (at the beginning of the war we had a maid who came from Friesland and I had spent a holiday at her parents' farm, so that seemed quite plausible to me at the time; potatoes were a very scarce resource). In fact, as I learned later, he was immured in the roof space above our parents' bedroom. The bedroom ceiling was redecorated to seal and hide the access. Food and excrement were passed in cocoa tins through a pipe, ostensibly an air ventilation duct above the washbasin in their bedroom.
>
> Whilst there he had jaundice and wrote a first draft of his book Nuclear Forces...Our house was searched on at least two occasions. I remember one evening mother, Andrée and I

[108] MP, L. Rosenfeld to C. Møller, 27 Apr 1944.

[109] MP, L. Rosenfeld to C. Møller, 22 Jun 1944.

[110] RP, "Copenhague: Møller (1943–1955)", C. Møller to L. Rosenfeld, 5 Jul 1944. RP, "Correspondance particulière: Hulthén and Serpe".

[111] Rozental, (1971), p. 6. Email correspondence J. L. J. Rosenfeld to A. S. Jacobsen 12 Jan 2005. A. Rosenfeld in email correspondence with A. S. Jacobsen, 20 Mar 2005. Peierls (1974).

> were having our evening meal when they came. They looked through the house, but I think that the fact that the table was laid for three and that there were no signs of any dishes having been hastily removed convinced them that there were just the three of us in the house (as indeed I also thought at the time), so they left. We were fortunate that they did not throw exploratory grenades through walls or ceilings, as they did in other houses.[112]

The reason for Rosenfeld's hiding is not clear. However, by the end of September 1943, Dutch Jewry as defined by the Nazis had been deported. From about this time (about the same time that the Germans decided to deport all Jews in Denmark), or even before, the Germans ordered re-examinations of cases where there was doubt about a person being Jewish or Aryan. J. Presser refers to a document of 5 July 1944 which among many others referred to the necessity of "unmasking the many full-Jews who, by all sorts of manipulations, have succeeded in passing themselves off as Aryans or part-Aryans in the Netherlands", and of exposing the "extensive swindle" that enabled several thousand people to "run about in the Netherlands as unrecognized Jews".[113]

Even if Rosenfeld was not subjected to this threat, he, as a healthy adult male about 40 years old, ran another fatal risk, *viz.*, to be sent to Germany to do forced labour in construction or in metal, ammunition, or aircraft factory. From 1942, all men, whether students, unemployed or employed, were in danger of this fate, and as the German need for workers increased "men were even picked up off the streets and from their homes during razzias". According to Inge Bramsen, about 330,000 non-Jewish men are estimated to have been in hiding, most of them in an attempt to avoid forced labour.[114]

---

[112] Email correspondence J. L. J. Rosenfeld to A. S. Jacobsen, 12 Jan 2005. In his obituary of Rosenfeld, R. Peierls also mentioned Rosenfeld's hiding during the war, Peierls (1974).
[113] Presser (1988), p. 300.
[114] Bramsen (1995), p. 4. van Walsum (1995), pp. 129–134. BPC, Y. Rosenfeld to M. Bohr, 15 Dec 1943, about that there were hardly any students left since they had all gone to Germany.

This last part of the occupation became by far the worst in the northern provinces of the Netherlands. While Belgium and the south of the Netherlands were liberated by November 1944, the north-west, which became known as *Festung* Holland, and which included Amsterdam, Rotterdam, Utrecht, and The Hague, was kept under German occupation throughout the winter 1944–1945 due to the failure of the Allies' Arnhem offensive 17–25 September. This was a particularly severe winter and was soon referred to as the "hunger winter". Following a general railway strike in September, the transportation of food to the western part of the country was stopped by the Germans for three months. And because of the Allied presence around Arnhem, transport of coal from the mines in the south was impossible too. There was no more electricity or gas. In order to cook and keep warm, people burned wood, often acquired by plunder and pillage. As a result, the people in the western part of the country suffered from hunger and cold, and between 15,000 and 20,000 died of malnutrition. There were large hunger marches where those who could, went begging in the countryside. Matters grew even worse due to the increased desperation of the Germans in their demands for forced labour and their retaliations against armed resistance.[115] Rosenfeld described this particularly grim time to Møller when they had resumed contact in August 1945.

> Now I could write at length about the horrible time we have had since September last year until the liberation in May, but I feel I would rather forget about these things, or rather (because I can never forget them) not let them appear in the foreground of my consciousness again. The most frightful was not the famine, the cold or the dark; it was the inhuman terror, which the Germans imposed upon us.
>
> In this winter we have seen the best and the most disgusting sides of man, but what takes precedence over everything else is that we have learned to appreciate personal freedom. "Les droits de l'homme", human rights are not merely beautiful

[115]Bramsen (1995), p. 3. Arblaster (2006), p. 230. Pais (1997), pp. 61–64.

> phrases which you learn in school; they have become for us vivid, precious realities. I only hope that people will not too quickly forget this lesson.[116]

A few days later Rosenfeld added,

> We have now already achieved relative prosperity to such a degree...that we no longer think about the time when we ate cabbage in water three times a day and had to break doors and chairs in order to get wood to cook it. Luckily we could keep Andrée and Jean healthy thanks to the excellent help we had from a church organization that provided for all children in town. Yvonne and I had become terribly thin (it is a strange feeling which I could call "skeletal consciousness"), but we have already regained the desirable rounding.[117]

On 6 September 1945, Rosenfeld informed Martin Strauss that

> I was interested in the proceedings of your seminar over [the] logic of science, although I am rather out of practice of logical thinking after all these years of a regime which I need not describe to you!
>
> On the whole, the Dutch physicists living in the Western provinces (which had to endure the hunger and terror winter) have emerged from the hell in good condition: this is the case, in particular, for Kramers, Mrs. Ehrenfest and myself, as well as for van Dantzig, who, as a Jew, had to live underground. As for Casimir, who had moved earlier to Eindhoven, he was liberated already in the autumn of 1944.[118]

It seems as if Heisenberg's name was indirectly involved in saving Pais' life after Pais had been arrested and thrown into prison in the spring of 1945. In April 1945, he was released after his girl friend had shown a high Nazi official a copy of a letter from Kramers to

---

[116] MP, L. Rosenfeld to C. Møller, 27 Jul 1945. Originally in Danish.
[117] MP, L. Rosenfeld to C. Møller I, 1 Aug 1945. Originally in Danish.
[118] RP, "Correspondance particulière", L. Rosenfeld to M. Strauss, 6 Sep 1945.

Heisenberg in which Kramers asked for Heisenberg's help to save Pais' life (Heisenberg apparently never received the letter).[119]

Møller and Bohr naturally expressed immense relief as soon as they heard that Rosenfeld and his family were alive and well, and these feelings were of course reciprocated on the part of the Rosenfelds.[120] To the Rosenfelds' great relief they also found all of their family in Belgium well, despite the fact that Rosenfeld's mother had lived in the basement for three months. "[I]t is a rare happiness for a family in these countries to be able to celebrate the liberation without mourning any of its members", he wrote to Møller.[121] And to Chandrasekhar who sent parcels to the Rosenfelds, he wrote in early 1946:

> Yvonne and I have been touched more than we can say by your kind letter of 1 February and the parcels which have just arrived. I need not say that the contents of these parcels is exceedingly welcome to us because it substantially helps in restoring our external appearance as civilized beings, but it is of still more value to us as a token of friendship.
>
> When we look back on these nightmare years, we can deem ourselves extraordinarily lucky. We have suffered no loss of life in our family and the material losses are not very heavy.... My mother's house was badly shaken by a V1 in Liege, but the damage is comparatively slight. Yvonne and I managed to live rather quietly in Utrecht with the children; only the last war winter (from September 1944 to our liberation in May 1945) was most horrible. We were cut off from the rest of the world, and so had no coal, no gas, no electricity and no sufficient food. And we were defenceless against

[119]Pais (1997), p. 121. Kramers approached Heisenberg for help with respect to colleagues and friends in other situations too, but whether it had any effect is unclear. Dresden (1987), p. 458. Rechenberg (2005), p. 182 note 94.

[120]RP, "Copenhague: Møller (1943–1955)", C. Møller to L. Rosenfeld, 11 Jul 1945. RP, "Bohr (1940–1948)", N. Bohr to L. Rosenfeld, 5 Sep 1945. The Rosenfelds seemed to have received the first news about Bohr and his brother, Harald Bohr, through Heisenberg. BPC, Y. Rosenfeld to M. Bohr, 15 Dec 1945.

[121]MP, L. Rosenfeld to C. Møller II, 1 Aug 1945. Originally in Danish.

> the ruthless terror of the Germans. When we were liberated, we were on the verge of starvation. Still, we had been able to save the children, whose health was not seriously impaired, so that after a few weeks' vacation in the country they were quite all right again. Also Yvonne and I, who had reduced to a one-dimensional form, recovered very rapidly.
>
> I have been able to carry on theoretical work all the time, even during the hunger months. I then started writing a book on nuclear forces, which is now nearing completion.[122]

Rosenfeld seems to have gone into hiding at just the right time. During the occupation he had kept a low profile politically. Simultaneously he seems to have pinned his faith on having close contact first and foremost to Bohr and Møller and later even to Heisenberg. In addition, he had contact with a Danish official in Holland. He could not just disappear without anybody noticing it. As evidenced by so many other cases, this, however, would have been absolutely no guarantee for his life. Some bureaucracy may very well have been involved too due to his Belgian citizenship and the poor communication between administrative authorities in Belgian and Holland in those years. Mentally, as already mentioned, it was important, that Rosenfeld had been able to continue his work. In this way he even added to his merits while taking his mind off the hardships he and his family went through. Thus, apart from publishing extensively, he worked on his book *Nuclear Forces* while hiding in the attic.

After the war, Rosenfeld spoke very little if at all about what he experienced in those dark years. As he told Møller in the summer 1945, he hid the terrible experiences in the back of his mind. Rosenfeld seems not to have suffered from guilt to be alive, like Casimir and many other Dutchmen, Jews and non-Jews alike.[123] Rosenfeld seems to have been able to deal with the existential issues arising

[122]RP, copy of letter from L. Rosenfeld to S. Chandrasekhar, 1945. Original in CP.
[123]Bramsen (1995). Pais (1997), pp. 68, 77. Casimir reports such a feeling of guilt and deplored that he had not done more to help people in need, Casimir (1983), pp. 191–192.

from these events and to manage his life afterwards. Probably contributing to this was the fact that he had not lost family members. In addition, he possessed such a clear stand against Nazi Germany and such clear ideas of the injustice done to mankind in general. His book *Nuclear Forces* was dedicated to his close friend and colleague, Jacques Solomon, who had been executed by Gestapo in Paris in 1943 as a result of his active participation in the French Resistance. Rosenfeld immediately engaged with renewed energy in political issues and wrote extensively in left-wing periodicals, as we shall see in the next chapter. A topic of particular concern to him became the social responsibility of scientists following the bombings of Hiroshima and Nagasaki with nuclear weapons in August 1945.

Chapter 5
Cold War and Political Commitment

This chapter analyzes Rosenfeld's political commitment during the developments on the Left in the early Cold War, including how it affected his career opportunities and his friendship with Bohr. Bohr and Rosenfeld were politically active on very different levels and in different ways. They did not occupy the same place in the political spectrum either and they were active on opposite political and ideological fronts of the Cold War. In the summer of 1950, they clashed over how to tackle the Russians and the international control of nuclear weapons. Thus, while Rosenfeld developed into Bohr's strongest supporter in physics he became simultaneously his critic in politics.

In the contests over ideology and cultural values embodied by science in the early years of the Cold War, Rosenfeld struggled to maintain an independent and critical balance between points of view which were by and large perceived to be mutually exclusive. He can be characterized as progressive left-wing; he was not a member of the Communist Party, yet supported communist-led organizations. He exhibited knee-jerk reactions against the politics of the United States and was clearly favorably inclined towards the Soviet Union in the dichotomy of Cold War politics. On the other hand, Rosenfeld fiercely opposed the ideological imperatives from Moscow which immediately aspired to become the monopoly leftist opinion. In particular, as we shall see in greater detail in the next chapter, unlike communist physicists and philosophers obeying the ideological decrees from

Moscow, Rosenfeld allowed Bohr's epistemology of quantum physics to have an impact on his Marxist ideology. The majority of communist physicists interested in this subject during the Stalin era opposed Bohr's complementarity doctrine on the ground that it could not, they claimed, be combined with Marxist–Leninist realism and materialism. Rosenfeld's individual points of view stood little chance of being recognized in either camp in the predominantly black-and-white pictures that were painted of communism and anti-communism by 1948 and the growing emphasis on the inescapable need to make a choice between them.

Dividing the World

By the attack on Hiroshima and Nagasaki with nuclear weapons in August 1945, the Americans achieved a speedy victory over Japan which surrendered unconditionally. As a side effect, the atomic bomb impressed and probably even intimidated the Soviet Union. In the long run this infernal technological innovation unmatched in the history of military technology turned out to be a motivation for peace and not war — at least concerning nuclear war, a scenario foreseen and promoted by Bohr. However, the immediate victory of the Allies did not bring the expected peace. The power vacuum in Europe in 1945 was soon filled and controlled by the military powers of the United States and the Soviet Union. Both countries sought national security through spheres of influence, but by different means and on the background of an uneven power balance. The US came out of the war as the strongest military power and the richest economy in the world as well as the world's leader in science and technology. It was intact and it possessed a new awesome weapon, the atomic bomb. The Soviet Union had a victorious army, but it faced enormous reconstruction work and had lost more than 27 million of its citizens during the war; in comparison there were hundreds of thousands of American victims. Stalin imposed buffer states around the Soviet Union by means of territorial demands in the negotiations with the British and the Americans and later by military power followed by purges. The US used more sophisticated bargains such as the Marshall Plan seeking to promote democracy through economic recovery. By 1947 Europe was divided and its politics polarized; states, people, and movements aligned themselves with either Moscow or Washington.[1]

The development of atomic weapons had made science, politics, and war inextricably linked. No longer could scientists claim that pure science was divorced from technology and political power. The atomic bomb was on the conscience of mankind but in particular on the nuclear physicists' conscience and posed a problem that had to be faced. As Robert Oppenheimer remarked to an audience at the Massachusetts Institute of Technology (MIT) in November 1947: "In some sort of crude sense which no vulgarity, no humor, no overstatement can quite extinguish, the physicists have known sin; and this is a knowledge which they cannot lose".[2] Many scientists became politically engaged. They felt the need for international cooperation and initiatives to avoid the use of the new weapon, particularly with the advent of the hydrogen bomb in 1952, since this weapon's destructive capability would mean nuclear obliteration of the world. The social responsibility of scientists was highlighted and the political attitudes of scientists and their cooperation became important to the governments of many nations, not least the US and Britain. Scientists became involved in scientific and non-scientific organizations with the aim of influencing political developments. In articulating political claims on how the new technology should be used — for military or peaceful purposes — they continued the line from the 1930s attempting to accommodate politics to science.[3]

In the United Nations, the US had from the beginning posed suggestions for controlling nuclear weapons. In June 1946, they put before the Atomic Energy Commission of the UN the Baruch Plan, after Bernard Baruch, which had a number of political advantages for the US. It advocated the freezing of atomic weapons development, which would secure the status quo balance with the atomic monopoly

[1] Gaddis (1997), pp. 4–48. Badash (1995), pp. 70–73.
[2] Oppenheimer (1955), p. 88.
[3] Many sources deal with this period and scientists' political engagement. The following are highlighted as sources used here. Elzinga and Landström (1996). Jones (1988). Petitjean (2008a). Petitjean (2008b). Edgerton (1996). Wang (1999).

of the US. Moreover the system of surveillance and inspection suggested by the Baruch Plan would have given the US access to the military installations of the USSR. Soon after the presentation of the Baruch Plan, the US began test explosions of atomic bombs at Bikini Atoll in the Pacific, as a result of which the Soviets challenged the honesty of American diplomacy. Soviet counter-proposals suggested plain abandonment and destruction of atomic weapons. Thus negotiations within the fledgling United Nations on international control of atomic weapons were beset with steep opposition between the two superpowers and there was little hope for a successful outcome in that forum. In the spring of 1948, the activities of its Atomic Energy Commission came to a halt.[4]

Besides national security and the threat of nuclear war, a growing anti-communism in the West also contributed to shape the Cold War atmosphere. During the Second World War, many leftists had taken an active part in fighting fascism. Indeed, in 1945 it proved of great advantage to the communist parties that they had dominated resistance movements during the war. Even in countries far from social revolution such as Belgium, Denmark, and Norway, the communist parties scored percentages as high as 10–12% of the votes; in France they emerged as the largest party of all in the 1945 elections. The influence of the European communist movements therefore peaked in 1945–1947.[5] There was an optimistic and a celebratory mood on the Left in 1945 and a hope that the broad alliances created in the resistance movements could be extended into the future, as well as a hope of extending into peacetime the new apparent friendliness with the Soviet Union in the alliance which had defeated Hitler.[6] However, the rise of communism in Western Europe, combined with the actions of the Soviets in Eastern Europe created fear on the part of Western governments that their liberal democratic foundations were under attack. Already during the war the British Government began to consider the USSR and communism the next major threat

[4]Badash (1995), pp. 68–70. Jones (1988), p. 81.
[5]Hobsbawm (1995), p. 166. Thing (1993), Vol. 2, p. 752.
[6]Jones (1988). Horner (1996). Petitjean (2008b), p. 255.

to British security, a point of view that became clearly enunciated in Winston Churchill's "Iron Curtain" speech in March 1946 (although at that time he no longer represented the Government). This speech in turn contributed to widening the rift between the United States and the Soviet Union.[7] In March 1947, the Truman Doctrine was announced invoking the policy of containment, according to which America committed itself to provide military, economic, and diplomatic support to any free democratic nation that became victim of aggression and pressure from totalitarian regimes. The doctrine was immediately demonstrated in practice in Greece where a civil war was being fought after the communist-led resistance had liberated the country. The Greek Governmental army was backed by Britain and the US and Britain restored the hated Greek monarchy. When the British military and economic assistance ended in February 1947, the US took advantage of that situation by taking responsibility for economic and military assistance to Greece. The Left saw this as a clear example of American imperialism and felt betrayed by Stalin. The Soviets never actively supported the military efforts by the Greek Communist Party for reasons connected with the sphere of influence division in Europe. In June 1947, the Marshall Plan was announced.[8]

The Soviet Union was also not inclined to extend the cooperation with its allies. The Comintern had been dissolved in 1943, but as a response to the Truman Doctrine and the Marshall Plan, Stalin announced the creation of a new communist organization, the Cominform (Communist Information Bureau), in September 1947. Contrary to Comintern which had been an international organization, the Cominform was meant to be a tool in the Cold War representing the communist bloc dominated by the USSR. The creation of the Cominform was Stalin's demonstration of a Europe that was divided in two.[9]

---

[7]Ullrich (2007), pp. 4–5. Wang (1999), p. 21. A. Kojevnikov, Cold-War Mobilization of Science in the Soviet Union, manuscript.
[8]Gaddis (1997), pp. 43, 50.
[9]*Ibid.*, p. 46.

Stalin had made Andrei A. Zhdanov Secretary of the Central Committee and responsible for the Party ideological line and Zhdanov created a two-camp rhetoric distinguishing between the war-mongering capitalist countries led by the imperialist, anti-democratic US and the anti-imperialist and democratic camp of peace-loving progressive forces led by the Soviet Union. The latter included the so-called people's democracies in Eastern Europe as well as Indonesia and Vietnam. It was claimed that the anti-imperialist camp was supported by the working classes, the democratic movements, and the fraternal communistic parties in all countries. However, in this new strategy by Stalin, class-struggle played no role. The "struggle for peace" constituted an opposition purely between states. The strategy was received by the West as confirmation of an aggressive foreign policy on the part of the USSR. Indeed, with the creation of the Cominform, communist parties were meant to act as extensions of Russian foreign policy. In addition, the period from 1947 until Stalin's death in 1953 saw a new wave of terror within the Soviet Union and its satellite states.[10] In order to prevent Czechoslovakia from participation in the Marshall Plan, the communists took power in 1948. The Berlin crisis simultaneously demonstrated that the US would not tolerate further expansionism by the Soviets in Europe. In September America allied itself militarily with Western Europe in the Western European Union, the precursor to the NATO agreement of April 1949. Thus, by 1948 international tensions much akin to war had come to dominate European politics. A genuine fear of nuclear war now prevailed in the West.

The general climate in international relations grew increasingly distrustful, tense, and hostile. In late August 1949, the Russians tested their first atomic bomb, the news of which came as a shock to the American public and officials. It was expected that the Russians would have atomic bombs eventually but not so soon.[11] The nuclear

[10]Thing (1993), Vol. 2, pp. 753–755. Ullrich (2007), p. 6. Werskey (1988), p. 292. Kojevnikov (2004), p. 184. Kojevnikov (2011).

[11]Badash (1995), p. 77. In 1948, Blackett estimated that the Russians would have atomic bombs within a period of five years. Blackett (1948), p. 46.

arms race was intensified; President Truman went ahead with the hydrogen bomb project. In October the same year, the communists gained power in China. Thus, overnight a fundamental shift in the balance of power seemed to have taken place; the communist world appeared almost to have doubled in size.[12] In December 1949, nuclear physicist Klaus Fuchs, who had worked at Los Alamos during the war, was arrested as a Soviet spy by the British Secret Service and in March 1950 he was sentenced to fourteen years in prison. In the summer of 1950, the electrical engineer Julius Rosenberg and his wife Ethel were arrested in the US, convicted in March 1951 for conspiracy to commit espionage and executed in June 1953.[13] The existence of atomic spies created a wave of hysteria and fed anti-communism in the US. The US experienced purges of communists, sympathizers, and American leftists, who were made scapegoats for the loss of the so-called "atomic secret." In Britain, an anti-foreigner tone began to be expressed in the British Press when Italian-born atomic physicist Bruno Pontecorvo, on the verge of taking up a position at the University of Liverpool, vanished to Moscow and was suspected of having given away secret information to the Soviets. However, Rudolf Peierls, who at this time served as President of Atomic Scientists' Association and had appointed Fuchs knowing he was a left-winger, continued to express his sympathy for free exchange of information internationally and his dislike of the attempt to impose strict barriers against this.[14]

Up until the summer of 1950, Europe was the central arena of the Cold War. This changed dramatically on the advent of the Korean War by which the Cold War was globalized thus involving the US more directly. It had the effect that the American leaders no longer distinguished between communists in different parts of the world and in America, but instead considered communism to be a unified and universal threat, the spread of which they were determined to stop.

---

[12] Gaddis (1997), p. 55.

[13] Hoffmann (2009), p. 416. Jones (1988), p. 42. Wang (1999), p. 262. Badash (1995), p. 106.

[14] Jones (1988), pp. 42–43. Rodian and Garrity (1950). Peierls (1950), in RP, "Manchester (1948–1958): Political (1948–1958)."

Science, Socialism, and Peace

Rosenfeld kept a low profile during the war, at least on the surface. However, he emerged from the war as a full-blown socialist, and in contrast to the 1930s he was now going visible. He came forward with a much more concrete program of political action compared to the intellectual movement which he had followed in the 1930s.[15] He shared the optimism among leftist intellectuals that the time had come to unite the Left in the creation of a better world based on reason, critical reflection, and socialism now that fascism and Nazism had finally been defeated.[16] He became a regular correspondent for several left-wing periodicals and newspapers such as the Belgian *L'Éclair* (The Flash) and the Amsterdam based weekly *De Vrije Katheder* (The Free Pulpit), and he wrote letters to the editor in *The Manchester Guardian. De Vrije Katheder* was an underground paper founded in November 1940 which had both communists and progressive non-communists on its editorial board. The fate of this journal is symptomatic for the period. As the Cold War rapidly intensified, the editorial board came under severe pressure from the Communist Party to change its profile so as to make it conform to the party line. As a result, the editorial board agreed to stop the journal altogether in 1950. Under the conditions of the Cold War there was no longer a place for a journal with the aim to bridge the gap between communists and progressive non-communists.[17] In this journal Rosenfeld wrote critical analyses of the political situation in the Netherlands and of Dutch imperialism in Indonesia. After his visit to Dublin, Ireland, in May 1946, where he gave a series of physics lectures, he published his impressions of Ireland's economical and political system in the same journal.[18]

As soon as the war was over, Rosenfeld was anxious to renew his contact with Bohr and learn what the "grand danois" had to say

[15]A similar development seems to have characterized the American Progressive Left. See Wang (1999), p. 6.
[16]Horner (1996), p. 132.
[17]Van den Burg (1983), pp. 377–378.
[18]RP, "Utrecht (1940–1947), Political (1945–1947), Lettres de Hollande ("L'éclair", Bruxelles)." Van den Burg (1983). Rosenfeld (1946b).

about the world situation.[19] Bohr's sixtieth birthday was coming up in October 1945. Among Rosenfeld's articles in *De Vrije Katheder* was a homage to Bohr in celebration of this event, which was also published in an English version.[20] It was not sheer unreserved adoration for Bohr which came to the fore in that article. In it Rosenfeld described Bohr's social character and his optimism on behalf of the future as a product of a privileged and protected life in the small democratic country of Denmark.

> His whole attitude, in the laboratory as in the family circle, towards the problems of science as towards those of society, always expresses without effort the same serenity, the same tranquil force, the same desire for universal comprehension, and above all the same radiant, unshakable optimism... He looks at mankind with unbounded sympathy and confidence; in fact, his warm belief in the good will of men is not sufficiently tempered by a realization of the harshness of contemporary social conditions. His subtle intuition into the human soul cannot here, any more than his open-mindedness, counterbalance the lack of direct experience; in a small democratic country, of relatively uniform and rather high social standard, he grew up in a purely intellectual environment, whose complacent liberalism had not yet been shaken by the tumultuous development of industrial civilization, and his naively idealistic representation of social relations is still under the spell of Spencerian individualism.[21]

In other words, Rosenfeld saw Bohr as "a bit naive in political questions", as he stated in the interview conducted by Thomas Kuhn and John Heilbron in 1963.[22] Apart from Bohr's sixtieth birthday,

---

[19] Rosenfeld (1945b), in RP, "Utrecht (1940–1947): Political (1945–1947)."
[20] Rosenfeld (1945c), in RP, "Utrecht (1940–1947), Political (1945–1947)." Rosenfeld (1979j), pp. 314, 323.
[21] Rosenfeld (1979j), pp. 314, 323. Rosenfeld's viewpoint was not shared by J. Kalckar, according to whom Bohr often commented on world politics and took a clear stand. Kalckar also mentioned Bohr's serenity, Kalckar (1967), pp. 234–235.
[22] Kuhn and Heilbron (1963c), pp. 7–8.

there was another occasion for writing about Bohr in the newspapers in the fall of 1945, *viz.*, his return to Denmark and Europe after having participated in the Manhattan Project. Nuclear power and the atomic bomb was a most urgent topic for Rosenfeld, as for most physicists in the second half of 1940s.[23] As a physicist in the occupied countries, Rosenfeld was not exposed to the same dilemmas faced by physicists in the allied countries in connection with their work related to nuclear weapons.[24] Upon the news of the bombings of Hiroshima and Nagasaki Rosenfeld wrote to Møller: "I need not say that the topic of the day is the atomic bomb. All the same I am sorry that nuclear physics should get into prominence among the public just by such an infernal invention!"[25] In November, Rosenfeld published in *L'Éclair* and in the Dutch "Independent weekly of socialist politics and culture" *De Baanbreker* (The pioneer), a conversation with Bohr during the latter's birthday celebration about the new world situation after the bomb.[26]

Bohr was very concerned with the consequences of atomic threat to the world; from 1943 and for the rest of his life this issue was on his mind.[27] In the first letter from Bohr to Rosenfeld after the war, Bohr touched upon the issue about the atomic bomb and enclosed his recent article prompted by the new situation, "Science and Civilization" published in *The Times* 11 August 1945.[28] It may have been difficult for the Rosenfelds to appreciate Bohr's abstract idea that in the long run the development of the atomic bomb could be beneficial for mankind and peace: "You see it was a peculiar matter I was involved in", Bohr wrote, "but I have the greatest faith that the development will contribute crucially to promote a peaceful relation between all nations".[29]

---

[23]Jones (1988). Edgerton (1996), pp. 16–17. Wang (1999), pp. 9–13.
[24]See for example Schweber (2000).
[25]MP, L. Rosenfeld to C. Møller, 14 Aug 1945. Originally in Danish.
[26]Rosenfeld (1945a; 1945b), in RP, "Utrecht (1940–1947): Political (1945–1947)".
[27]Aaserud (2005). Aaserud (1999).
[28]Bohr (2005b).
[29]RP, "Bohr (1940–1948)", N. Bohr to L. Rosenfeld, 5 Sep 1945.

Less than a month after the atomic bombs had fallen on Hiroshima and Nagasaki, Rosenfeld held a series of six lectures at Utrecht University for students of all faculties in which he informed them about the scientific principles behind the atomic bomb and the development of nuclear physics leading up to it. Even the largest lecture theater was too small for the enormous audience for these lectures.[30] The lectures were based on the Smyth report, the official American document, written by the Princeton physicist Henry D. Smyth, and defining what information about the atomic bomb could be given to the public.[31] Under the title *De ontsluiting van de atoomkern* (The opening of the atomic nucleus), the lectures were published in 1946.[32] Rosenfeld concluded the lectures by discussing the wider societal implications of the existence of nuclear power and the bomb. He began by outlining Bohr's thoughts on the matter. Bohr had, according to Rosenfeld, with "his usual optimism" expressed "the opinion that with the atomic bomb there is a possibility of avoiding any future war".[33] The background of Bohr's reflections — which was not known to the public — was the documents he had produced in connection with his confidential discussions with statesmen already during the war. These statesmen included President Franklin Roosevelt, Prime Minister Winston Churchill, Sir John Anderson, who from 1941 had been given responsibility at ministerial level for the British atomic bomb project, code-named "Tube Alloys", and the British ambassador in the US Lord Halifax. However, until his Open Letter to the United Nations in 1950, Bohr did not mention these confidential discussions, nor his continuous confidential approaches to statesmen after the war in his publications; he merely provided a general account of the implications of the atomic bomb for the future relations among nations, and this was most likely also the only information he gave to Rosenfeld at the time.[34] When Rosenfeld later learned about Bohr's confidential activities it added

[30]Hooyman (1979).
[31]Smyth (1945). Aaserud (2005), p. 56.
[32]Rosenfeld (1946a).
[33]*Ibid.*, p. 123.
[34]Aaserud (2005), pp. 4–60. Aaserud (1999).

to his picture of Bohr being naive in political matters: "He was trying to assume that all those statesmen were just as honest as he was and that they were seriously analyzing the (epistemological) aspects of politics. He was very naive and inexperienced. In his youth he had lived in this euphoric atmosphere of bourgeois hegemony, in an environment of this ruling class which was very satisfied with itself. Especially in the small agricultural country like Denmark, what political experience could he have got? He was very much out of touch with the times".[35]

There was general agreement among scientists about the necessity of building and organizing an effective supervision worldwide of the production and use of the new destructive weapon.[36] However, opinions differed as to how that was to be done. In September 1945, for example, Einstein proposed for the first time the creation of a world federal government in which the security of nations was founded upon law, an idea he returned to frequently and which for a short while he perhaps hoped would be fulfilled by the United Nations.[37] In his lecture at Utrecht University, Rosenfeld referred to Bohr's argument for creating a new relation between nations, an open world in which barriers that until then had served as a necessary protection of national interests had to be broken down. And, according to Bohr, here there was something to be learned from the international cooperation of scientists. The close links between scholars from all countries could benefit the discussion about the control of nuclear energy. Scientists could contribute crucially and influence the spirit in which the discussions took place. In particular, a free and unrestricted exchange of information about the realizable scientific progress was an essential precondition for the efficiency of international control. Such an exchange could strengthen the cultural connections and the mutual insight within and among nations, helping to create an atmosphere of trust and goodwill. The scientists also had

[35]Kuhn and Heilbron (1963c), p. 7. According to Rosenfeld, Bohr's brother, the mathematician Harald Bohr, was much more astute and had much more acute judgment in political matters than Niels Bohr. *Ibid.*, p. 8.
[36]Jones (1988). Edgerton (1996), pp. 17–18. Wang (1999), pp. 9–13.
[37]Baratta (2004), Vol. 2, pp. 301–314.

an important mission to fulfil in enlightening the public about the dangerous perspectives of the new weapon. So far Rosenfeld fully shared Bohr's thoughts and echoed them again nine years later.[38] However, Rosenfeld emphasized that it would take more than scientists' expressions of goodwill if Bohr's idea was to work out under the current social relations in the world. Nor was he reconciled with Bohr's idea that peace might come from the atomic bomb.[39] In later years, that is in retrospect, Rosenfeld viewed the development of the bomb in a different light and considered the arms race a more or less necessary development in order to achieve a power balance in the world. However, he maintained the view that it had been unnecessary to throw the bombs on Hiroshima and Nagasaki.[40]

When voicing his own opinions in the fall of 1945, Rosenfeld excused himself, "I know that I move outside the field of my profession in this". He justified his move, however, by referring to declarations made in connection with the recent reopening of Utrecht University, which announced reform in order to ensure more contact between students and professors and preventing students from developing into what we today call nerds, characterized by political apathy and ignorance of the social dimension of science.[41] Rosenfeld further pleaded the authority of objective and rational scientific argumentation in dealing with social problems even though they were so much more complex than those pertaining to inanimate nature.[42] Next he emphasized that the state of affairs pertaining to the atomic bomb was not at all controlled by the scientists. The so-called secret of the atomic bomb, was, as was well-known, no *scientific* secret. All relevant scientific principles and facts were known to the physicists already at the beginning of the war, and in the recent official publications further procedures were clearly laid out. The secret involved technical design and procedures, and it was not the scholars who hid the secret, but rather men with industrial and financial interests, who

[38]Rosenfeld (1946a), pp. 124–125. Rosenfeld (1954).
[39]Rosenfeld (1946a), p. 125.
[40]RP, "Supplement D", radio broadcast with Rosenfeld in Danmarks Radio, 6 Mar 1972.
[41]Molenaar (1994), pp. 37–38.
[42]Rosenfeld (1946a), p. 126. See also Rosenfeld (1954).

had been given the fabrication of the atomic bomb in their hands. Even though governments controlled the uranium mining and probably still had control over the use of the new weapon, Rosenfeld saw a danger in a capitalistic system in which the real powers in control of nuclear energy were the big businesses that merely followed their own interests. "Do not misunderstand me. I do not want to paint these men in the trusts as monsters, although for my part I cannot forget that it was these people who released all the horrors of fascism on the world".[43] The way to prevent capitalist interests from taking over decision-making about nuclear energy was to build a socialist economy driven by the interests of the people. That this could be done successfully could no longer be doubted, Rosenfeld claimed. "Perhaps this vision of the future seems too radical for you", Rosenfeld suggested to his audience, but he ended nevertheless with invoking the Marxist imperative: "Society will be a socialist world federation, or it will not be at all".[44] Hence Rosenfeld believed strongly at this time that capitalism was to be blamed for war (and for fascism). Unless socialism was invoked in the world at large, the world would simply not survive. Only global socialism would guarantee peace. And a precondition for scientific progress was peace. Science and its peaceful applications could therefore only thrive under socialism. A counter-argument could be made that the idea that science would only thrive under socialism had just been thoroughly falsified during the war; capitalist systems were more than capable of sustaining and advancing science and its associated technologies. Rosenfeld's argument, however, concerned *peaceful* applications of science. Only under socialism would scientists' work benefit the people, he claimed.[45] In 1954, he clarified that "I am not so naively optimistic, however, as to suggest that the emergence of socialism would automatically secure conditions under which the delicate tree of science would thrive and yield its fruits in abundance".[46] Of course, patience

[43]Rosenfeld (1946a), p. 128. Originally in Dutch.
[44]*Ibid.*, p. 130.
[45]See also Rosenfeld, lecture notes in RP, Supplement "Science and Peace".
[46]Rosenfeld (1954).

and unrelenting effort were required. Whereas some physicists would participate in the production of atomic weapons because they saw it as a secondary evil compared with "the overwhelming danger and evil" coming from the policy pursued by the Soviet Union, Rosenfeld was strongly convinced that the moral responsibility of the scientists demanded that they did not take part in science with military ends.[47] Furthermore, capitalism had outlived itself, Rosenfeld sincerely believed, when he discussed the political situation in 1949 in a long letter to his younger colleague, the Polish physicist Wladyslaw Opechowski who had previously worked in the Netherlands, and who Rosenfeld later called "my best friend".[48] The crisis of capitalism Rosenfeld described as one of decadence. Like most leftists at the time he sincerely believed that capitalism was no longer adaptable to the conditions of a continuingly changing world.[49]

The End of the Popular Front

Rosenfeld's ideological baggage and political conviction can be described as progressive socialist and non-communist; he was what Perry Anderson has called an intellectual freelance.[50] His political views bore some resemblance to Trotsky's position, for example that of the permanent revolution. "Only a European and then a world federation of socialist republics can be the real arena for a harmonious socialist society", Trotsky said in his Copenhagen speech in 1932.[51] The united front against fascism and the far right, which Rosenfeld advocated, was also a strong feature in Trotsky's position. However, Rosenfeld would never side with the so-called Trotskyists. And only on one occasion did Rosenfeld mention Trotsky by name. The commotion that this brought about probably taught him not to do it again. The incident shows again how difficult it was to mount a

[47]RP, "Manchester (1948–1958): Political (1948–1958)", J. D. Cockcroft to L. Rosenfeld, 8 Apr 1952. The American physicist E. Teller was motivated by the Soviet challenge in promoting the American atomic weapons program. Hewlett and Holl (1989), p. 35.
[48]RP, Supplement "O", copy of letter from L. Rosenfeld to J. Maddox, 18 Jun 1971.
[49]RP, Supplement, draft of letter from L. Rosenfeld to W. Opechowski, 30 Dec 1949.
[50]Anderson (1976), p. 44. See also Wang (1999), pp. 5–6.
[51]Trotsky (1932a), p. 10. See also Chapter 3.

critical defence of one's chosen camp at the time, an activity Rosenfeld nevertheless gradually increased with ambivalence the following years.

In the spring of 1946, Rosenfeld submitted an article to the periodical of the Dutch Communist Party, *Politiek en Cultuur* (Politics and culture). It was entitled "De socialistische student in de maatschappij" (The socialist student in society). The article can be seen as a contribution to a discussion that had already been going on for half a century in intellectual circles in the Netherlands. Changing social and economic circumstances and changing career opportunities for students made universities and other higher education institutions debate their role in society. The question was: should they educate students to become scientists and scholars or rather future leaders of society?[52] Rosenfeld's article reflects his general and sincere concern with education, both at university level and in secondary school, and its role in society. He even made comparative studies of the education systems in the different countries he had lived in.[53]

From a leftist perspective, the topic of the article was very close to the one that Trotsky had been asked to talk about with representatives of the student council in Copenhagen in 1932. And, in fact, Rosenfeld explicitly referred to that interview in the article.[54] Rosenfeld believed that the current post-war crisis led the Dutch intellectual middle class, from where students were predominantly recruited, to seek refuge in unrealistic spiritual movements, and he was afraid that the current political apathy could function as the substrate of a lurking fascism.[55] His article was meant as a call to arms to socialist students against this movement. The "detrimental influence of the present situation on the youth", was close to Rosenfeld's heart especially in those early post-war years.[56] Rosenfeld's

[52]Baneke (2008), p. 208.
[53]Rosenfeld (1936), p. 135. Rosenfeld (1949). See also RP, "Correspondance particulière: Kemmer" about "Physics teaching in England".
[54]RP, Supplement "Science and society", Rosenfeld, manuscript "De socialistische student in de maatschappij". Neergaard (1932). See also Chapter 3.
[55]RP, Supplement "Science and society", Rosenfeld, manuscript De socialistische student in de maatschappij.
[56]L. Rosenfeld to R. Peierls, 4 Jun 1948, in Lee (2009), Vol. 2, item [461], pp. 148–149.

article was also meant as a call for Popular Front policy. He expressed satisfaction with the opportunity the article gave him to address not only moderate socialists but also full-blooded communists and reiterated that it was about time that democrats of all nuances spoke with one voice.[57] People are in fact pre-disposed to seeking cooperation, he stated; this tendency was only disturbed by the short-term policy of some politicians. It should be possible to unite forces on the Left instead of fighting each other. He recalled how, in a Marxist club in Zurich sixteen years earlier, democrats of all shades, including Social Democrats and communists, had engaged in "Homeric discussions"; for uniting did not mean agreeing about everything, Rosenfeld emphasized. The fight between opinions through discussion is a healthy phenomenon because "truth has no greater enemies than, on the one hand, schematizing dogmatism, blind to the manifold and frequently antagonistic aspects of reality and, on the other hand, the slack spirit of compromise which, in the delusion of an enchanting artificial consensus, side-steps the core of the problems".[58] A completely honest and frank strife between opposed opinions would forge unity through a dialectical process, a unity he considered crucial in order to overcome the social problems of the post-war era.

However, a strong unity can only arise between people who have learned to know and appreciate each other. Students and intellectuals in general, including scientists, must therefore show solidarity with the workers. Rosenfeld referred to the advice given by Trotsky to the student council in Copenhagen in 1932 in such matters spicing it with his own views. The way for the average student to find his place in a socialist society was to repudiate his class and social type. It was the students' destination to take intellectual leadership in society, or at least in the transition to socialism intellectuals could have such a leading function. The noblest task of the socialist student was to contribute to the dissolution of the artificial distinction between

[57]Rosenfeld's viewpoints were in tune with those of the Russian physicist S. Vavilov, the new president of the Soviet Academy of Sciences in 1945. Kojevnikov (2011).

[58]RP, Supplement "Science and society", Rosenfeld, manuscript De socialistische student in de maatschappij.

intellectual and manual labor. The way to overcome this separation was to approach the worker, but in a modest, not a patronizing, manner. In particular, it was Rosenfeld's opinion that the intellectual and the scientist should become a mediator between the workers and the treasure chambers of science to which the worker otherwise had no access. One important way to accomplish this was through the history of science.

In his article for *Politiek en Cultuur*, Rosenfeld not only praised Marx and Lenin, but also referred to Trotsky as the great revolutionary. At the same time he distanced himself from the so-called Trotskyists, whom he considered to constitute a sectarian division of the Left. Given Rosenfeld's aim to address all wings with the task of uniting the forces on the Left, the reference to Trotsky may have seemed a natural thing for him to do. However, the editor of *Politiek en Cultuur*, Eva Tas, immediately censored the references to Trotsky. In the ensuing correspondence about the article Rosenfeld argued that it was an impossible task to try to create unity among the Left under the current conditions of which Tas' response was a symptom. However, as his article was meant as a positive contribution to the fight for Marxist democracy, he agreed to leave Trotsky's name out in order to avoid misunderstandings. Rosenfeld attempted to explain his reference to Trotsky by his belief that Trotsky, who had been assassinated in Mexico in 1940, could now be regarded as a historical figure. Apparently that was considered a naive thought at the time. In the end, the article was never published.[59] From this incident Rosenfeld may have learned not to speak out about his position to just anybody. However, he outlined his position to Opechowski, "one of the rare people, whose judgment I value", a few years later.[60] Rosenfeld then deplored to Opechowski the new situation with the

[59]RP, "Centre intellectuel français d'Utrecht 1945–1947", E. Tas to L. Rosenfeld, 18 Apr 1946. RP, Supplement "Science and society", Rosenfeld, manuscript "De socialistische student in de maatschappij" and draft of letter from Rosenfeld presumably to E. Tas, 20 Apr 1946. I am grateful to Leo Molenaar and Internationaal Instituut voor Sociale Geschiedenis (International Institute of Social History), Amsterdam for checking the periodical for Rosenfeld's article. It was not found.

[60]RP, Supplement, Draft of letter from L. Rosenfeld to W. Opechowski [not dated]. Originally in French.

"Western epigones of Marxism, for whom October is nothing but an historical event, the anniversary of which is celebrated by falderal; for whom Lenin is an archangel, Stalin a pope, and Trotsky Beelzebub!"[61]

Rosenfeld's optimism on behalf of a united Left was further tested in 1947 when the Dutch Labor Party refused to collaborate with the communists and instead entered a coalition with the conservative Catholic Party. In other European countries such as Denmark, Belgium, Norway, France, and Italy, communists were in government at the time.[62] Rosenfeld did not trust the Catholic Party an inch, because, as he stated, they supported the Franco regime in Spain and compromized when it came to fascists and former collaborators with the Germans.[63] His critical letters to the editor in *The Manchester Guardian* about Dutch imperialist policy in Indonesia were noticed among the Left in Holland, and he was soon invited to write in yet another Dutch left-wing weekly, *De Groene Amsterdammer* (The green Amsterdammer).[64]

The History of the Social Relations of Science

Before he left Utrecht in 1947 to take up a position in Manchester, Rosenfeld gave a farewell lecture which was published in *De Vrije Katheder* under the title "Dynamisch denken in de wetenschap" (Dynamic thinking in science).[65] It was a materialist analysis of the social conditions that best serve science when its development is otherwise affected and seems determined in a detrimental way by historical process. Once again Rosenfeld took as his starting point in the current direction taken by Dutch society in the ruins of war. Rosenfeld regarded Dutch society with its many religious fractions

[61][de nos épigones occidentaux du marxisme, pour qui Octobre n'est plus qu'un événement historique dont on célèbre l'anniversaire par les flonflons pour qui Lénine est une archange, Staline un pape, et Trotzky Belzébuth!] *Ibid.*
[62]Thing (1993), Vol. 2, p. 752. Arblaster (2006), p. 232.
[63]Rosenfeld (1947), Geyl (1947), in RP, "Utrecht (1940–1947), Political (1945–1947)".
[64]RP, "Utrecht (1940–1947), Political (1945–1947)", *De Groene Amsterdammer* to L. Rosenfeld, 29 Aug 1947. See also De Vrije Katheder to L. Rosenfeld, 5 Sep 1947.
[65]Rosenfeld (1948a).

or pillars — the Protestants, the Calvinists, and the Catholics — as declining (like all capitalist states at the time), as a result of which people clung to spiritual movements.[66] It was Rosenfeld's strong conviction at this time that science could not thrive in a religious society. Several years later, in his correspondence with the East German historian of science Friedrich Herneck, Rosenfeld's reflections about variants of belief and to what extent each interfered with the rational thinking of intellectuals in different countries and belonging to different religious denominations such as Catholicism, Protestantism, and Judaism, were more nuanced.[67]

Sketch of Rosenfeld by a Dutch student in 1947.[68]
Courtesy of The Niels Bohr Archive.

In the article for *Vrije Katheder*, Rosenfeld grouped people and scientists into social types according to the categories static *versus* dynamic and dialectical *versus* metaphysical (in Engels' meaning of the word as the opposite of dialectical) ways of thought and attitudes to life. He saw the groups as being shaped by the ideological

[66]About the fractionalized Dutch society, see Somsen (2008), p. 232–233.
[67]RP, "History of science folder 7", copy of letter from L. Rosenfeld to F. Herneck, 3 Jan and 9 Mar 1959.
[68]RP, "Liège 1922–1940: Utrecht".

bias of the society they were part of, whether its economy was based on farming or commerce. Each time a new society had emerged as a result of the historical process, science and technology would undergo yet another blossoming period. This view was connected to the conviction that the purpose of science was not only to try to find answers to the big existential questions, but it was also to control the forces of nature in order to make them benefit society. According to Rosenfeld, curiosity was a crucial condition for science, and he argued that curiosity would thrive when a new social order arose. He exemplified this with Simon Stevin and Christiaan Huygens who benefited from the ideology of the Dutch trading tradition, and the French encyclopaedists who drew on the ideology of the Enlightenment.[69] Ideology had a strong impact on cultural life, art, and science. Since science, like any other human activity, is a social product, the way for science to escape decline, stagnation, and death due to a given ideology, was by upholding a dynamical and dialectical way of thinking in its scientific practice. As Rosenfeld had made clear in his 1945 honorary essay for Bohr, the "Copenhagen spirit" epitomized that kind of scientific thinking. The "Copenhagen spirit ... is *par excellence* that of a complete freedom of judgment and discussion". Bohr, who personified the Copenhagen spirit, was in Rosenfeld's view a "mind always alert, always open to doubt, to possible improvement, attentive to the least objection, never resting until it has completely enveloped the object of his search and dominates all its aspects".[70]

Rosenfeld shared his burning interest in the history of the social relations of science and the belief in the importance of its social and political function with *The Visible College*, that is, the scientific intellectuals who had dominated the Social Relations of Science Movement in Britain in the 1930s, including J. G. Crowther, P. M. S. Blackett, Joseph Needham, and John D. Bernal. Rosenfeld wrote to Crowther in September 1945:

---

[69]See also RP, Supplement "Science and Society", L. Rosenfeld to W. Opechowski, 15 May and 30 Dec 1949.
[70]Rosenfeld (1979j), p. 313.

> I have been much interested in your books on British and American men of science, especially (as you may easily imagine if you remember our past conversations) on account of the fundamentally sound historical method which you adopt. By the way, one should like to have for this method some less dusty name than "dialectical or historical materialism". I recently chanced upon a little Pelican book on Greek science by Benjamin Farrington, in which the same method is applied with great success. Doubtlessly there must be quite a lot of people who are at the same time interested in problems of the history of science and convinced that the only sensible way of tackling such problems is the above-mentioned one. I wonder whether there is any club or society providing, so to say, a meeting-place for these people (who may belong to a wide variety of professions and branches of learning). If (as I believe) there isn't, don't you think it would serve a useful purpose to found one? It goes without saying that I mean a society free from any dogmatism, and in particular from the unconditional admiration of everything sovietical which seems to be de rigueur nowadays in certain circles (at least on the Continent).[71]

A venue such as the one suggested by Rosenfeld for scientific intellectuals interested in social history of science was established in 1946 by Needham under the auspices of UNESCO and the International Academy for the History of Science, *viz.*, a Commission for the History of the Social Relations of Science. When Rosenfeld was asked to chair it he justified the need for such a Commission to the Portuguese historian of science and representative of UNESCO, Armando Cortesao, in the following way:

> Unfortunately, history of science is in a state of utter neglect and needs not only coordination and encouragement, but

[71] J. G. Crowther Archive, University of Sussex Library, Special Collections, Brighton, L. Rosenfeld to J. G. Crowther, 14 Sep 1945.

> building up from the very foundation. In fact, notwithstanding the existence of an International Academy of the History of Science and of active specialized journals like "Isis", there is practically no deeper cooperation between workers in this field. Everyone pursues his own line of research, there is no general outlook, no clear recognition of problems and, of course, no planning at all... it remains true that this branch as a whole is nowhere a fully recognized subject of study, comparable, e.g. with theoretical physics or embryology or medieval history. In few other fields of knowledge is there such a display of dilettantism and unscientific approach.[72]

During the 1950s, Rosenfeld contributed as much as his other commitments allowed him to the institutionalization of the history of science in Britain and later in Denmark.[73] As President for the Commission for the History of the Social Relations of Science, Rosenfeld collaborated closely with the young British Marxist historian of science, Samuel Lilley.[74] There was one major product of the commission, *viz.*, an anthology which appeared in a special issue of *Centaurus* in 1953, edited by Lilley. The anthology listed contributions from a variety of mostly left-leaning scholars including Anton Pannekoek, Benjamin Farrington, V. G. Childe, R. J. Forbes, Needham, R. Taton, S. F. Mason, Richard H. Shryock, Lilley, Bernal, and Dorothea Waley Singer. Rosenfeld wrote the introduction.[75] During the cultural Cold War of the 1950s, this leftist movement in the history of science was efficiently curbed by opposing rightwing movements such as the Society for Freedom of Science (SFS), especially following the Lysenko affair (see further below). Even though Lilley never held a formal position in history of science he continued nevertheless writing and teaching the social history of science. In the

[72] RP, "Union Internationale d'Histoire des Sciences: Commission for the History of the Social Relations of Science (1946–1956)", L. Rosenfeld to A. Cortesao, 23 Oct 1946.
[73] Jacobsen (2008). RP, "History of Science: British Society for the History of Science", "Union Internationale d'Histoire des Sciences".
[74] Jacobsen (2008). Enebakk (2009). RP, "Union Internationale d'Histoire des Sciences: Commission for the History of the Social Relations of Science (1946–1956)".
[75] Lilley (1953).

1960s, the social history of science experienced a revival on the initiative of Robert S. Cohen among others.[76]

In early 1947, Rosenfeld again complained about the "terrible spiritual demolition, which idealism spreads among the students" in this "all too holy country", that is, the Netherlands, this time in a letter to Bohr.[77] The correspondence between Bohr and Rosenfeld from this time is characterized by Rosenfeld's blunt outbursts of his political opinions and Bohr's avoidance of political issues. Rosenfeld was a close friend of Claude Morgan, the editor of the "quite excellent" French intellectual, left-wing weekly *Les Lettres Françaises*, another underground paper founded during German occupation.[78] In the late fall of 1947, Morgan informed Rosenfeld about the plan by the French left-wing journalist and writer Yves Farge to write a book on the history of the atomic bomb. Rosenfeld immediately spotted that the manuscript was not historically accurate with respect to Bohr's role in the decision to use the atomic bomb and was worried that such misrepresentations of the historical facts should be spread. Via Morgan, he informed Farge about the "truth", as he saw it, about Bohr's role.[79] Subsequently, Rosenfeld expressed his concern about the matter to Bohr and urged Bohr either to publish the letter he had written to Roosevelt during the war or at least write a small piece explaining the events preceding the "disastrous decision about using the bomb".[80] Bohr answered Rosenfeld that at the present time he felt it his "duty to keep the utmost discretion, which is of absolute necessity in order not to harm the great cause".[81] Bohr reasoned that for the time being the current crisis, "so serious for humankind", was much more important than saving his own reputation.[82] However, he asked Rosenfeld to keep him informed about what and how much Rosenfeld thought Morgan and Farge knew. As

[76]Mayer (2004). Mayer (2000). Enebakk (2009), pp. 588–593.
[77]BSC, L. Rosenfeld to N. Bohr from Utrecht, 1 Feb 1947. Originally in Danish.
[78]BSC, L. Rosenfeld to N. Bohr, 5 Dec 1947.
[79]*Ibid.*, RP, "Political 1948–1958", C. Morgan to L. Rosenfeld, 17 Nov 1947. RP, "Copenhague: Bohr (1940–1962), Open Letter", Y. Farge to L. Rosenfeld, 25 Nov 1947.
[80]BSC, L. Rosenfeld to N. Bohr, 5 Dec 1947.
[81]BSC, N. Bohr to L. Rosenfeld, 16 Dec 1947. Originally in Danish.
[82]*Ibid.*

the case of Robert Jungk's book *Brighter than a Thousand Suns: A Personal History of Atomic Scientists* (1958) was going to show with respect to the infamous Bohr–Heisenberg meeting in Copenhagen in 1941, the role of the great physicists in the war proved an extremely delicate matter in the public sphere during the Cold War.[83] Bohr published an extract of his 1944 letter to Roosevelt as part of his Open Letter to the United Nations in 1950 when there was no longer hope of achieving agreement in that forum about the prevention of a nuclear arms race.[84]

In his political activism, Bohr may be characterized as an elite statesman of science comparable to the American physicists Vannevar Bush, James B. Conant, and Robert Oppenheimer;[85] at least in the sense that Bohr addressed governments, governmental organizations such as the UN, and heads of state rather than supporting grass-root movements when he wanted to make his influence felt. However, contrary to Bush and Conant, Bohr operated alone and independently of the interests of the American military.[86] In contrast to these elite scientist-administrators and to Bohr's solitary route, Rosenfeld's placed his political activism within concerted attempts through scientists' organizations aiming at persuading governments and public opinion of the desirability to end the atomic arms race, to further the role of science's peaceful applications in society and for public welfare, and to promote international relations and cooperation. Rosenfeld became involved in and was a driving force in several such organizations.

Scientific Internationalism

After the war Rosenfeld learned about the activities of the British scientists on the progressive Left from Crowther, who had developed into an enterprising and unsurpassed master in navigating in the

[83]The book was originally published in German in 1956 with the title *Heller als tausend Sonnen: Das Schicksal der Atomforscher* Jungk (1956). See also Carson (2010), pp. 398–412.
[84]Bohr (2005a).
[85]Badash (1995), p. 104.
[86]As for Conant, Oppenheimer, and Bush, see Wang (1999), pp. 6–7, 14.

corridors of British science policy.[87] Like their British counterparts, Dutch scientists were very active in discussing and deciding the role of science in society. During the German occupation of the Netherlands, there was much underground reflection and discussion among Dutch scientists about the future organization of scientific research and higher education and on the future role of scientists in society, and Rosenfeld took an active part in these negotiations. These political projects resulted in publications and measures taken after the war.[88] Similar academic activities took place in Denmark, involving predominantly left-wing researchers who were also active in the resistance movement, and who were strongly inspired by the British Social Relations of Science Movement and by Bernal's writings.[89]

In the summer of 1946, Rosenfeld became a driving force in the establishment of the Dutch Verbond van Wetenschappelijke Onderzoekers (VWO) (Federation of scientific researchers). He became Vice-President of the board. It was dominated by leftist forces and had among its objectives the democratic re-organization of higher education and research. Recruitment from all parts of society was to be improved and research should be regarded as a useful investment for society and not merely as cultural luxury. The aim was to strengthen the social position and influence of the scientific worker. Research was to be viewed as a profession rather than a hobby.[90] In this organization Rosenfeld collaborated with the Delft physicist and former communist Johannes Martinus (Jan) Burgers, the communist astrophysicist Marcel Minnaert, and "a Marxist of the old guard from the heroic time", the astronomer Anton Pannekoek (more of whom in the next chapter), among others.[91]

---

[87]J. G. Crowther Archive (Brighton), University of Sussex Library, Special Collections, Rosenfeld to Crowther, 31 Oct 1945. Crowther (1970), pp. 226–227.

[88]Rip and Boeker (1975), p. 462. Molenaar (1994), pp. 37–38, 90–91. Somsen (2008), p. 239. Rosenfeld (1946a), pp. 123–130. Later when in England, Rosenfeld published an article, Rosenfeld (1949), which was an outcome of his previous reflections on these issues while in the Netherlands.

[89]Knudsen (2010), pp. 216–222. Knudsen (2005), pp. 328–339. Nielsen (2008).

[90]Rip and Boeker (1975), p. 463. Molenaar (1994), pp. 47–71. RP, "Verbond van Wetenschappelijke Onderzoekers (1946–1948)".

[91]J. Burgers was the initiator, secretary and chairman of the Committee on Science and its Social Relations (CSSR) of the International Council of Scientific Unions (ICSU),

In the same summer, Rosenfeld was elected as the Dutch delegate to the board of the World Federation of Scientific Workers, which held its inaugural meeting in London on 20–21 July 1946. This non-governmental organization was a progressive leftist response to the questions of international social relations of science and of post-war reconstruction. Although the Russians refused to join it, the WFSW quickly gained a reputation as a communist front organization dominated by British and French communists and fellow travellers.[92] The French nuclear physicist and prominent member of the French Communist Party Frédéric Joliot-Curie was President, Bernal Vice-President, while Crowther served as Secretary-General.

The board of WFSW in 1946. Rosenfeld Album.
Courtesy of The Niels Bohr Archive.

see Somsen (2008). As regards to M. Minnaert, see Molenaar (2003). RP, Supplement, L. Rosenfeld to F. Herneck, 2 Dec 1969. In this letter, Rosenfeld promotes Pannekoek's book. See also RP, "History of science: XIXe siècle", L. Rosenfeld to S. G. Brush, 7 Oct 1966.

[92]Petitjean (2008b). Werskey (1988), pp. 276–277. RP, "World Federation of Scientific Workers (1947–1955)".

Based on the view that scientists should not, and in reality could not, isolate themselves from the major political trends of the day, the aim of the WFSW was to work for international solidarity, the protection of scientists' interests as professionals, and the mobilization of the rank and file of the international science community.[93] Bernal wrote the charter for the first general assembly in Prague in 1948. The ideology and practice of the organization were thus derived from Bernal's writings from the 1930s, particularly his book *The Social Function of Science* (1939). Bernalism, as this body of thought has since been called, was indicative of a new politics of science which linked the development of science and technology for public welfare to socialism and the classless society. A central issue was the planning of science from a leftist perspective. To Bernal, the power of planned science with particular goals in mind had been clearly demonstrated by the centrality of science and technology in the victory of the allied countries over fascism.[94]

It was considered of utmost importance by the board members to have Bohr's name associated with the World Federation of Scientific Workers, and it became Rosenfeld's mission to win Bohr's support for their cause and for their various activities. Rosenfeld hardly succeeded in this, however. In 1947, he appealed to Møller to persuade Bohr to give his consent and support to the WFSW in the shape of "a brief statement to be printed in a small propaganda book in which it is explained what the WFSW is and wants".[95] The Danske Forening til Beskyttelse af Videnskabeligt Arbejde (The Danish organization for the protection of scientific work) had joined the WFSW the year before.[96] After Møller's approach, Bohr replied to Rosenfeld that he had forgotten about Joliot's invitation to write a piece for the leaflet "due to the pressure of work" and asked Rosenfeld "what answers

[93]Petitjean (2008b). Elzinga (1996b), p. 93.
[94]Elzinga (1996a), p. 3, 9. Horner (1996), pp. 132, 139–140.
[95]MP, L. Rosenfeld to C. Møller, 1947.
[96]MP, copy of letter from C. Møller to L. Rosenfeld, 14 Nov 1946. Knudsen (2010), p. 219.

have been received upon similar approaches to others".[97] Bohr was extremely cautious and did not easily fall victim to manipulation in political matters. Besides, he was engaged on the opposite front, so to speak, and had developed a trustworthy relationship with the new American ambassador to Denmark, Josiah Marvel.[98] Bohr did not spell out his own political standpoint in detail to Rosenfeld in their correspondence, but Rosenfeld drew his own conclusions. In September 1947, Crowther asked Rosenfeld if he could "do all you can to persuade Bohr to come to the Rutherford Meeting? He has refused so far".[99] The meeting in question was a commemoration of Rutherford ten years after his death, arranged by the WFSW which was to take place in Paris 7–9 November that year. Rosenfeld responded to Crowther "With regard to Bohr's attitude towards the WFSW (and the world in general) I was at first dismayed to see to what extent he had succumbed to the current propaganda. But in the course of some conversations I am glad to say I was able so far to reverse his attitude as to get from him a (still rather vague) assurance that he would come to Paris after all. I am watching a favorable moment to try to press the matter still further".[100] On this occasion Rosenfeld succeeded; Bohr wrote a tribute, but due to pressure of work he ended up sending Møller in his place to read it.[101]

In the meantime, the Manchester physicist, socialist, and war hero P. M. S. Blackett sent Rosenfeld a telegram on 5 November 1946 asking whether he would consider accepting the chair of theoretical physics at Manchester University, a position with "only light teaching duties".[102] D. R. Hartree had left to become Plummer

---

[97] RP, "Copenhague: Bohr (1940–1948)", N. Bohr to L. Rosenfeld, 3 Jun 1947. See also Rosenfeld to Bohr 19 Jun 1947.
[98] Aaserud (2005), p. 67.
[99] RP, "World Federation of Scientific Workers (1947–1955)", J. G. Crowther to L. Rosenfeld, 17 Sep 1947.
[100] RP, "World Federation of Scientific Workers (1947–1955)", Copy of letter from L. Rosenfeld to J. G. Crowther, 24 Sep 1947.
[101] Aaserud (2007), p. 111. Bohr (2007a).
[102] RP, "Manchester 1: Appointment 1947", P. M. S. Blackett to L. Rosenfeld, 5 Nov 1946. See also Blackett to Rosenfeld, 8 and 11 Nov 1946.

Professor of mathematical physics at Cambridge in 1946, and Blackett needed a new theorist for his laboratory. The first candidates that had occurred to Blackett as possible replacements had been Casimir and Møller. He conferred with Bohr about it, who, however, expressed doubt that either of these physicists would accept the offer and suggested instead that it might be worth trying a younger candidate. Bohr was also reluctant to let Møller go.[103] When exactly Rosenfeld, who was not exactly a younger candidate, came into the picture is not clear. However, Blackett and Rosenfeld who knew each other well from the 1930s most likely met at the inaugural meeting of the WFSW in London in late July 1946, where Blackett in his capacity as President of the British Association of Scientific Workers gave the opening address.[104] In late November, three physicists had been nominated and were listed in the following order, H. Frolich, N. Kemmer, and Rosenfeld, of which the first two were British and younger candidates. Blackett, however, preferred Rosenfeld. The Deputy Vice-Chancellor of the university contacted Bohr asking his opinion about the nominations. Bohr suggested that Rosenfeld, "due to his extensive and fruitful researches in many fields of atomic theory and his exceptional wide knowledge and great gifts for lucid representation... should be considered in the first instance".[105] By February 1947, the University of Manchester had settled for Rosenfeld and he accepted the offer.[106] The news was announced in *The Manchester Guardian* and in *Nature*. Bohr, Møller, Peierls, Crowther, and other colleagues from abroad such as John Wheeler expressed pleasure at Rosenfeld's new appointment. His Dutch colleagues congratulated him, but deplored that he was now leaving their country.[107] He began his new job in the early fall of 1947.

[103]BSC 27, P. M. S. Blackett to N. Bohr, 14 Mar and 11 Apr 1946. Bohr to Blackett, 1 and 15 Apr 1946.
[104]Horner (1996), p. 138.
[105]BGC, the Deputy Vice-Chancellor of the University of Manchester to N. Bohr, 19 Nov 1946. Copy of letter from N. Bohr to the Deputy Vice-Chancellor, 10 Dec 1946.
[106]RP, "Manchester 1: Appointment 1947", P. M. S. Blackett to L. Rosenfeld, 11 Dec 1946. Blackett to Rosenfeld, 11 Feb 1947. BSC, Rosenfeld to Bohr, 1 Feb 1947.
[107]RP, "Manchester 1: Appointment 1947". RP, "Copenhague: Bohr (1940–1948)".

Personnel at the institute in Manchester 1947, including Blackett and Rosenfeld. Rosenfeld Album. Courtesy of The Niels Bohr Archive.

Rosenfeld was now geographically united with the group of left-wing scientists in Britain that had been so influential in the 1930s and with whom he had so much in common and collaborated with in several political organizations. Like Joliot-Curie and Blackett, Bernal was widely admired for his courage, intelligence, and militant initiatives.[108] Rosenfeld also admired Bernal, but soon he began criticizing "Bernal and his friends" for their uncritical adoration of the Soviet line in scientific and cultural matters. Rosenfeld was also of the opinion that British left-wing intellectuals were much more bourgeois and had much less contact with the working class than left-wing intellectuals in France and Italy.[109] This was reflected in the British Left's preoccupation with the analysis of the social and international function of science. Bernal's book *The Social Function*

[108]About Joliot-Curie, see for example Blackett's description of him and his leader and oratory skills in Nye (2004), p. 170.

[109]RP, Supplement "Science and Society", L. Rosenfeld to W. Opechowski, 15 May 1949.

of Science (1939) was never translated into French and the social function of science remained of less interest to French intellectual scientists, who were more concerned with defending rationalism and theoretical approaches such as dialectical materialism. Their political commitment lay in the sympathy for the politics of the proletariat and the USSR rather than in science policy.[110]

On the other hand, Rosenfeld was quite fond of British *manners* and he severely regretted the Americanization of Europe after the war, especially as he witnessed Bohr adopt American social conventions.

> When I saw Bohr again after the interruption of the war years, and later on observed how he more and more imitated the American manners, I was sorry for the change from the natural distinction of the British manners he had formerly adopted. I do not at all think that the American manners are less formal than the British ones, they are just a different kind of formality — which is "democratic" only in appearance. I never touched this point with Bohr, of course. As to the "traditions" of Danish life Bohr was fond of, they were partly again the special kind of formality of the higher bourgeoisie, but mainly the cultural aspects of Danish (and European) life and conversation, which one misses so much in America.[111]

Rosenfeld maintained his very formal manners admitting in 1962 to the philosopher of science Wolfgang Yourgrau that he was "too old, or too old fashioned, or both, to relinquish my European customs".[112] Thus he would not address colleagues by their first name in their correspondence. This happened only with very few very close colleagues and friends, including Delbrück, Chandrasekhar, and Møller. Rosenfeld refused to provide the first names of for example Bohr and Heisenberg in a paper he had submitted for the journal

[110] Elzinga (1996a), p. 13. Petitjean (2008b), p. 253.
[111] RP, Supplement, L. Rosenfeld to D. Danin, 16 Oct 1973.
[112] RP, "Correspondance particulière", L. Rosenfeld to W. Yourgrau, 18 Jul 1962. Yourgrau had proposed that they used first names when addressing each other. Yourgrau to Rosenfeld, 12 Jul 1962.

Physics Today in the summer of 1969, because, as he maintained, it was not customary in the early twentieth century to provide first names. He wrote to the editor R. Hobart Ellis Jr.: "if we are to remain good friends, better take my text as it is".[113] Hobart Ellis Jr. did not give in as a result of which Rosenfeld replied "I see that I am too old-fashioned to deal with American editors" and withdrew the paper altogether.[114]

"The Soviet Crisis"

On behalf of Blackett who was prevented from attending, Rosenfeld participated in the World Congress of Intellectuals at Wroclaw (formerly Breslau) in Poland on 25–28 August 1948 as a member of the British delegation.[115] The purpose of the conference was to let "scientists, writers and artists... gather to discuss what contribution they can bring to the free, creative and peaceful cooperation of the peoples of the world".[116] The Wroclaw meeting can be seen as constituting the first seed in the dawning World Peace Movement, which was immediately perceived as a Soviet ploy by the British Government.[117] From the perspective of the Left, the meeting was another marker of the end of the Popular Front between leftist and liberal intellectuals with the advent of Cold War.[118]

Whereas the Danish delegation presented only three members including the ageing writer Martin Andersen Nexø, the British delegation constituted one of the largest, counting 42 members, in fact only exceeded by the Polish of 53 members. Besides Rosenfeld, the delegates from Britain included Bernal, Crowther, the biochemist J. B. S. Haldane, the biologist and first Director of UNESCO Julien

[113] "Physics Today Archives" Box 19, corresp. 48–70, Centre for History of Physics, AIP, L. Rosenfeld to R. H. Ellis Jr., 29 Jul 1969.
[114] "Physics Today Archives" Box 19, corresp. 48–70, Centre for History of Physics, AIP. L. Rosenfeld to R. H. Ellis Jr., 8 Aug 1969.
[115] RP, "Manchester (1947–1958): Cultural World Congress for Peace, Wroclaw (1948)", A. Slonimski to L. Rosenfeld, 5 Jul 1948.
[116] RP, "Manchester (1947–1958): Cultural World Congress for Peace, Wroclaw (1948)", Invitation to Rosenfeld by the French–Polish Organization Committee.
[117] Ullrich (2007), pp. 6–7. Deery (2002), pp. 461–462.
[118] Horner (1996), pp. 142–143. Pinault (2000), p. 411.

Huxley, the historian A. J. Taylor, and the mathematician Hyman Levy. From the Soviet Union the writers Alexander Fadeyev, Ilja Erenburg, and Dawid Zaslawski, among others, took part but no natural scientists. From France appeared the nuclear physicist Irène Joliot-Curie, the biologist Eugene-Marcel Prenant, the painter Pablo Picasso, and the journalist and writer Yves Farge in a relatively large delegation of 27 members. From Brazil came, among others, the physicist Mario Schönberg, with whom Rosenfeld became close friends, and the United States presented a delegation of 30 members.[119]

Like many delegates Rosenfeld was deeply disappointed by the vitriolic language of the Soviet delegates dismissing any idea of the meeting being non-partisan and an attempt to bridge the two camps. Instead they promoted the line created by the Cominform.[120] On the back of a leaflet about the congress Rosenfeld jotted down his impressions of the congress in brief notes. The notes suggest that his expectations for the meeting were not met. Here for the first time Rosenfeld clearly witnessed the Russian Cold War rhetoric; he considered the congress somewhat a farce; all communication was clearly in a deadlock. There were, according to Rosenfeld, "many dialectical contradictions, but very little synthesis". The congress had been announced as cultural, but was "conducted as political". The political thesis made for the "greatest paradox of all". As he expressed it: intellectuals tricked into political affair, knowing it, furious about it, but voting the resolution. What was this resolution? Disclosing danger of American imperialism: protection of fascism, economic penetration and enslavement. Protesting against encroachments of freedom of thought and expression and frustration of science and culture in decaying capitalism". Rosenfeld considered the political thesis convincing, but it "failed to convince anybody not previously

[119]RP, "Manchester (1947–1958): Cultural World Congress for Peace, Wroclaw (1948)". About M. Schönberg, see Freire Jr. (2005), p. 7. RP, M. Schönberg to L. Rosenfeld, 5 Apr 1949.

[120]RP, "Manchester (1947–1958): Cultural World Congress for Peace, Wroclaw (1948)" and RP, Supplement "Science and Society", L. Rosenfeld to W. Opechowski, 15 May 1949. Ullrich (2007), pp. 7–8. Horner (1996), p. 143.

convinced, since no effort was made in that direction. Russians found this quite natural and were shocked at opposition". There was "no discussion, but invectives" at the congress and "only pitiful caricature", according to Rosenfeld.[121] The resolution gave rise to a lot of debate especially among the British delegates. Some, for instance Taylor, opposed the resolution due to its "over-simplification". Others signed the statement and then withdrew their signatures. Rosenfeld agreed that the resolution was a simplification.[122] He was not that happy about it, partly because of the process of its negotiation and partly due to its simplicity and one-sidedness directed as it was only against the United States. But he signed it anyway. As he wrote to Crowther in December 1948:

> So far as I am concerned, I can say that the only reason why I signed the Wroclaw resolution is that it seemed to me to embody an accurate statement of the menace to peace and culture arising from the blind development of economic forces dominated by motives of self-interest. This conclusion has in no way been "dictated" or "inspired" to me by anybody (least of all by the Russian intellectuals who, in my opinion, made a very poor show at Wroclaw) but is the result of independent and impassionate observations that I have endeavoured to carry out in a scientific spirit. So long as nobody can convince me, by the same method, that I am wrong, I have no choice but to act on this conclusion.[123]

Subsequently he had serious doubts about it, which the need for such a justification also reflects. The resolution was about the denunciation of American and colonial imperialism on behalf of the oppressed cultures that constituted the victims of it, a fight which Rosenfeld considered "as the most urgent task;" so he signed. Meanwhile, that act caused great disappointment in his younger colleague Opechowski

[121] RP, Supplement "Science and society", Rosenfeld notes "Wroclaw".
[122] RP, "Manchester (1947–1958): Cultural World Congress for Peace, Wroclaw (1948)".
[123] RP, "Manchester (1947–1958): Cultural World Congress for Peace, Wroclaw (1948)", draft of letter from L. Rosenfeld to J. G. Crowther, 11 Dec 1948.

when he heard it, since Opechowski could no longer agree as regards the meaning of the words "freedom" and "science", when employed by one of the driving forces behind the resolution, Bernal. Opechowski's response initiated a longer discussion between Rosenfeld and Opechowski that continued throughout 1949 about the current political situation from a radical leftist perspective.[124] In the best of all worlds, Rosenfeld conceded to Opechowski, the resolution should have been based on careful, rational reflection. In particular, Rosenfeld would have liked the resolution to include a warning about the dangers of dogmatism and intolerance in the cultures of the very same oppressed masses. He saw similar tendencies to dogmatism and intolerance taking root in Soviet society and called it "the Soviet crisis". Opechowski called it plainly "the horrors of the Soviet regime".[125] The reason for the Soviet crisis, as Rosenfeld saw it, was a fast decay of Marxist thinking plus a "falsification of the political and ideological situation which has followed the death of Lenin".[126] However, "the Soviet crisis (the existence of which is rather stupidly denied by the Communists) is, if you ask me", Rosenfeld wrote to Opechowski, "a crisis of growth, not a crisis of decadence", which he believed characterized capitalism.[127] In seeing such dangers in the development of Soviet society, Rosenfeld clearly felt at odds with the communists. They ironically refused to see the current development, even though it was so conspicuous from the very perspective of Marxist theory, Rosenfeld claimed. A solution to the Soviet crisis lay, according to Rosenfeld, in a "clear recognition of [the] historical situation".[128] In the end, Rosenfeld concluded that the Wroclaw resolution should be seen as a political act, not as a scientific analysis of a given situation as he had first explained to Crowther. He seems to have struggled with the same problems with scientific politics as his former Dutch colleagues; scientific analysis can guide in decision making, but it

---

[124] RP, "Correspondance particulière: Opechowski" and RP, Supplement "Science and society".
[125] RP, "Correspondance particulière", W. Opechowski to L. Rosenfeld, 6 Oct 1949.
[126] RP, Supplement "Science and Society", L. Rosenfeld to W. Opechowski, 15 May 1949.
[127] RP, Supplement "Science and Society", L. Rosenfeld to W. Opechowski, 30 Dec 1949.
[128] RP, Supplement "Science and Society", Rosenfeld notes "Wroclaw".

cannot make the decision for you. Allegedly, Crowther should have dismissed Rosenfeld's hesitation at the Congress by stating that he had "stayed in Holland for too long".[129]

Apart from the main resolution, the offshoots of the Wroclaw meeting were the formation of a Permanent International Committee of Intellectuals in Defence of Peace and National Peace Committees of Intellectuals that would be responsible for organizing the peace movements within their respective countries.[130] The International Committee included Crowther (of course!). Rosenfeld was prevented from participating in the first meetings of the British Cultural Committee for Peace, but he wholeheartedly supported the initiative, as he informed Crowther in December 1948.[131] Thus, despite being critical about Soviet domestic policy, Rosenfeld saw no other choice but to support the dawning Peace movement. He refused to "howl with the wolves", as he wrote to Opechowski, because he saw the cry for "freedom" in the other camp as pure hypocrisy. Indeed, anti-communism was already in full force in America by the spring of 1949 with its effective political repression of the liberal-left, including the State Department's strict visa policy affecting both American and foreign scientists.[132]

Despite his critical position towards the domestic politics of the Soviet Union, Rosenfeld was, on the other hand, hurt by the characterization of the Soviet Union as a fascist state. Fascism was the enemy of science, humanity, and free thought, not communism! It was of course a tragic mistake when the Russians took refuge in similar methods, but according to the exegesis of Rosenfeld and most of the Left at the time, the Russians took their precautions in defense. Rosenfeld excused the Soviet Union because the intellectual level in that socialist state was not yet mature enough. At this point,

[129]RP, Supplement "Science and Society", drafts of letters from L. Rosenfeld to W. Opechowski, 15 May and 30 Dec 1949. Somsen (2008).
[130]Ullrich (2007), pp. 7–8.
[131]RP, "Manchester (1947–1958): Cultural World Congress for Peace, Wroclaw (1948)", J. G. Crowther and L. Golding to L. Rosenfeld, 14 Oct 1948. Draft of letter from L. Rosenfeld to J. G. Crowther, 11 Dec 1948.
[132]RP, Supplement "Science and Society", Wroclaw notes. Wang (1999), pp. 252–253, 278–279. Badash (1995), p. 104.

Rosenfeld took this understanding of the situation to its extreme; if the Russians employed methods reminiscent of fascism, such mistakes could be blamed on their enemies who put them under pressure, he argued![133]

Following Wroclaw, the first World Peace Congress was held in Paris 20–25 April 1949. At this congress, the World Peace Council (WPC) was created which took the name "Partisans of Peace", making it clear that action, not talk, was needed. As a striking symbol of peace, the WPC chose a white dove designed by Pablo Picasso. A permanent Executive Bureau was created and chaired by Joliot-Curie, who was also president of the World Federation of Scientific Workers.[134]

The cultural Cold War can be said to have begun in earnest on 24 June 1947 when, after a time with increasing ideological measures taken against cultural institutions in the Soviet Union, Zhdanov gave an important speech which led to new measures in the restriction of the right of philosophers, film directors, artists, writers, and even musicians to govern intellectual activity by their own means and regulations. A series of demands to socialist scholarly work proper was invoked. Near the end of his speech Zhdanov mentioned specific issues in science including the following sentence about quantum mechanics "The Kantian vagaries of modern bourgeois atomic physicists lead them to inferences about the electron's possessing "free will", to attempts to describe matter as only a certain conjunction of waves, and other devilish tricks".[135] According to Loren Graham, Zhdanov actually opposed the Party's intervention in the natural sciences, but immediately after his death in 1948, controls in scientific fields including physics, genetics, cosmology, structural chemistry, and physiology were tightened. Science, like art and literature, should serve the people, it should be chosen by the Soviet leadership, and it should be planned and should rest on a materialistic basis and practical orientation.[136] The most striking and famous

[133] RP, Supplement "Science and Society", Wroclaw notes.
[134] Ullrich (2007), pp. 8–9. Deery (2002), pp. 449–451.
[135] Zhdanov (1947), p. 43. Quoted from Graham (1987), p. 325.
[136] Graham (1987), pp. 15, 174, p. 325. See also Ivanov (2002), pp. 317–318.

example of Stalin's renewed denial of the autonomy of Soviet scholarship was his decision in July 1948 to put T. D. Lysenko in charge of Soviet biology and prohibiting genetics as known in the West.[137] Thus, simultaneously with the Wroclaw conference, Soviet biology became officially Lysenkoist. While according to genetics heredity is embodied in the genes, Lysenko believed heredity could be changed more rapidly by the influence of the environment upon the organism. This could be demonstrated by the process of vernalization, he claimed, that is, the freezing of winter seeds and their planting in the spring, in order to increase wheat yields.[138] The political step to place Lysenko in charge of Soviet biology meant that Marxist scientists were faced with the choice between modern genetics and their political commitment. The Lysenko controversy therefore served to disillusion many liberals both as regards the USSR and as regards the political program of left-wing scientists, and to isolate the latter. Bernal took a position favorable to Lysenko and critical of genetics in 1949. For Bernal classical genetics was bourgeois genetics.[139] The Lysenko affair was utilized quite effectively by the ideological right, embodied by the Society for the Freedom of Science, for example, in counter-attacking the Left. The result was eventually that a victorious right-wing view on science came to set the agenda with respect to the role of science in society, notably in the field of history of science. According to that view, there was a close connection between liberal capitalism and the production of scientific knowledge. It contrasted a free market in scientific ideas linked with individual initiative to rapid technological progress with the rigidities of state planning.[140]

The implications of the new emphasis on ideology, widely known under the name *Zhdanovschina* for quantum physics and Bohr's complementarity in particular became Rosenfeld's paramount concern (see next chapter).

---

[137] Werskey (1988), pp. 293–294. The political "games" in which scholars were enrolled in the post-war milieu is well-described in Kojevnikov (2004), pp. 183–207.
[138] Jones (1988), pp. iii–iv, 16–37, 55.
[139] *Ibid.*, p. 31.
[140] *Ibid.*, Mayer (2004). Enebakk (2008).

The Maverick

The Manchester chemist and philosopher Michael Polanyi, who despite their political disagreements was also a close friend of Blackett, seems to have found in Rosenfeld a congenial colleague in matters of philosophy of science.[141] In October 1949, Polanyi invited Rosenfeld to take part in a discussion group on the philosophy of science that included M. G. Evans, Max Newman, and Dorothy Emmet. It is more than likely that Rosenfeld participated when Polanyi and Evans invited scholars such as Jean Piaget and Ilya Prigogine, the latter another Belgian scientist with Russian ancestors.[142] Prigogine and Rosenfeld became collaborators in later years, and Rosenfeld found Piaget's developmental cognitive psychology of children in close agreement with his own views on the dialectical development of science.[143] As regards Polanyi, while assuring Rosenfeld of the "profound respect in which I hold you", he at the same time remained puzzled by Rosenfeld's "great respect for dialectical materialism" and he was "wholly opposed to the political action" of some of the "Science for Peace" circulars Rosenfeld sent him during the years.[144]

Science for Peace was a British organization of scientists and scholars founded in 1951 with Rosenfeld as one of its driving forces. Other supporters included apart from the usual left-wingers such as Bernal, Joseph and Dorothy Moyle Needham, and Haldane, also Rosenfeld's more liberal-minded friends and colleagues, Max Born (who lived in Edinburgh after the war) and the historian of medicine Charles Singer. Born agreed to support the organization when he

---

[141] Nye (2004), pp. 36–41.

[142] RP, "Manchester 7: Epistemology 1947–1958", M. Polanyi to L. Rosenfeld, 8 Oct 1949. Rosenfeld mentioned that he knew Emmet in a letter to D. Danin on 5 Aug 1971, RP, "Supplement". Scott and Moleski (2005), p. 219. Prigogine was invited for a seminar in Manchester later in October 1949 and asked to meet Rosenfeld. RP, "Correspondance diverse (1947–1958)", I. Prigogine to L. Rosenfeld, 24 Oct 1949. Rosenfeld apparently met Piaget for the first time in Manchester in 1955, RP, "Supplement U", L. Rosenfeld to J. Piaget, 10 Feb 1968.

[143] See for example Rosenfeld (1979y), pp. 618–622. Rosenfeld (1979e), pp. 651–654.

[144] RP, "Manchester: Science for Peace (1951–1956)", M. Polanyi to L. Rosenfeld, 7 Feb 1952. RP, "History of Science 1: Review of Bernal's Science in History (1955–1956)", M. Polanyi to L. Rosenfeld, 8 Dec 1955.

felt assured that it was not under communist influence.[145] Indeed, it was stressed in the organization's pamphlets that support for the organization came from scientists "of diverse political and religious beliefs".[146] The first supporter listed in the pamphlets was always the liberal Scottish nutrition physiologist and agricultural scientist, Lord Boyd Orr, who received the Nobel Peace Prize in 1949 for his scientific research into nutrition and for his active efforts to find means to improve man's health with an eye to secure peace and promote cooperation between nations.[147] Thus, Science for Peace appeared to be a more liberal-left initiative compared with the communist-controlled WFSW and the WPC. The motivation behind the initiative was the danger of a third world war and the conviction that weapons are not made without the intention to use them, nor so nuclear weapons. The organization opposed nuclear weapons as weapons of mass destruction and as a "perversion of scientific effort".[148] Thus, scientific activity was seen as morally neutral but with the potential to be used for good or evil purposes. Since nuclear weapons were a product of scientific technology the organization appealed to scientists to recognize their moral responsibility. As Rosenfeld expressed it: "a very little honest reflection will make anybody realize that the decision to put the results of scientific investigation at the service of either war or peace is not itself a matter of science, but of social ethics".[149] They saw it as the scientist's duty to inform the public and Governments about "the destructiveness and misery of modern war", and, on the other hand, that there were other much more beneficial sides to science. The international character of science was stressed; "it is a world wide republic of the mind. Scientists form one fraternity, united in a common attempt to understand nature and a common

[145]Schirrmacher (2007).
[146]RP, "Manchester Science for Peace (1951–1956)", *Science for Peace Bulletin*, 1 June 1951.
[147]Jahn (1949).
[148]RP, "Manchester Science for Peace (1951–1956)", Observations on the M.R.C. Report "The Hazards to Man of Nuclear and Allied Radiations" by a Sub-Committee of Science for Peace.
[149]RP, "Manchester Science for Peace (1951–1956)", L. Rosenfeld, preface, in pamphlet "Atomic Challenge".

concern for human betterment".[150] The organization wanted to work for the removal of all barriers that restricted the free intercourse of scientists throughout the world. Still, when Rosenfeld sent material on Science for Peace to Polanyi, Bohr, and others, they seemed to have conceived them as yet another communist agitation.

It is clear that Rosenfeld and Polanyi belonged on opposite sides in the political spectrum. Polanyi had founded the anti-Marxist organization, Society for Freedom of Science (SFS) in 1941, together with the Oxford zoologist John Baker and the professor of plant ecology, Sir Arthur G. Tansley. The SFS consisted of scientists and scholars (including the Belgian historian of science Jean Pelseneer, whom Rosenfeld knew well)[151] uniting in opposition to socialism and the very idea, claimed by Bernal and Crowther *et al.*, that science ought to be planned with the needs of society in mind. Members of this organization favored the view that science is essentially a pursuit of truth for its own sake, an enterprise whose sole object is to understand nature. One criterion of the purity of a science was seen as military irrelevance. Unlike Bernalists, members of SFS also drew a sharp line between science and technology. Technology was recognized as having a legitimate social function, but science should be self-regulating. Members of this society attempted to show that the history of science and the history of war barely intersected and that science's only external link was with philosophy.[152] In 1953, Polanyi became involved with the Congress of Cultural Freedom Committee on Science and Freedom, an anti-communist organization with the purpose of combating totalitarian threats to freedom of critical thought and to isolate the Left.[153]

Rosenfeld was to represent WFSW at the Paul Langevin and Jean Perrin Commemoration in Paris in June 1948. As the proud French

[150]RP, "Manchester "Science for Peace (1951–1956)", "Observations on the M.R.C. Report "The Hazards to Man of Nuclear and Allied Radiations" by a Sub-Committee of Science for Peace.
[151]RP, "History of Science 5: Correspondance particulière: Pelseneer".
[152]Elzinga (1996a), pp. 9–10. Mayer (2004), p. 58. Werskey (1988), pp. 281–285. Edgerton (1996), pp. 18–19. Jones (1988), pp. 21–22.
[153]Scott and Moleski (2005), p. 222. Jones (1988).

tradition prescribes, the remains of the two famous scientists were to be laid to rest in the Panthéon.[154] On this occasion Crowther sought to make the most of Rosenfeld's philosophical bent as well as his close connection with Bohr's institute. He asked Rosenfeld to speak about the philosophical, moral, and historical aspects of the organization of scientific research "with illustrations from the development of research in theoretical physics".

> In other words, could you make your discourse thoroughly highbrow, so that there can be no doubt of the WFSW's keen interest and deep thought on these aspects. You could, without mentioning the Society for Freedom of Science, attack its position indirectly and by implication. For instance you might discuss the Copenhagen School as an example of planned research, under definite leadership, at the most abstract and advanced level... I have discussed these suggestions with Bernal, who is quite in agreement with them.[155]

Rosenfeld did not use the Copenhagen School as an example of planned research in his talk, as suggested by Crowther. In fact, Rosenfeld did not uncritically adopt the planning imperative in Bernal's thought. Even though Rosenfeld found there was a need for planning with respect to the emancipation of the field of history of science from dilettante scientist-historians' random approach, in science itself he stressed the importance of scientists being able to choose their own research topics. This was intimately connected with the importance of curiosity, creativity, and spontaneity in scientific practice.[156] In continuation of this view, Rosenfeld disliked Bernal's

---

[154] RP, "World Federation of Scientific Workers (1947–1955)", J. G. Crowther to L. Rosenfeld, 19 Mar, 4 and 18 Oct, 1948.

[155] RP, "World Federation of Scientific Workers (1947–1955)", J. G. Crowther to L. Rosenfeld, 4 Oct 1948.

[156] RP, "Manchester (1947–1958): Prix Francqui (1949)", typewritten speech of thanks. RP, Supplement, draft of letter from L. Rosenfeld to W. Opechowski, 30 Dec 1949. RP, "Union Internationale d'Histoire des Sciences: Commission for the History of the Social Relations of Science (1946–1956)", L. Rosenfeld to A. Cortesao, 23 Oct 1946.

reduction of science to technology and the thesis that science developed merely as a response to social needs, consisting of solutions to technical problems. Rosenfeld agreed that the needs of society may initiate scientific development, but this in itself did not automatically ensure scientific answers to social problems. He believed that upon being prompted by social needs science develops its own autonomous knowledge, building on natural laws that are true. At this point the scientist is no longer influenced by society in his scientific work, but refers only to nature. Scientists were materialists by necessity in this sense, according to Rosenfeld. Thus, science *also* developed according to its own internal logic and that should be enforced by freedom of research, according to Rosenfeld. One could only hope that science's own internal development would in the end provide answers to social needs. The two aspects of scientific development, that is, those due to its relations with economic and social development, on the one hand, and those due to its internal logic, on the other, should be seen as complementary, he argued. "Science depends on form of economy and social order: stagnates or grows with society, as a part of it. But this relationship must not be understood as a direct causal one. Scientific ideas develop according to internal logic, but must be activated by contact with problems raised by social needs".[157] This meant that Rosenfeld did not entirely dismiss all the values represented by the SFS opposing Bernal's idea of the planning of science. From the perspective of Marxist thought, Rosenfeld could justify his position by emphasizing Engels's *dialectical* view on the development of science in contrast to Lenin's emphasis on *materialistic* aspects in accounting for the historical development of science. Engels' view reflected a strong faith in the progress of the sciences, whereas Lenin's viewpoint gave a more fortuitous picture of how the sciences developed depending exclusively on historical process (more about this difference between Lenin and Engels in the next chapter).[158]

[157]RP, Supplement, notes on "Success and failure in the growth of science".
[158]Graham (1987), pp. 38, 45. Sheehan (1993), pp. 133–134.

On the other hand, Rosenfeld attacked the SFS severely even if indirectly for its protagonists' loud assertion that there was no connection between pure science and politics. Rosenfeld gave historical examples from Victorian Britain, suggesting that class differences between the industrial bourgeoisie of Manchester and representatives of the old English aristocracy in the Royal Society of London might have played a key role in the Royal Society's rejection of the first important paper on the discovery of the law of conservation of energy by the Manchester physicist James Prescott Joule. If such a thing could happen, "the freedom of the scientist" could be seriously questioned.[159] Another example was the recent development in Germany where scientists out of an idealized self-understanding as being engaged only in pure science, and being above and beyond politics, in fact ended up legitimating the Nazi ideology.[160] In this way, Rosenfeld intended to demonstrate how the notion of social neutrality pleaded for by members of the SFS was dangerous and deceptive.

In 1954, Rosenfeld was asked to review Bernal's book *Science in History* (1954) for the magazine *Horizons*, which was the literary organ of the World Peace Movement.[161] This was a journal with at least a million subscribers among the politically conscious laity in no less than 17 countries.[162] Bernal's book aimed at a lay audience rather than historians of science. The purpose of the book was to trace the effect of science on economic, social, and political history and to illuminate how scientific progress was produced by emerging social classes and sustained by reactionary elements in society.[163] It was generally favorably received, even among historians of science, and even by those who did not otherwise share his political ideas. He was met with admiration for his courage to attempt at making the great synthesis and it was acknowledged that his perspectives on

[159]Rosenfeld (1979z), p. 881. Emphasis in original.
[160]*Ibid.*, p. 887. The same view is expressed in Elzinga (1996a), p. 10. As for Heisenberg's position see Carson (2010), pp. 161, 174.
[161]This section is largely taken from the paper Jacobsen (2008). RP, "History of Science 1: Review of Bernal's *Science in History* (1955–1956)", G. Scharffe to L. Rosenfeld, 8 Dec 1954.
[162]RP, "Review of Bernal", G. Scharffe to L. Rosenfeld, 8 Dec 1954 and 8 Dec 1955.
[163]Bernal (1954), pp. vii, xii.

history ought to stimulate further research.[164] Among its lay readers, the book became immensely popular; it appeared in many editions and was translated into numerous languages.[165]

Rosenfeld, however, wrote a devastating critique of Bernal's book, as if it had been a book for historians of science.[166] His critique appears somewhat surprising given that Rosenfeld and Bernal were on the same side in the grand cause and the fact that their historical methods were after all very similar; they agreed about the social and political function of history of science. However, Rosenfeld criticized Bernal for, among other things, collectivizing scientific achievement too much. Intimately connected with this, Rosenfeld strongly opposed what he conceived as the ready-made Marxist–Leninist inspired framework that Bernal wanted to fit the story into. As Rosenfeld put it in a letter to Herneck later, his criticism foremost concerned "the harmful tendency to dogmatism which can only discredit the cause of historical materialism".[167] Of course, it was not less important in Rosenfeld's eyes that this dogmatism clearly influenced Bernal's view on modern physics. In fact, that was Rosenfeld's strongest concern.[168]

Bernal was completely taken by surprise by Rosenfeld's attack. He complained to Rosenfeld and to the editors of the journal that he found the review completely out of line with its place of publication. In Bernal's view Rosenfeld discussed pedantic scholarly details, losing sight of the grand purpose of the book: to use history of science as a tool for promoting the greater political cause. Bernal's responding letter to Rosenfeld ended in a sigh: "I could have wished it had fallen into the hands of someone with less erudition and more sense of proportion".[169] Of course Bernal feared deeply what harm an unfavorable review in such a widespread journal could do to the book and

[164]See for example Williams (1957), p. 472.
[165]Ravetz (1981), p. 394.
[166]Rosenfeld (1979q).
[167]RP, "History of science 7: Herneck", Copy of letter from L. Rosenfeld to F. Herneck, 31 Aug 1961.
[168]RP, Supplement "Evolution of scientific thought and history of science: notes", copy of letter from L. Rosenfeld to B. Farrington, 10 Jan 1956.
[169]RP, "Review of Bernal", J. D. Bernal to L. Rosenfeld, 15 Dec 1955.

to the World Peace Movement more widely. He resented this greatly and asked the editors to turn down the review or to allow him to respond to it in the same issue.[170] In the end the editors decided against publication of Rosenfeld's review on grounds similar to the objections raised by Bernal, but Rosenfeld did not seem to regret this very much or he had foreseen this could happen.[171] Anyhow, he had already made arrangements for the review to be published in the Danish based history of science journal *Centaurus*, if it should be rejected. This being the case, Bernal relaxed due to the fact that in *Centaurus*, with its much narrower spectrum of readers, the review would receive much less notice.

Among his countless book reviews, this review was among the few that Rosenfeld selected many years later for publication in his *Selected Papers*. He was proud of the review and sent it to his acquaintances in the history of science milieu in Britain including Benjamin Farrington, Crowther, Charles and Dorothea Waley Singer, Born and Polanyi. He later bragged to Herneck in East Berlin about the fact that the review had been so critical that a so-called progressive journal would not publish it.[172] Rosenfeld's strong critique of Bernal was welcomed by some, not surprisingly Polanyi, Born, and the Singers. Crowther found that Rosenfeld had "raised a number of points that have also worried me".[173] Farrington, however, like Bernal regarded Rosenfeld as a maverick who betrayed the grand cause.[174] He expressed great surprise at some of Rosenfeld's criticism

[170]BP, J. D. Bernal to P. Biquard, 16 Dec 1955. Biquard to Bernal, 19 Dec 1955. Bernal to C. Morgan, 21 Dec 1955.
[171]RP, "Review of Bernal", G. Scharffe to L. Rosenfeld, 1 Mar 1956. BP, L. Rosenfeld to J. D. Bernal, 12 Mar 1956.
[172]RP, "History of science 7: Herneck", Copy of letter from L. Rosenfeld to F. Herneck, 31 Aug 1961.
[173]RP, "Review of Bernal", J. G. Crowther to L. Rosenfeld, 17 Dec 1955. B. Farrington to L. Rosenfeld, 5 Jan 1956. M. Polanyi to L. Rosenfeld, 8 Dec 1955. A. Clow to L. Rosenfeld, 6 Jan 1956. D. W. Singer to Rosenfeld, 15 Dec 1955. RP, "History of Science: C. and D. Singer 1947–1953", D. W. Singer to Rosenfeld, Cornwall, 29 Apr 1955. RP, "Correspondance particulière", M. Born to Rosenfeld, 28 Mar 1957.
[174]RP, "Review of Bernal", B. Farrington to L. Rosenfeld, 5 and 18 Jan 1956. RP, Supplement "Evolution of scientific thought and history of science: notes", copy of letter from L. Rosenfeld to B. Farrington, 10 Jan 1956.

and his reaction is again an example of the wall Rosenfeld ran into when attempting to raise criticism of the acts of fellow travellers. Farrington's reaction can be described along the line: if you are not unconditionally for us, you must be against us.[175] It was clear that it had been impossible to keep the research program in the history of the social relations of science clear of dogmatism and Soviet ideology as Rosenfeld had wished for since 1945. This combined with the generally increasing marginalization of the Left both politically and culturally, partly as a result of the effect of the Lysenko affair, resulted in a rapid and effective denigration of the social history of science program at least until the mid 1960s.[176] Rosenfeld blamed his fellow Marxist scholars partly for this development. When he criticized the British historian of science Arthur Rupert Hall for failing to take the social relations of science into account in his book *The Scientific Revolution 1500–1800* (1954), suggesting that "[a]bove all, [Hall] seems to be hampered by some remnant of Cantabrigian snobbishness which leads him to grossly underestimate the role of craftsmen in the development of science", he added that "I readily admit that this relationship is by no means straightforward or ubiquitous and that in suggesting this, some self-styled "Marxists" may have muddled up the issue".[177]

Anglo-American Science Policy and Anti-Communism

Like Blackett, Rosenfeld was well-informed about the scientists' political discourse in America and the activities of the Federation of American Scientists through the journal *Bulletin of Atomic Scientists*.[178] Rosenfeld may also have followed closely how Blackett rapidly came to oppose the conservative turn of Anglo-American nuclear policy during these years, which culminated in the publication in 1948 of his book *Military and Political Consequences of*

[175] RP, B. Farrington to L. Rosenfeld, 5 Jan 1956.
[176] Werskey (1988), pp. 277–279. Enebakk (2008).
[177] Rosenfeld (1955–1956).
[178] Blackett (1948), p. viii.

Atomic Energy. Blackett had served as a member of the MAUD Committee, which was the initial British atomic bomb project during the war. From August 1945 until the early spring of 1948, Blackett served as a member of the Advisory Committee on Atomic Energy to the British Government along with James Chadwick, Cockcroft, Peierls, Charles Darwin, and others. However, during the first years after the war, Blackett's views quickly diverged from those of his colleagues in the committee. After the committee was abolished in early 1948 Blackett decided to go public about his opposition. In his book Blackett laid out his views about military strategy that had been carried out by British and American air forces during the war as well as his opposition to the British development of atomic weapons. Simultaneously with the book's publication in November 1948 Blackett was awarded the Nobel Prize in physics "for his work on cosmic radiation and his development of the Wilson method", which brought the book even more publicity; it was received as quite controversial and criticized by many, especially American, scientists.[179]

In the same year, the *Bulletin of Atomic Scientists* published an open letter to Einstein from four Russian physicists (Sergei Vavilov, A. N. Frumkin, Abram F. Joffe and N. N. Semyonov) expressing their objection to Einstein's idea of a World Government that should secure continuing world peace. Einstein had announced his idea in an open letter to the United Nations in October 1947. If necessary, he suggested, the UN should go forward with the idea even if the Russians would not participate. The proposal was dismissed by the Russian physicists because they saw it as surrendering the Russian and poorer countries' independence to a foreign capitalist monopoly and imperialism which, however, were what they had freed themselves from.[180] "I have to say that I support the Russian side a hundred percent in this discussion", Rosenfeld wrote to Bohr on that occasion and called Einstein's response arrogant and dilettantish.[181] Rosenfeld may have worried that Bohr would support

[179]Nye (2004), pp. 86–90.
[180]S. Vavilov *et al.* (1948). Einstein (1948). Baratta (2004), Vol. 2, pp. 441–444.
[181]RP, L. Rosenfeld to N. Bohr, 1 Mar 1948. About the Russian physicists' open letter to Einstein, see Vucinich (2001), p. 91.

Einstein and the Movement for World Federal Government. This movement enjoyed support among some prominent Danish scientists and scholars such as the physiologist Poul Brandt Rehberg and Piet Hein, and internationally by Sir John Boyd Orr who served as President of the movement 1948–1951.[182] Bohr supported the movement in as much as he had contributed his "Science and Civilization" article from *The Times* (1945) as a foreword to the report *One World or None* (1946) and also contributed a foreword to a pamphlet about the Danish part of the organization, Een Verden (One World), in 1949.[183] Bohr was in America at the time, when Rosenfeld wrote to him and Rosenfeld took the opportunity to express his concern about the American scientists whom he considered "hopeless" and "dangerous" because, as he saw it, they were letting themselves be exploited as propaganda instruments by reactionary forces in an ideological preparation for war against the Soviet Union. Rosenfeld seems to have sensed from abroad how the American science policy was gradually but effectively being transformed these years by the rise of the Cold War. A number of American scientists collaborated with military leaders and high-level public officials. For American scientists in general, however, the late 1940s was a crucial period during which they realized the extent and limit of their political power. Increasing political repression accompanied the post-war revival of anti-communism and eventually put an end to the liberal-left politics of science which had briefly enjoyed support in the US immediately after the war. Liberal-left scientists eventually found little room to manoeuvre outside the Cold War consensus. Within only a few years they had to retreat from their commitment to an international world order and international cooperation on scientific research and instead concentrate on countering the threat to their own political freedom posed by the American loyalty-security system. By the spring of 1949, even before the announcements of communist victory in China, the

---

[182]Baratta (2004), Vol. 2, p. 515.

[183]Knudsen (2010), pp. 269–283. Masters and Way (1946), with contributions, besides Bohr's foreword, by J. R. Oppenheimer, H. Bethe, A. Einstein, H. Urey, E. P. Wigner, L. Szilard, among others. Bohr (1949).

Soviet acquisition of the atomic bomb, and the onset of the Korean War, a qualitative change had taken place in American political culture and the tolerance of the American Left had disappeared. America was committed to the Cold War and domestic anti-communism was already overwhelming. The American scientific community contributed to this Cold War political consensus through its general failure to speak out. Simultaneously, the content and direction of scientists' research agendas were swayed heavily by the technological needs and funding support of the American military.[184] Back in Europe Rosenfeld did not yet recognize the difficulties which the American scientists were facing, and he was not prepared to excuse their adaptation to the US political situation. Furthermore, whereas the communist coup d'état in Czechoslovakia in the spring of 1948, which prevented this country from participation in the Marshall Plan, was seen as an appalling Moscow manoeuvre by many observers in America and Western Europe, Rosenfeld saw it as a demonstration of the European people being strong and wise enough to reject "American Fascism", no less![185] Like the majority of the Left at this time, Rosenfeld adopted the communist account of the events, according to which reactionary American forces attempted to win Czechoslovakia for the Western Bloc.

During the last years of the 1940s and during the McCarthy era many American scientists were refused passports by the State Department and entry of foreign scientists with left-wing ties into the United States was likewise severely restricted. Like several British scientists, for example the physicist E. H. S. Burhop and even Polanyi, Rosenfeld was refused admission into the US in the 1950s.[186] In Rosenfeld's case, the Central Intelligence Agency (CIA) and the Federal Bureau of Investigation (FBI) were concerned about his association with the WFSW, which "has been cited by the Congressional Committee on Un-American Activities... as "another

[184] Wang (1999), pp. 6–8, 250–253.
[185] RP, L. Rosenfeld to N. Bohr, 1 Mar 1948.
[186] Jones (1988), p. 79.

international communist front organization" which seeks to win scientists to the communist cause".[187] In addition, a "confidential source abroad advised that the name of L. Rosenfeld, 23 Rue de la Rogie, Liege, appeared in the address book of the British atomic scientist, Dr. Klaus Fuchs at the time of his apprehension in February 1950".[188] If anything, this suggests that Rosenfeld and Fuchs had not been in contact since the 1930s when Rosenfeld lived in Liège. Rosenfeld objected to the US immigration laws when turning down invitations from American colleagues to attend conferences in the US on the ground that he was denied a visa. He even refused to apply because of the intrusive inquiries about his general political beliefs and his opinion about US foreign policy as a condition for entry.[189] In January 1957, when Rosenfeld applied for visa in order to attend a conference on the role of gravitation in physics at the University of North Carolina, Chapel Hill, he was declared a "current member of communist party" by the FBI and denied entry.[190] However, on this occasion the physicist Cecile DeWitt-Morette, who was a member of the Institute for Field Physics where the conference was going to take place, got a visa for Rosenfeld "simply by calling up the U.S. Attorney General by phone. I did not know the Attorney General, I did not even know his name. I found his phone number by calling Directory Assistance. Giving my name with a self-assured tone of voice I was quickly transferred to him by his secretary. He assured me that he had the authority to reverse a consul decision and would do so right away. I then asked for an emergency visa so that Rosenfeld would arrive in time for the conference", which took place on 18–23 January.[191] In March the same year when Rosenfeld again applied

[187] FBI HQ file 105-56664. File 21 May 1956 was based on information "sent [by] CIA on March 30, 1951 . . . the purpose of the request was the sponsoring of the captioned individual to attend the Congress on Theoretical Physics at the University of Washington". U.S. Department of Homeland Security, U.S. Citizenship and Immigration Services.
[188] *Ibid.*
[189] RP, "Manchester (1947–1958): Political (1948–1958)", R. E. Marshak to L. Rosenfeld, 24 Sep 1954. A. Roberts to L. Rosenfeld, 22 Dec 1955. Wang (1999), p. 278.
[190] FBI file 25 January 1957.
[191] C. DeWitt-Morette in email to A. S. Jacobsen, 21 and 22 Jul 2010. DeWitt-Morette knew Rosenfeld well through the summer school Les Houches she founded in the French Alps in 1951. Rosenfeld was among the lecturers in 1952 and 1957. *Europhysics News*

for visa to attend the Seventh Annual Rochester Conference in High Energy Nuclear Physics in mid April, he was granted permission to attend the conference. By then he had acquired an official Belgian Chargé de Mission passport which he had been advised to get by the American physicist R. E. Marshak. By the second half of the 1950s, the altered political conditions meant that the anti-communist hysteria was gradually on the retreat. This, combined with the Federation of American Scientists' documentation that a great number of foreign scientists seeking to enter the United States experienced some sort of difficulty in obtaining a visa, resulted in a modification of the visa policies that made it easier for foreigners to enter the country.[192] After he settled in Copenhagen in 1958, the American Embassy at Copenhagen vouched for Rosenfeld and he no longer had trouble acquiring a visa.[193]

Belgian Recognition

In November 1948, Rosenfeld received the exciting news that his long-term friend, the astrophysicist in Liège, Pol Swings, intended to promote him for the Francqui Prize of 250.000 francs.[194] This prestigious prize had been institutionalized in 1932 by one of the most powerful men in Belgian public life, the industrial magnate, director of the Société Générale, and philanthropist Emile Francqui. Since then it has been given each year to a Belgian savant who is particularly distinguished through his (or her, presumably) scientific or

(1999), p. 68. RP, "Manchester (1947–1958), 4: Ecole des Houches (1952, 1957)". About the conference, see *Conference on the Role of Gravitation in Physics*, WADC Technical Report 57-216, ASTIA Document No. AD 118180. I am grateful to Cecile DeWitt-Morette and John J. Stachel for drawing my attention to this information.

[192]FBI file 22 March 1957. RP, "Manchester: Political (1948–1958)", R. E. Marshak to L. Rosenfeld, 24 Sep 1954 and 3 Jan 1956. W. E. Meyerhof to Rosenfeld, 1 Mar 1955. A. Roberts to Rosenfeld, 22 Dec 1955. Wang (1999), pp. 278, 286.

[193]FBI file September 1966. Unfortunately the information about Léon Rosenfeld originally kept in the Danish Security and Intelligence Service (PET) has been destroyed. Juridical Executive L. Sørensen to A. S. Jacobsen, 20 Oct and 8 Dec 2008.

[194]RP, "Manchester (1947–1958): Prix Francqui (1949)", P. Swings to L. Rosenfeld, 17 Nov 1948.

scholarly career, provided there is a suitable candidate.[195] The prize is recognized worldwide as a supreme distinction. When it was inaugurated, it was the world's second highest scientific award after the Nobel Prize. The second to receive it in 1934 was the astrophysicist and cosmologist George Lemaître, and Rosenfeld and Chandrasekhar were asked to review and assess Lemaître's candidacy.[196] When Rosenfeld was nominated there was one other candidate, namely the professor of mathematics and physics at Liège, Florent-Joseph Burau, but Rosenfeld was awarded it for that year, 1949.[197] The lodge-like ceremony, in which Yvonne was not even allowed to take part, took place 23 June 1949 at the University Foundation in Brussels. In his speech of thanks, Rosenfeld took the opportunity of expressing his indebtedness to his Belgian education, which had impressed upon him the Belgian traditions and attitude to life which he summarized as "this spirit of tolerance so largely widespread among our working populations of Wallonia", in other words a "respect for the human person and aversion for all things outraged, dogmatic and savage". This spirit of tolerance, he stated, "is the heart of science" and transmitted to the youth it is the foundation of the future.[198]

It is likely that both Léon and Yvonne were anxious to return to Belgium.[199] In 1952, an opportunity arose when Rosenfeld was asked if he would be interested in a position at Brussels as the successor to the professor of general and mathematical physics M. Henriot, who was seriously ill.[200] The offer may also have been tempting for Rosenfeld because Blackett began negotiating a move to Imperial College in the same period.[201] After Rosenfeld had inquired about

[195]No women seem to have received it, at least not prior to 2000. Halleux *et al.* (2001), Vol. 2, pp. 60–61.
[196]RP, copy of letter from L. Rosenfeld to S. Chandrasekhar, 18 Feb 1934. Original in CP.
[197]RP, "Manchester (1947–1958): Prix Francqui (1949)", P. Schwings to L. Rosenfeld, 24 Jan 1949.
[198]RP, "Manchester (1947–1958): Prix Francqui (1949)", typewritten speech of thanks. J. Willems to L. Rosenfeld, 13 Jun 1949.
[199]See for example BPC, Y. Rosenfeld to M. Bohr, 13 Jan 1946.
[200]RP, "Manchester (1947–1958): Université de Bruxelles (1954)", J. Géhéniau to L. Rosenfeld, 10 Sep 1952.
[201]Nye (2004), pp. 155–156.

the administrative conditions regarding the position, it was arranged that Rosenfeld should visit the University and give a seminar at the Centre de Physique Nucléaire on 15 January 1953. This would coincide with an initial meeting about CERN in which Rosenfeld was to participate anyway.[202] A year later, nothing had been decided yet, but Swings informed Rosenfeld about salaries at the Free University. Swings also told Rosenfeld that both he and the university administrator Jean Willems "wished highly for his return". Finally, Swings hinted at some difficulties which the physicist Max Cosyns had experienced at the Free University in Brussels due to his militant communist attitude. Swings therefore advised Rosenfeld to stick to the position he had earlier explained to Swings, *viz.*, that although he was Marxist this did not mean that he did not detest "Stalinist imperialism"![203] In March 1954, the chair was announced vacant and Rosenfeld applied for it. It took another year until twenty-two out of twenty-five commissioners that were appointed to advise the Rector of the Free University in the matter voted for Rosenfeld. The final decision now lay in the hands of the Rector. On his request, Rosenfeld sent some documents "which will give you an idea about the movement 'Science for Peace' ",[204] to which the Rector responded that he had read them with "real interest". However, simultaneously he cancelled his meeting with Rosenfeld in April 1955.[205] After this there are no more letters about the position in the file and Rosenfeld did not get the position. It may therefore be inferred that contrary to when he was offered a position at Manchester, where the fact that he was a leftist may even have been to his advantage from Blackett's point of view, on this occasion Rosenfeld's political commitment decided the matter to his disadvantage. In the meantime, Blackett's successor in Manchester had been announced to be Samuel Devons,

[202]RP, "Manchester (1947–1958): Université de Bruxelles (1954)", J. Géhéniau to L. Rosenfeld, 23 Nov 1952. BSC, Rosenfeld to Bohr, 30 Jan 1953.
[203]RP, "Manchester (1947–1958): Université de Bruxelles (1954)", P. Swings to L. Rosenfeld, 29 Jan, 11 Feb, and 5 Mar 1954.
[204]RP, "Manchester (1947–1958): Université de Bruxelles (1954)", copy of letter from L. Rosenfeld to the rector of the University of Bruxelles, presumably 15 Feb 1955.
[205]RP, "Manchester (1947–1958): Université de Bruxelles (1954)", E. J. Bigwood to L. Rosenfeld, 11 Mar 1955.

which made Rosenfeld's position there even more satisfying from a scientific point of view than before, since he brought a "big van de Graaf-accelerator in his pocket", as Rosenfeld wrote to Bohr.[206] Rosenfeld stayed in Manchester until 1958.

Rosenfeld versus Bohr

As we shall see in the next chapter, Rosenfeld criticized the line taken by the Communist Party in matters of science and culture, but he did not hesitate to throw in his lot with the Russians and the communists when he sensed they were being treated unfairly on the world political scale. Rosenfeld's political alliance with the Russians also meant that he did not accept without reservations Bohr's idea that it would take an Open World, in which all information on technology, science, and social relations was shared between nations, to prevent a catastrophic war with atomic weapons.

In March 1950, the international committee of the World Peace Council convened in Stockholm where they launched a world campaign for signatures to an appeal against nuclear weapons. The Appeal was written by Joliot-Curie and stated:

> We demand prohibition of the atomic weapon, this terrible weapon for mass destruction of men.
> We demand establishment of a strict international control in order to secure the accomplishment of this prohibition.
> We consider that the government which first uses the atomic weapon against whichever country not only commits a war crime, but also a crime against humanity, and we hold that it should be treated as a war criminal.
> We ask all honest men all over the world to sign this appeal.[207]

[206]RP, "Manchester (1947–1958): Université de Bruxelles (1954)", J. Géhéniau to L. Rosenfeld, 7 Mar 1954. Rosenfeld to Monsieur le Président du Conseil d'Administration de l'Université Libre de Bruxelles, 16 Mar 1954. J. Géhéniau to L. Rosenfeld, 22 Jan 1955. E. J. Bigwood to L. Rosenfeld, 21 Jan 1955. BSC, L. Rosenfeld to N. Bohr, 1 Feb 1955.

[207]Quoted from *BCW*, Vol. 11, p. 189.

It was soon signed by millions of people all over the world, most notably in Russia. Both Léon and Yvonne signed the appeal.[208] The wording of it was clearly manipulative. People were confronted with the option of signing the appeal, or, if they refused, being viewed as warmongers and "enemies of peace". Even the former Nazi Pascual Jordan signed the appeal it was reported by Hamburg newspapers on 7 July 1950. However, he withdrew his signature three days later when he realized its context.[209] The British Government immediately conceived the appeal as directly serving the interests of Soviet foreign policy and as an attempt to conjuring up what it claimed was a non-existent threat from the West. The British Government was of the opinion that the proposals offered by the Stockholm Appeal ought to have been covered by UN resolutions, which, however, had been vetoed by the Soviets.[210]

From the summer of 1949, Rosenfeld and Bohr took up their collaborative work again on the measurability of charge and current distributions in quantum electrodynamics. They seemed to have quickly found the tone from the 1930s. Rosenfeld's stays in Copenhagen also led to "heated political discussions" with Margrethe Bohr about the Greek civil war and the British role in it; Rosenfeld subsequently sent Margrethe "some documents" to ensure her that the League of Democracy in Greece was not a communist enterprise but was comprised of "decent liberal Englishmen".[211] During the second half of the 1940s, Bohr did not take part in the public debate on specific issues raised by the advent of atomic armament, but he continued approaching high officials in Britain and the US, such as Anderson, Baruch, and George Marshall through the American ambassador to Denmark, Josiah Marvel. Bohr's conviction about the necessity of introducing openness between states was only strengthened with time. Marvel encouraged Bohr to go forward with his idea about openness as the key to the promotion of confidence and mutual

[208]This is clear from the letter RP, "Manchester (1948–1958): Political (1948–1958)", C. Morgan to L. Rosenfeld, 18 Jul 1950.
[209]Beyler (1994), p. 438.
[210]Deery (2002), pp. 450–451. Ullrich (2007), pp. 9–11.
[211]BSC Supplement, L. Rosenfeld to M. Bohr, 17 Jan 1950.

trust between East and West. As Bohr believed that an initiative in this direction would have to come from the US, he attempted to persuade the American Government to take steps towards making openness a paramount issue. The State Department took Bohr's proposal seriously although it was concluded that the time was not ripe for carrying it out.[212] About the same time as the announcement of the Stockholm Appeal in March 1950, Bohr decided to make his hitherto confidential reports public. He began writing an open letter to the United Nations and it was published on 12 June 1950. In it Bohr included the letter he had sent to President Roosevelt in July 1944 and passages from a second memorandum of March 1945 to Roosevelt.[213]

Bohr appeared to be unaware of or he ignored the Stockholm Appeal. However, on the day of publication of the Open Letter, the Danish communist daily, *Land og Folk* (Land and People), interpreted the Open Letter as expressing support for the Stockholm Appeal and as a request to others to sign it. According to *Land og Folk*, the Soviet Government supported the Stockholm Appeal and the newspaper inferred from this that the Russians would then also support Bohr's Open Letter. However, as a response to *Land og Folk*, Bohr published a short announcement the following day in which he denied that he would sign the Stockholm Appeal, giving as reason that it did not mention anything about an open world.[214]

Léon and Yvonne got the news about Bohr's Open Letter on the morning of 14 June through the leading article in *The Manchester Guardian.*[215] Léon immediately called Niels and Yvonne sat down to write a letter to Margrethe Bohr. In her letter, Yvonne expressed regret that physicists had gained a "reputation as the men who worked on the destruction of humanity!" Therefore she

[212] Aaserud (2005).
[213] Bohr (2005a).
[214] *Land og Folk* (1950). Professor Niels Bohr's answer through Ritzau's Bureau, *BCW*, Vol. 11, p. 189.
[215] *The Manchester Guardian* (1950). RP, "Copenhague: Bohr (1940–1962), Open Letter".

was enthusiastic about the "very good idea" of Margrethe's husband and she believed that the Open Letter would no doubt "influence a lot of researchers" due to Bohr's stature and authority.[216] After their telephone conversation, Bohr sent copies of his Open Letter to Rosenfeld as well as "some copies of an answer which I considered it correct to address to a question which the communist daily 'Land og Folk' directed to me".[217] The latter was Bohr's refusal to sign the Stockholm Appeal. This did not go down well with Rosenfeld. He was horrified and extremely disappointed by Bohr's answer to *Land og Folk* and his refusal to sign the appeal. Rosenfeld saw Bohr's move as a rejection of the peace initiative by the communist bloc which he himself supported. In a letter to Bohr, Rosenfeld gave vent to his feelings that were in accordance with common opinions among the Left at this particular time.

> In my opinion you have thereby destroyed any possibility for your so honest and inspiring effort to have any practical impact. Now the communists can use your own words as support of the accusation that your demand about free information is a hidden attack against the Soviet Union! I was downright flabbergasted when I read this response and I am still completely in despair about it.
>
> Of course I understand very well what you really mean and where you are heading. I certainly do not need to say with how much sympathy, admiration, and emotion I follow your endeavors. However, I also try to understand the Russians and the communists, and therefore [I] have perhaps become more sensitive towards certain phrases and sentences, which for you sound innocent, but which have a provoking effect on others. You presume preconditions, which you treat as given and true at all times and for all people; the Russians, on the other hand, emphasize that the value of all such principles is limited in time and is only valid within a certain class. Now,

[216]BPP j. nr. 120, Y. Rosenfeld to M. Bohr, 14 Jun 1950.
[217]BPP j. nr. 120, N. Bohr to L. Rosenfeld, 16 Jun 1950.

> it is not, in the first place, about whether they are right in this or not. In order to achieve a rapprochement with them, you have first of all to try to make yourself acquainted with their way of thought and not express any direct or indirect assessment of their behavior out of criteria, the applicability of which they deny...
>
> In the meantime, the real situation is that the Russians and the communists are too suspicious to approve of your proposal. I do not think that the Americans will ever agree either: everything points in the opposite direction. It is easy to say that the Russians fear free information because it will reveal the bad state of the Soviet Union and allow the Russian people to compare it with the blessing of democracy. In reality the Americans are less interested than the Russians in allowing the propagation of free information about the condition of the world. Because then such questions as the treatment of the Blacks in the United States itself and in South Africa, the concentration camps in Greece, the cruelties of the French in Indo-China, and so many other cases of misery and oppression, for which the responsibility lies in American imperialism. On the other hand, you would get to hear about the enormous material and mental progress which the people's societies of Hungary, Poland, Yugoslavia, [and] China can achieve as soon as they have freed themselves from this imperialism. The greatest support for the American policy and the continuous tyranny of capitalism — namely lies and deceit — would thereby be overthrown. How can you then be surprised that your negotiations with American statesmen fail?[218]

Bohr had hoped to reach the public with his letter, but Rosenfeld assured him that he should not be too optimistic about that either. People apart from intellectuals would find it too abstract and

[218]BPP j. nr. 120, L. Rosenfeld to N. Bohr, 21 Jun 1950. Originally in Danish.

rational, he declared. The masses could relate much more easily to the simple message in the Stockholm Appeal. And the intellectuals had no political power since they had no connection with the masses, according to Rosenfeld. He added that Yvonne "shares my feelings".[219]

Léon and Yvonne in the French Alps in 1952. Rosenfeld Album. Courtesy of The Niels Bohr Archive.

[219] *Ibid.*

This was not at all the tone in which Rosenfeld normally addressed Bohr, and he also ended his letter with an apology for his outburst and opposition. Bohr, however, must have been well aware of Rosenfeld's standpoint from previous letters and conversations. Nevertheless, Rosenfeld's reaction came as a surprise for Bohr, he stated, but he appreciated Rosenfeld's sincere honesty in this matter.[220]

In the meantime, North Korean troops crossed the 38th parallel on 25 June. Bohr considered his Open Letter even more relevant in the light of these events. Despite the misunderstandings he expected from both sides, Bohr hoped that his Open Letter would contribute to bring in new arguments in the efforts to overcome this crisis in the international situation. It was his hope that he could with the Open Letter raise understanding of the fact that progress would only be possible if extensive concessions were made from all sides.[221] Rosenfeld was not yet ready to make concessions on behalf of the Soviet Union in this matter. "[T]he violent events of the last few days show very clearly what immediate threat the American imperialism means to [the mental development of all humanity]!" he wrote to Bohr.[222] Bohr's Open Letter did not evoke much response; it was mentioned with approval in the press but then ignored, overshadowed by the Korean War.[223]

Bohr and Rosenfeld's relationship suffered seriously for a couple of years from this clash. Their second joint paper had just come out in *Physical Review* and Bohr spoke about continuing their joint work.[224] However, a plan to attend a conference together in Bombay, India, later in 1951 was canceled on Bohr's part and he asked Rosenfeld to represent him at the meeting. Rosenfeld was greatly disappointed.[225] He was also disappointed when Bohr turned down an offer to be the

---

[220]RP, "Copenhague: Bohr (1940–1962), Open Letter", N. Bohr to L. Rosenfeld, 28 Jun 1950. Carbon copy in BPP j.nr. 120.
[221]*Ibid.*
[222]BPP j.nr. 120, L. Rosenfeld to N. Bohr, 30 Jun 1950.
[223]Aaserud (2005), p. 81.
[224]BPP j.rn. 120, N. Bohr to L. Rosenfeld, 21 Jul 1950. RP, "Copenhague: Bohr (1940–1962)", N. Bohr to L. Rosenfeld, 12 Aug 1950.
[225]RP, "Copenhague: Bohr (1940–1962)", N. Bohr to Rosenfeld, 30 Nov 1950. BSC, L. Rosenfeld to N. Bohr, 4 Dec 1950, 25 Aug 1951. RP, "Manchester (1947–1958), 4: Bombay (1950–1951)".

main speaker at a conference in 1951 on the question of atomic energy organized by the British Association of Scientific Workers. Rosenfeld tried to make Bohr change his mind about it: "Your authority among them is tremendous and from your point of view such a conference would afford you an excellent opportunity to present your views to a really intelligent and sympathetic audience, which at the same time is quite influential in scientific and technical circles in this country".[226] But to no avail. In the summer of 1951, Rosenfeld complained to Frédéric Joliot-Curie about the "insurmountable psychological barrier" that separated Bohr and himself.[227] On that occasion Joliot conferred with Rosenfeld about the possibility of persuading Bohr to take a leading position in a meeting between scientists from East and West arranged by the WFSW. Bohr had previously been invited by Joliot-Curie to take part in the Second World Peace Congress, which was originally planned to take place in Sheffield, England, in November 1950. When the congress was cancelled and moved to Warsaw due to the British Government sabotaging it, Bohr was again invited by Joliot-Curie. Each time Bohr declined the invitation on the ground that for him a precondition for positive results about international understanding and common security was openness.[228] In light of his previously failed attempts, Joliot-Curie, who was strongly attached to this idea, asked Rosenfeld how to approach Bohr about participating in the new initiative by the WFSW.[229] The meeting should be of scientists of such repute that governments would pay attention to their conclusions on nuclear warfare and international liaisons. It was to be held in May 1952. Rosenfeld answered Joliot-Curie that "my impression is that Bohr, to speak confidentially, would be close-fisted".[230] Rosenfeld assessed that it would be best

[226]BSC, L. Rosenfeld to N. Bohr, 11 Dec 1950.
[227]RP, "World Federation of Scientific Workers (1947–1955)", F. Joliot-Curie to L. Rosenfeld, 2 Jun 1951.
[228]Aa. Bohr informed Rosenfeld about these invitations in January 1952. BSC, Aa. Bohr to L. Rosenfeld, 7 Jan 1952. Deery (2002). Ullrich (2007), pp. 26–33.
[229]RP, "World Federation of Scientific Workers (1947–1955)", F. Joliot-Curie to L. Rosenfeld, 2 Jun 1951.
[230]Archives Joliot-Curie, Fonds Frédéric Joliot-Curie, Paris, F121, L. Rosenfeld to F. Joliot-Curie, 15 Jun 1951. Originally in French. P. Biquard to L. Rosenfeld, 20 Jun 1951.

to discuss the matter with Bohr "by word of mouth" and volunteered to approach Bohr and try to persuade him on behalf of Joliot-Curie at a conference in Copenhagen on quantum physics in the beginning of July 1951. However, Bohr expressed greater resistance towards the project than Rosenfeld had imagined. Rosenfeld interpreted Bohr's resistance as sheer disappointment about the lack of enthusiastic reception the Open Letter had enjoyed among communists: "I have had the impression that Bohr is affected profoundly by the small echo which his Open Letter to the United Nations has awakened in the communist circles, which have effectively (not without reason) given him a rather glacial reception".[231] He advised Joliot-Curie that should he have any hope of success with Bohr, he better make a bigger deal out of Bohr's open world.

In his capacity as general secretary of the meeting in question — Rosenfeld gladly accepted when Joliot offered him that position — Rosenfeld attempted to make it up with Bohr in December 1951.[232] Whereas his previous letter had been in English he now wrote in Danish again, although less perfectly than he usually was capable of. "Both Joliot and Bernal (who has just been in Russia and knows something about the opinion of the Russians) attach the greatest importance to the idea that you should take the leading position at this meeting... Joliot and Bernal [find] that the question about an "open world", which is so close to your heart, should constitute the foundation for the discussions. First of all an attempt should be made to find practical means to re-establish the "open world", at least within science."[233] Rosenfeld stressed that a first preparatory meeting would not imply any obligation for participants to continue if it should turn out that the antagonisms between them were greater than expected. However, he believed it was definitely possible to reach a common program at such a meeting. He continued slightly

[231] Archives Joliot-Curie, Fonds Frédéric Joliot-Curie, Paris, F121, L. Rosenfeld to F. Joliot-Curie, 19 Jul 1951.
[232] Archives Joliot-Curie, Fonds Frédéric Joliot-Curie, Paris, F119, F. Joliot-Curie to L. Rosenfeld, 22 Oct 1951. L. Rosenfeld to F. Joliot-Curie, 29 Oct 1951.
[233] BSC, L. Rosenfeld to N. Bohr, 12 Dec 1951. Emphasis in original. Originally in Danish.

in despair or slightly impatient, it may seem, in trying to penetrate Bohr's armor.

> I merely want to point out to you that this proposal is the first positive reaction to your Open Letter. Whereas you harvest only empty politeness or vicious criticism from other sides, here you have a concrete suggestion which aims at the realization in practice of the "open world" we wish for.
>
> In this connection I must add some more personal remarks. I still maintain the critical considerations which I communicated to you at the time when the Open Letter came out. But the criticism concerned only the external circumstances of the publication of the letter, not the content of the letter itself.[234]

Rosenfeld urged Bohr to seize the more opportune situation which he considered existed now for working for an open world. He enclosed some documents about "Science for Peace" in order to give Bohr an idea of the atmosphere surrounding the problem in England.[235] Rosenfeld stressed as further encouragement that by this new initiative British scientists had already expressed their willingness to participate in the meeting. He hoped this would encourage Bohr. Rosenfeld ended his letter to Bohr December 1951 by suggesting that should Bohr be very busy he could let his son, Aage Bohr, make the arrangements. They would surely welcome a visit by Aage in Manchester. At the institute in Manchester, Aage "would even get to see a piece of 'open world', in which some communists, a Titoist Yugoslav, a few Catholic South Americans and even an Irishman live in a harmonious and enjoyable community", Rosenfeld jokingly ended the letter.[236] Hence, in a way he did exactly the opposite of what he had advised Joliot-Curie to do when approaching Bohr. As Rosenfeld did not receive a reply over Christmas he seems to have

---

[234] *Ibid.* Emphasis in original.
[235] Probably *Science for Peace Bulletin*, No. 1, June 1951 with "Contributed Views, by Professor L. Rosenfeld" on p. 7. L. Rosenfeld, preface, in pamphlet "Atomic Challenge", in RP, "Manchester (1948–1958): Science for Peace (1951–1956)".
[236] BSC, L. Rosenfeld to N. Bohr, 12 Dec 1951. Originally in Danish.

become impatient and asked Møller just after New Year if Bohr had received his letter.[237] Aage Bohr responded on behalf of his busy father a week later. They were not prepared to joke about the open world: "It has to be emphasized that the "open world" is not a mere polite way of speaking, but a concrete proposal... about different nations through mutual agreement immediately giving complete access to all information not only about scientific questions, but also about all other questions, which are of significance for the relation between the countries, as the only way to create real mutual trust between nations, and in that way make possible common measures to secure peaceful relations in the world", Aage Bohr wrote.[238]

Niels Bohr declined the invitation because he was of the opinion that until all sides, including "people like Joliot and Bernal" and "the circles to which they are linked", agreed to the proposal of openness, such a meeting would not lead to positive results. Aage declared that it was news to him and his father that Joliot would be positively inclined towards the question of openness, "since hitherto he has not at all clearly declared where he stands in this matter".[239] Bohr stood firm on his demands and conditions for participating, and he continued to do so in the years to come, which may also have been connected to his wish that the United States or the United Nations should take the lead in the initiative for an open world not the Peace Movement. He declined the invitation to sign both the Russell–Einstein Manifesto, which was issued on 9 July 1955 in London, the content of which resulted from negotiations among Bertrand Russell, Joseph Rotblat, Born, and Joliot-Curie, and the Mainau Declaration, drafted by Born and Otto Hahn, which was circulated at a conference of Nobel Laureates in Lindau, Germany, on 15 July 1955.[240] The Mainau Declaration included signatures by

---

[237] MP, L. Rosenfeld to C. Møller, 3 Jan 1952.

[238] RP, "Copenhague: Bohr (1940–1962)", Aa. Bohr to L. Rosenfeld, 7 Jan 1952.

[239] *Ibid.*

[240] Joliot-Curie managed, despite Russell's reluctance to the idea, to put his stamp on the content of the text by insisting on a call for a conference of scientists. This idea became a pillar of the Russell–Einstein Manifesto and led to the first Pugwash Conference in Nova Scotia in 1957. Einstein signed the final draft, but died on 18 April 1955 before the Manifesto was announced at a press conference on 9 July 1955. The signatories of the

non-communist Nobel Laureates from the West only. Both declarations were appeals against the use of nuclear weapons prompted by the American detonation of a particularly powerful hydrogen bomb, the Castle Bravo Test in the Bikini Atoll in March 1954. The test explosion got out of control and turned out to be about a thousand times more powerful than the Hiroshima bomb. Radioactive fallout spread hundreds of miles downwind and produced the first human casualties of the hydrogen bomb era. It shocked by producing global environmental consequences and produced genuine fear that a third world war using thermo-nuclear weapons would extinguish the civilized world.[241]

Bohr gave grounds similar to those that he had given Joliot-Curie the previous years when he turned down the invitations to sign the Russell–Einstein Manifesto and the Mainau Declaration; there was no explicit talk of openness in these initiatives and he doubted whether such declarations would have the desired effect anyway.[242] Bohr instead set his hopes high for the international Geneva conference on the peaceful uses of atomic energy on 8–20 August 1955, sponsored by the United Nations (see next chapter).[243]

Rosenfeld's and Bohr's opposed political positions clearly reflect the wider Cold War context in terms of their political engagement on different fronts and levels. Their clash reflected the struggle between governmental organizations and the peace movement to obtain people's allegiance. The question was with whom should the responsibility for preventing war and preserving peace reside, with the UN, which was then officially waging the Korean War, or with the World Peace Council? Since the World Peace Council desired peace but

---

Russell–Einstein Manifesto included, apart from Russell, Einstein, Born, Joliot-Curie, and Rotblat, Linus Pauling, L. Infeld, H. Yukawa, C. F. Powell, H. J. Muller, and P. W. Bridgman, nine of who had received a Nobel Prize. The Pugwash movement became an international forum where problems could be discussed informally and is still very viable. BSC, A. Einstein to N. Bohr, 2 Mar 1955. B. Russell to N. Bohr, 8 Mar 1955. Bohr to Russell, 23 Mar 1955. Butcher (2005), pp. 10–11.

[241]Gaddis (1997), p. 225.

[242]BSC, N. Bohr to B. Russell, 23 Mar and 20 Apr 1955. N. Bohr to M. Born 20 April 1955.

[243]BSC, N. Bohr to M. Born, 20 Apr 1955. *Bulletin of the Atomic Scientists* (1955), p. 274. Gaddis (1997), pp. 206–207.

on Soviet terms and the US dominated resolutions in the UN in their own favor, the choice was also between the two camps.[244] As we have seen, Bohr continued to emphasize the importance of the United Nations.

In the long run, Bohr and Rosenfeld did not let politics stand in the way of their friendship. Besides, the times changed into a less tense period. Coinciding with a thaw in the East–West intellectual relations the psychological barrier between them dissolved and there was room for mutual concessions about weaknesses or wrongdoings of their respective camps.[245] After a visit to Copenhagen in the summer 1954, Rosenfeld reported to his friend and colleague, the now East German philosopher of physics, Martin Strauss: "As for Copenhagen, I just want to tell you that I could observe a very interesting development there in the general political atmosphere. There was a general indignation against the politics of the United States... I am especially pleased to report that Bohr himself now takes a very firm position".[246] Strauss replied that he was glad to hear that Bohr made good progress with regard to politics. He hoped that Bohr would let the outside world know about this as well.[247]

Despite their disagreements on politics Rosenfeld's admiration for Bohr was undiminished. Bohr reciprocated the feelings and paved the way for Rosenfeld to take up a position in Copenhagen. In the spring of 1956, Bohr informed Rosenfeld about the new funding at the institute in Copenhagen from the Ford Foundation to strengthen international collaboration. Half a year later, this information from Bohr's side was turned into a more concrete invitation for Rosenfeld to take up a chair at the new Nordic Institute for Theoretical Atomic Physics, NORDITA, as it came to be called. Rosenfeld accepted the proposal sometime during the autumn of 1957, when he could feel

[244] Deery (2002), pp. 450, 462.
[245] RP, "Copenhague: Bohr (1940–1962)", N. Bohr to L. Rosenfeld, 31 Mar 1954. Bohr to Rosenfeld, 22 Jan 1955.
[246] RP, "Correspondance particulière", Copy of letter from L. Rosenfeld to M. Strauss, 9 Jul 1954.
[247] RP, "Correspondance particulière", M. Strauss to L. Rosenfeld, 3 Oct 1955.

absolutely sure that the new position had really materialized.[248] The Rosenfelds settled in Copenhagen in 1958. This may be interpreted as if Rosenfeld was Bohr's choice of a successor when he retired, at least with respect to the epistemology of quantum physics.

[248]RP, "Copenhague: Bohr (1940–1962)", N. Bohr to L. Rosenfeld, 9 May and 10 Oct 1956. M. Bohr to Y. and L. Rosenfeld, Dec 1956. N. Bohr to L. Rosenfeld, 23 Jan and 9 Aug 1957. RP, Supplement Box 5 "R", copy of letter from L. Rosenfeld to S. Rozental, 21 Oct 1957.

Chapter 6
Bohr's Cold Warrior

For a better understanding of where Bohr and Rosenfeld found common ground in the Cold War, we need to take a closer look at Rosenfeld's efforts from about 1949 defending Bohr's ideas on quantum theory against left-wing accusations of them for being idealistic and positivistic, and this is the topic of this final chapter. Rosenfeld's controversy with the American physicist David Bohm about the latter's hidden variables interpretation of quantum mechanics in the early 1950s is central here. In the quantum controversy Rosenfeld and Bohm took opposite standpoints, aligning themselves with the two giants Bohr and Einstein, respectively, with whom they were personally acquainted. However, their clash in the 1950s also has to be understood within the context of the Cold War Marxist disputes. Their opposed alignment in the quantum controversy tied in with their different interpretations of Marxism and their different standpoints in philosophy of science. The Marxist dispute in the quantum controversy seems to have culminated around 1957–1958 coinciding with the Soviet physicist Vladimir A. Fock visiting Bohr in Copenhagen and Rosenfeld being appointed to the first chair at the Nordic Institute for Theoretical Atomic Physics (NORDITA), Copenhagen. At that time, the world experienced a renewed openness between nations as a result of Stalin's death, at least on an intellectual level. However, First Secretary of the Communist Party of the Soviet Union, Nikita Krushchev clearly demonstrated that there was a limit to this openness when he commanded the Red Army to crush the people's revolution in Hungary in 1956. This event led to many on the Left denouncing their belief in the communist cause.

In the 1930s, Rosenfeld fought what he conceived as reactionary views on quantum physics epitomized by Jeans–Eddington idealism. From the late 1940s, there was a new battle to be fought, now with the opposite sign, namely against the harmful ideological decrees from Moscow pertaining to culture and science.[1] Quantum physics and particularly Bohr's complementarity were under attack once more by the changing Party ideology in Moscow. In May 1949, Rosenfeld expressed his concern to Bohr about "the different misunderstandings which appear when trying to blend complementarity and all kinds of mysticism (whether it concerns idealism à la Eddington and others or the Russian pseudo-Marxism). These many 'isms' are probably much too trivial for you, but I feel, after all, that one can no longer be satisfied with ignoring all the nonsense. More than ever the spirit of science needs to be defended against attacks from all sides".[2] Bohr was probably relieved that Rosenfeld took affair so that he did not have to engage in these discussions which he considered trivial, not new, and a waste of time.[3] In the interview with Kuhn and Heilbron in 1963, Rosenfeld described a conversation with Bohr about the Russian critique. According to Rosenfeld, Bohr asked him

> "Why are those Russians dissatisfied?" And I tried to explain, "They accuse you of being a positivist", and so on. Then he said, "Is that it? But those things are so trivial; they are not of interest to physicists. Physicists are beyond that point and that is not the thing that we are interested in. We are struggling with real problems, not with those trivial statements about our living in an external world".[4]

On the other hand, the complementarity interpretation of quantum theory continued to occupy Bohr. In the late 1940s, Bohr made contributions to the consolidation of his quantum epistemology with renewed emphasis on measurability based on the quantum postulate, that is, the existence of the quantum of action.[5] Bohr and

[1]This chapter is a further elaboration of Jacobsen (2007).
[2]BSC, L. Rosenfeld to N. Bohr, 31 May 1949. Originally in Danish.
[3]RP, "Supplement 1971–1974", L. Rosenfeld to D. Danin, 7 Jan 1972.
[4]Kuhn and Heilbron (1963c), p. 12.
[5]Bohr (1996d), pp. 333–335.

Rosenfeld resumed their analysis of the measurability of electrical charges and currents that constitute the sources of the electromagnetic field, a work in progress that had been shelved since the 1930s and now had to be completely rewritten in light of the recent work in quantum electrodynamics done independently by Julian Schwinger, Sin-Itiro Tomonaga, and Richard Feynman. Bohr and Rosenfeld's work concerned to what extent measurements of electric charges are limited by quantum mechanical fluctuation phenomena. They examined once again complicated idealized experiments to see if they would yield the same conclusions as the new formalism of quantum electrodynamics. They concluded that they had found complete consistency of Schwinger's formalism and the commutation relations of charge and current components with their analyses of complex idealized experiments.[6] This second Bohr–Rosenfeld paper did not attract nearly as much attention as Bohr's paper "Discussion with Einstein on Epistemological Problems in Atomic Physics" that appeared in the book *Albert Einstein: Philosopher — Scientist* (1949).[7] In this paper, Bohr gave his version of his discussions with Einstein about quantum mechanics in the late 1920s and early 1930s. These new papers by Bohr as well as Einstein's position in the quantum controversy — many physicists who found quantum mechanics difficult found it comforting and encouraging that Einstein had also found it difficult — contributed to make the old discussion resurface.[8]

The Relationship between Science and Philosophy

In a Russian context, the decade 1947–1958 has been called "the age of the banishment of complementarity".[9] Following the new Soviet ideological line in science, art, and literature, *Zhdanovschina*, Bohr

[6]Bohr and Rosenfeld (1979b). See also N. Bohr to W. Pauli, 15 Aug 1949, in von Meyenn (1993), Vol. 3, pp. 684–689. BGC, "Stern, A. W. 1949–1962", N. Bohr to A. Stern, 16 Aug 1949.
[7]Bohr (1996b).
[8]See for example Bohm (1952), de Broglie (1958). The Einstein factor continued to be an argument for work in this field, see for example Bell (1982).
[9]Graham (1987), p. 328.

and Heisenberg were attacked both for being positivistic and for building their ideas on quantum theory on speculation. The orthodox interpretation was criticized for stating that the wave function is an expression of the information available to the observer of the behavior of quantum systems rather than an expression of the actual objective behavior of quantum systems. Hence this interpretation of the wave function was conceived by the Soviet critics as subjective and idealistic. Party philosophers and physicists understood complementarity to say that the behavior of atomic systems, and hence matter, was dependent on an experimental context and therefore ultimately on the presence of an observer. Quantum mechanics seemed to have lost its objective character by this interpretation. If complementarity was mentioned by the Russian physicists, it was therefore usually presented simply as synonymous with Heisenberg's uncertainty relation. The fact that Heisenberg had stressed the importance of the principle of observability in the development of matrix mechanics was criticized and rejected as idealism and positivism. The materialistic view proper on the development of scientific theories was that they grew out of the response to new experimental facts; they were not based on some speculative principle.[10] Bohr's emphasis on measurability and the role of the observer in his interpretation was ascribed to his allegedly positivistic attitude. Besides, Bohr's popular exposition of his quantum philosophy from 1934 not only contained statements that could be disturbing for materialists as mentioned in Chapter 3; statements that could be interpreted as positivistic could also easily be found in that publication if that is what one was looking for.[11] To Party ideologists in Moscow, these features of the Copenhagen Interpretation implied that it was bourgeois, not

[10] See for example Blokhintsev (1953b).
[11] A. Grünbaum pointed out the following sentence to Rosenfeld in 1957: "We meet here in a new light the old truth that in our description of nature the purpose is not to disclose the real essence of the phenomena but only to track down, so far as it is possible, relations between the manifold aspects of our experience". Bohr (1934), p. 18. RP, "Epistemology 1955–1958", A. Grünbaum to L. Rosenfeld, 20 Apr 1957. Of course, this would not have been a problem for the logical positivists in the 1930s, but from the perspective of the leftist critique of neopositivism after the war, this could apparently be seen as a problem.

compliant with proletarian, that is, materialistic, science proper.[12] It serves to be mentioned that apart from the motivation in Marxist ideology, the Russian critique of the Copenhagen Interpretation did not introduce much new compared with the criticism in the West represented by Schrödinger, Einstein, de Broglie, and others.

In Rosenfeld's eyes, with this development, the Party line had reached a point where reason and political action parted. As he wrote to his close friend, the Polish physicist Wladyslaw Opechowski in December 1949:

> Alas, I already start to feel like "the old guard" and I see with sorrow a widening gap on the ideological level, between what I regard as the true Marxist tradition and what has currency under this name among the young people. In my time Marxism had nothing of this savage and dogmatic religion which it has become today. I know very well that you will protest loudly about this misunderstanding, etc., but there are after all facts staring you in the face. One of them, ..., is this ill-considered attack launched by coryphaei of the official philosophy in Russia against complementarity.... I regard this attack as inspired by a dogmatic and false conception of Marxism; it is the same attitude in the Party that you condemn. In my opinion the Party makes a disastrous mistake by taking this position pertaining to scientific and artistic questions; it is the same as showing ignorance about the essential element of spontaneity which characterizes scientific work as well as artistic creativity. Of course Marx and Engels understood that well; they passionately followed the progress of science in their own time. They did not try to force scientific ideas arbitrarily into some preconceived scheme, but with a rare mastery they made use of them [these ideas] to construct their vision of the world. They possessed neither the naïveté

[12] See Cross (1991), pp. 738–742 for a good description of the antipathy towards the Copenhagen Interpretation on materialist grounds amongst Party philosophers and physicists.

> nor the presumption to believe that they had reached final insights on everything and that they were capable of commanding scientific research *urbi et orbi.* They well-grasped and were profoundly imbued by that which comprises the soul of science, namely what I call having an open mind, that is always being ready to gladly accept every new acquisition of science without wishing to measure it against any pre-established scale. "Materialism" is only a word useful for polemical ends, that is to clear away the idealist muck but with no constructive value.[13]

Rosenfeld suggested that Bohr could in fact be regarded as an unconscious Marxist in the sense of his dialectical approach, his open-mindedness, and critical spirit. "The dialectical nature of this step ["this extension of traditional determinism which is called complementarity"] seems to me to be one of the most beautiful illustrations of Marxism that can be found. The fact that Bohr has used Marxism as M[onsieur] Jourdain used prose could only raise the convincing value of this example of the dialectical progress of science".[14] However, promoting Bohr as an unconscious Marxist stood little chance in matching the solid influence of *Zhdanovschina* and the renewed promotion of Vladimir I. Lenin's notorious and enthroned book *Materialism and Empirio-Criticism* among communist philosophers and scientists when the relationship between science and philosophy was at stake. This book was originally published in Russian in 1909 and translated into English and German in 1927. Alongside Engels' books *Anti-Dühring* and *Dialektik der Natur*, published posthumously in 1925, Lenin's book was the most influential publication in the early years of the Soviet Union and became influential among communist

[13]RP, Supplement, draft of letter from L. Rosenfeld to W. Opechowski, 30 Dec 1949. Emphasis in original.

[14]*Ibid.* In Molière's ballet comedy from 1670 *Le Bourgeois gentilhomme* (The Bourgeois Gentleman), Monsieur Jourdain is a foolish middle-aged bourgeois who aspires to become an aristocratic gentleman. To this end he takes philosophy lessons, among other things, in which he discovers to his surprise that he has been speaking prose all his life without knowing it.

scientists in the West too.[15] It was therefore either wishful thinking or a rhetoric trick when Rosenfeld later proclaimed to Friedrich Herneck that "Lenin's book has had no influence at all on the development of physics. It has remained unknown to virtually all physicists who are interested in the epistemological questions".[16] On the question whether Bohr was familiar with Lenin's book, Rosenfeld diplomatically told Herneck that he had himself of course read the book "carefully", however, only to reach the conclusion that it was dilettantish and useless; "I therefore considered it unnecessary to introduce it to Bohr".[17] According to Rosenfeld, "Lenin, while no doubt intending to follow Engels, in effect advocated mechanistic materialism with its metaphysical conception of determinism".[18] Ultimately, mechanistic materialism referred to the reductionistic, determinist, and positivist materialism prevailing in the nineteenth century. It also referred to the controversies Soviet Marxism underwent during the 1920s between so-called Deborinites, after Abram Moiseevich Deborin, and mechanist factions. The Deborinites favored dialectical over mechanistic materialism and they eventually came out as the winners of the controversy. They found support for their position in Engels' *Dialectics of Nature* when it was published posthumously in 1925.[19]

During *Zhdanovschina*, there were examples of Russian physicists who complied with the Party line whereas other physicists withdrew completely from the discussions. D. I. Blokhintsev and J. P. Terletskii were examples of the former; they followed suit in the renewed criticism of the orthodox interpretation as positivistic and idealistic, whereas Landau, for example, thought it wiser to avoid

[15] Josephson (1991), pp. 204, 226, 249–250. Graham (1987), pp. 25–26, 34–35. Lenin's book grew out of a complex Russian context connected with the political tumult prior to the Bolsheviks coming to power. However, the doctrines and positions upheld in the book, most notably against Ernst Mach, were taken at face value by Lenin's later followers and were not seen in this historical context. See Joravsky (1961), pp. 3–44.
[16] RP, Supplement Box 2, L. Rosenfeld to F. Herneck, 2 Dec 1969. Originally in German.
[17] *Ibid.*
[18] Rosenfeld (1979t), p. 482.
[19] Joravsky (1961), p. 215.

philosophical foundations altogether.[20] Blokhintsev was a specialist on solid state physics, statistical mechanics, and quantum field theory. In the second edition of his textbook on quantum mechanics, which was used widely at universities in the Soviet Union and in East Germany, Blokhintsev introduced a criticism of Bohr's views. Blokhintsev bolstered his indignation with the orthodox interpretation by mentioning that there was a whole section devoted to "The Liquidation of Materialism" in Pascual Jordan's polemical book from 1936, *Physics of the 20th Century*.[21] In the last chapter of the 1953 edition of his textbook, Blokhintsev literally echoed Zhdanov's criticism of the "crisis in the bourgeois philosophical thought" abroad, of which the Copenhagen Interpretation was seen as a clear manifestation. The tendency which it reflected was explained by ignorance of Lenin's book *Materialism and Empiriocriticism* among the bourgeois physicists abroad.[22]

Ideology aside, Blokhintsev was inspired by de Broglie's ideas and von Neumann's axiomatic foundation and statistical presentation of quantum theory in terms of the density matrix in putting forward an ensemble interpretation of quantum mechanics. According to this interpretation, the wave function *only* described the behavior of ensembles, that is, collections of particles, not individual particles. Following von Neumann, he rejected the possibility of hidden variables.[23]

Besides the more narrow concern from the point of view of research in quantum foundations, the new cultural political line in the Soviet Union may have widened the rift between physicists in the East and West in general, and between communist and non-communist physicists in the West, and it may have constituted an obstacle for attempts by Western physicists to bring about a rapprochement between physicists in America and Russia.[24] Rosenfeld

[20]Terletskii is sometimes spelled Terlezki. Terlezki (1952). Blokhintsev is sometimes spelled Blochinzew or Blokhinzev, especially in an East German context. Blochintsev (1953b).
[21]Jordan (1944). Blokhintsev (1953b), p. 552. Blokhintsev (1953a), p. 499.
[22]Blokhintsev (1953a), p. 498.
[23]*Ibid.* See also Kuzemsky (2008). Cross (1991), pp. 741–742. Graham (1987), pp. 329–336.
[24]RP "Correspondance particuliére: Mott", N. F. Mott to L. Rosenfeld, 27 Feb 1950. In this letter, Mott drew Rosenfeld's attention to a theoretical physicist, Corson, who was apparently quite willing to work for a rapprochement between America and Russia.

engaged actively in attempting to bring about such a rapprochement. However, contrary to how gently he thought the Russians ought to be handled with respect to their foreign policy and control of nuclear weapons, *viz.*, by being attentive to their way of seeing things and trying to understand them, in scientific, cultural, and philosophical matters he showed no mercy.

Whereas most communists followed the party line and some leftists of other variants denounced their belief in the cause due to a development they could not agree with, Rosenfeld followed neither of these paths. His faith in his own Marxist line was undiminished and he would not give in. He therefore sought to strengthen his own position and arm himself for verbal fight. He undertook a thorough study of Engels' writings intending to revive Engels' position, the one in the Marx–Engels' partnership dealing with the relationship between science and philosophy, in order to argue against the dogmatism, as he saw it, of the younger generation. In 1957, Rosenfeld suggested to a young Danish physical chemist, Jørgen Koefoed, that to understand and learn to appreciate the use of dialectical materialism in science he should read Engels' publications, of course keeping in mind that they were historical documents; not every word should be taken literally in those books.[25] Rosenfeld added that "it must be realized that it is impossible in principle to write a text-book about dialectics, since this would be to fix a mode of thought which is essentially flowing! It is exactly the same with complementarity (which is the modern form of dialectics): you cannot give a "definition" of it, but only understand what it is by re-thinking for yourself the typical cases in which it occurs".[26] Thus, Rosenfeld's conception of dialectical (and historical) materialism was in clear opposition to the way these topics were rigidly taught as part of an obligatory curriculum in the higher education system in the East bloc.

However, since Corson made references to de Broglie, Mott had doubts about him being a suitable candidate.

[25]In my paper Jacobsen (2007), I erroneously suggested Rosenfeld addressed the Danish physicist Otto Kofoed-Hansen in this letter.

[26]RP, "Epistemology 1955–1958", draft of letter from L. Rosenfeld to J. Koefoed, 20 Feb 1957. Emphasis in original. For Koefoed's original inquiry and his reply to Rosenfeld see RP, "Manchester 7: Thermodynamis, opticks, etc. Foundations of statistical mechanics", J. Koefoed to L. Rosenfeld, 13 Feb 1957. J. Koefoed to L. Rosenfeld, 4 Mar 1957.

Among contemporary Marxist thinkers, Rosenfeld turned to the Dutch astronomer and Marxist Anton Pannekoek. Pannekoek had played a pivotal role in the formation of European communism early in the century, and he had been a leader of the Comintern's Western European Bureau. However, he was also among the first to break with the authoritarian turn Marxism took in Russia in the late 1920s. He not only developed a forceful criticism of Marxism–Leninism and the organization model Lenin had founded for Russian society, but also enunciated an alternative Marxism in which he attempted to strip it of its concern with metaphysics and highlighted its importance as a critical method and as a method of social emancipation.[27]

Pannekoek and Rosenfeld knew each other and had corresponded at least since November 1943, but curiously enough it was only in 1949 they began to exchange their Marxist viewpoints. In connection with a correspondence about some issues in history of astronomy, Pannekoek's curiosity was raised concerning the incentive for Rosenfeld's interest in the societal background of science, because "usually it coincides with a socialist point of view", he wrote.[28] Pannekoek went on referring to "the real theory of Marx", in which all intellectual phenomena were viewed in consistency with the social development, at the same time emphasizing that this theory was "entirely different from the dogmatic scholasticism which the Russian C. P. imposes upon its followers under this name".[29] This view resonated entirely with Rosenfeld's position. Rosenfeld's answer, which we do not know, was received with great surprise by Pannekoek because of what it said about Marxism. He regretted that they had wasted years not exchanging ideas face to face, while Rosenfeld was still living in the Netherlands.[30] Pannekoek sent his booklet, *Lenin as Philosopher*, to Rosenfeld which had just been translated into English in 1948 (originally published in German in 1938 under the

[27]RP, Supplement, L. Rosenfeld to F. Herneck, 2 Dec 1969. See also RP, "History of science: XIXe siècle", L. Rosenfeld to S. G. Brush, 7 Oct 1966. Brendel (2001). Gerber (2005).
[28]RP, "Correspondance particulière", A. Pannekoek to L. Rosenfeld, 19 Jul 1949. Originally in Dutch.
[29]*Ibid.*
[30]RP, "Correspondance particulière", A. Pannekoek to L. Rosenfeld, 17 Aug 1949.

pseudonym of J. Harper), and which Rosenfeld swore by from then on. It contained a sharp critique of Marxism–Leninism.[31]

The renewed promotion of Lenin's old book overlapped with a general Soviet critique of neo-positivism.[32] Controversies between metaphysical and positivistic factions within Marxist circles dated back to the very origins of Marxism and positivism in the nineteenth century.[33] When the relation between science and philosophy was broached in a Marxist context, Rosenfeld referred to the following quote from Engels' *Anti-Dühring*:

> As soon as every single science is required to elucidate its position with respect to the universal relationship between things and to the knowledge of things, any special science of this universal relationship becomes superfluous. What then remains of all previous philosophy is the doctrine of thinking and its laws: formal logics and dialectics. Everything else is embodied in the positive science of nature and history.[34]

The quote was given in 1951 in Rosenfeld's review of a booklet containing two articles by the odd couple, Pascual Jordan and Klaus Zweiling; the latter was a former student of Max Born turned socialist politician, journalist, physicist, and philosopher. Whereas Jordan with his paper, "Das Plancksche Wirkungsquantum", in Rosenfeld's eyes "shows himself a faithful disciple of Bohr: he writes throughout *im Kopenhagener Geist*", Rosenfeld felt he had to protest against Zweiling's understanding of Marxism. Rosenfeld accompanied the Engels quote with the comment: "It seems to me that this single sentence adequately settles the question of the relations between science and philosophy".[35] He

---

[31] *Ibid.* Pannekoek (1948).

[32] Boeselager (1975).

[33] For the contradictions in the writings of Marx and Engels on metaphysics and positivism see Joravsky (1961), pp. 5–6. Boeselager (1975).

[34] Quoted from Rosenfeld (1951). Reference given to "F. Engels *Herrn Eugen Dührings Umwälzung der Wissenschaft* (Moscow: German edition), 1946, p. 29 (reviewer's translation)". The same quote is given in RP, "Supplement H", L. Rosenfeld to M. P. Herzog, 18 Dec 1970. The quote was kept in Rosenfeld's drawer. There is a discussion of this particular passage in Joravsky (1961), p. 5.

[35] Rosenfeld (1951), p. 324.

suggested that Engels' task in *Anti-Dühring* was the "elimination of philosophy" in science in the recognition that the increasing specialization of the scientific disciplines left philosophers irrevocably behind.[36] Wolfhard Boeselager seems to agree with Rosenfeld's interpretation of Engels. Engels' dialectical philosophy, the so-called science of the universal dynamic and developmental laws of nature, human society, and thought, was not about ontology.[37] According to David Joravsky, on the other hand, both positivistic and metaphysical Marxists turned to the above quoted passage to find support for their position. The quote appears self-contradictory in its claim of dialectics as the queen of science, on the one hand, whereas, on the other hand, Engels rendered a special science dealing with "the universal relationship between things" superfluous.[38] In contrast to Engels, Lenin interpreted the dialectic as ontology and upheld it as a philosophical system. Lenin maintained a close connection between materialism and modern science in his time. According to Boeselager, the "object of philosophy is for Lenin the ultimate explanatory principle of reality ... He wanted to establish an ontologically grounded theory with a scientific method".[39] Be that as it may, with respect to the old Marxist discussion between metaphysicians and positivists, we may characterize Rosenfeld as a Marxist with a positivistic inclination; he maintained that the only role left for philosophers in science was to make epistemological analyses of the basic assumptions underlying scientific activity, but then again only scientists were capable of doing that due to the advanced state of science, Rosenfeld claimed. Followers of Lenin, on the other hand, trumpeted a strong critique of neo-positivism and tended to have a metaphysical inclination.

Rosenfeld considered the renewed controversy about the relation between science and philosophy to be the heart of the matter in the new ideological decrees from Moscow with their damaging influence on science, as well as on Marxism. He expressed his view on the

[36]RP, Supplement, folder "Science and society". Sheet of notes entitled "Scientific synthesis and socialist philosophy".
[37]Boeselager (1975), pp. 25–26.
[38]Joravsky (1961), pp. 5–6.
[39]Boeselager (1975), pp. 33–34.

relation between science and philosophy to the British historian of science Dorothea Waley Singer in 1955 in the following manner by contrasting the philosopher Friedrich Hegel with Marx, whom Rosenfeld, however, considered to be a scientist.

> Every philosophy, whatever its original social background, contains potentially valuable elements, which, however, can only be put to effective use when applied to some concrete scientific investigation. As soon as this happens, however, the philosophy in question is bound to be coloured by the progressive outlook of science. By contrast, any philosophy degenerates into a barren dogmatism as soon as it loses the stimulation of scientific criticism. Thus, Hegel's outrageously reactionary system contained the germ of dialectical logic, which Marx transformed into a powerful method of analysis when he grappled with the concrete problem of capitalism. Nowadays, Marxism has sunk to scholasticism because it is divorced from scientific research: on this side it is not thought advisable to expose the decay of imperialism, while in Russia any serious social study is discouraged because it would immediately explode the official myth according to which Socialism is established there.[40]

Moreover, this damaging influence on science and Marxism, as Rosenfeld saw it, was not confined to Soviet territory; communists worldwide were complied to follow the new ideological decrees. As Max Jammer noted in 1974, the development in the Soviet Union partly sparked the renewed interest in the interpretation of quantum mechanics in the West.[41] Recent studies of this historical episode

[40]RP, "Supplement: Evolution of scientific thought and history of science: notes", L. Rosenfeld to D. W. Singer, 30 Dec 1955.

[41]Jammer (1974), pp. 250–251. When he prepared this book, Jammer consulted Rosenfeld on various aspects of it. See RP, "Supplement J", M. Jammer to L. Rosenfeld, 22 Mar 1971 and L. Rosenfeld to M. Jammer, 16 Apr 1971. Rosenfeld was critical of Jammer (1966). His criticism concerned Jammer's claims that Bohr had been influenced by William James and Søren Kierkegaard and that complementarity was merely an expression of a specific philosophical position. See Rosenfeld (1969a), and RP, "Supplement 1969–1971 A–K": L. Rosenfeld to M. Jammer, 8 Jan 1970.

have shown that the very idea of a Copenhagen Interpretation seems to some extent to be a construction of the 1950s. In particular, as recently argued by Kristian Camilleri, "[i]t was not only the defenders of the orthodoxy who gave the impression that a unified philosophical viewpoint lay behind the new quantum mechanics, but also their opponents. Nowhere is this more evident than in the case of the Soviet critique of quantum mechanics, which reached its crescendo in the early 1950s".[42] Rosenfeld certainly was against this term, as he stated many times, because it clearly suggests that there are alternative interpretations of quantum mechanics, which he stubbornly denied.

Even though communist scientists and philosophers in the West soon conformed to the renewed opposition against the orthodox interpretation of quantum mechanics, Bohr remained silent as regards this development. At least there is nothing in print from his hand about it. According to Rosenfeld, the normal procedure for Bohr when confronted by statements, which to him seemed crazy or obviously wrong and which he therefore violently disagreed with, was to keep silent.[43] In addition the topic was clearly explosive during the early the Cold War; it was impossible to separate completely the philosophical–ideological from the political. Meanwhile, criticism of the ideas of Bohr and Heisenberg from a communist perspective also reached the Danish press. Zhdanov's Party-ideological program from June 1947, in which the role of Soviet philosophy in the sciences was consolidated, was immediately reproduced in the Danish Marxist journal *Tiden: Tidsskrift for aktivt Demokrati* (*Time: Journal of Active Democracy*).[44] So were excerpts from an extremely radical speech by the Russian philosopher, A. A. Maksimov in January 1949, promoting Lenin's book *Materialism and Empirio-Criticism* and its significance for the sciences.[45] In this paper, the ideas of Bohr and Heisenberg were called "idealistic figments of the imagination", among other things, and it was asserted that all Soviet scientists were

[42]Camilleri (2009a), p. 35. Freire Jr. (2005), p. 28. Freire Jr. (2003), p. 582. See also Cross (1991). Freire Jr. (1997).
[43]RP, "Supplement 1971–1974", L. Rosenfeld to D. Danin, 7 Jan 1972.
[44]*Tiden* (1947–1948).
[45]Maksimov (1949).

obligated to fight such idealistic viewpoints.[46] Maksimov's paper seems to have initiated a dispute in the Danish press about dialectical materialism and modern physics. Shortly afterwards, the young communist physicist and politician, Ib Nørlund (Christian Møller's former assistant and Margrethe Bohr's nephew, see Chapter 4) reiterated Maksimov's viewpoints.[47] At the Institute for Theoretical Physics, the situation was carefully monitored. The physicist and historian of physics, Mogens Pihl, soon got involved in the dispute with Nørlund. This was also a normal procedure at the institute, according to Rosenfeld, that Bohr did not answer in person "but asked one of us to do it".[48] Of course this depended on the opponent he faced. In the following years the dispute between Nørlund, his supporters, and his opponents, gained momentum in the Danish press. Pihl attempted to reason with Nørlund at a meeting in the students' association, on the airwaves, and later in print in the Marxist intellectual journal *Dialog*. Nørlund apparently refused to budge.[49]

Whereas Bohr did not involve himself directly in the dispute with Nørlund, the professor of philosophy at the University of Copenhagen, Jørgen Jørgensen, was another matter. As a prominent member of the positivist movement, Jørgensen had organized the Unity of Science Congress in Copenhagen in 1936 in close collaboration with Bohr (Chapter 3). At this conference, Bohr's complementarity interpretation of quantum physics was discussed, and it was concluded that Bohr's interpretation was not inconsistent with logical positivism. Meanwhile, Jørgensen was also attracted to communism and during the 1940s he followed the new line of thought dictated from Moscow. He echoed the Russian critique of the ideas of Bohr

---

[46] *Ibid.*, 265–266.

[47] Nørlund (1949).

[48] RP, "Supplement 1971–1974", L. Rosenfeld to D. Danin, 7 Jan 1972.

[49] Nørlund (1954). This particular number of *Dialog* with Nørlund's contribution soon sold out completely. Pihl (1955), p. 19. In 1992 Bohr's assistant since World War II, S. Rozental, jotted down some comments about the dispute. Rozental Papers, NBA, Rozental, manuscript, 3 pp. (1992). To Rozental's regret, no consensus was reached between Pihl and Nørlund. In this connection, Rozental mentioned the Soviet physicist Fock as a Marxist scientist who combined dialectical materialism with Bohr's views. It is quite strange that Rozental did not mention the views of his close friend Rosenfeld.

and Heisenberg in his textbook of philosophy for first year students. All students at the University of Copenhagen took his introductory course in philosophy — filosofikum — and thus, during their first year of study, they were exposed to Jørgensen's criticism of Bohr's ideas![50] Sometime in 1953, a meeting was set up between Bohr and Jørgensen in the club of physics students called *Parentesen* (the Parenthesis) in order for them to discuss their disagreements. However, they could no longer come to terms with each other's positions.[51] Bohr also took part in a meeting in Selskabet for Filosofi og Psykologi (Society for Philosophy and Psychology) in 1958, where he confronted the philosophers Jørgensen, Alf Ross, and Bent Schultzer, among others, about his interpretation of quantum physics.[52] Rosenfeld was not directly involved in this local dispute, but it probably added to his concern about the public conception of complementarity.

Returning to Pihl, he was no fan of dialectical materialism, nor did he consider historical materialism relevant for his research in history of science. Nevertheless, in his examiner's report of former assistant to Bohr Aage Petersen's dissertation *Quantum Physics and The Philosophical Tradition* in April 1968, one of Pihl's first complaints was that Petersen had not included a chapter on the assessment of the relation between Marxist philosophy and quantum physics.[53] Although, according to Pihl, this chapter of the history of atomic physics had not been "elevating", he considered this episode of the European philosophical tradition interesting and significant. Even though Petersen stated that both Rosenfeld's and Heisenberg's views were of utmost importance for the topic he treated, neither Rosenfeld's materialistic standpoint nor his blending of dialectical materialism with complementarity was mentioned in the dissertation.[54]

[50]Jørgensen (1956), pp. 116–117. Faye (2010), pp. 42–44. Favrholdt (2009), pp. 385–387, 392–398, 412–413.

[51]I am grateful to Henning Refsgaard for information regarding this meeting.

[52]Favrholdt (2009), pp. 392–398.

[53]Mogens Pihl Papers, NBA, M. Pihl, Opposition ved Aage Petersens disputats, manuscript. For Pihl and Rosenfeld's divergent views on history of science, see RP, "History of Science 1: (1948–1951) (Manchester)", Pihl-Rosenfeld correspondence. A. Petersen's dissertation was published in 1968. Petersen (1968).

[54]Petersen (1968), p. 191.

Rosenfeld versus Bohm

As mentioned in the previous chapter, Rosenfeld was disappointed and critical of the American scientists' adaptation to their changing political and professional conditions in the late 1940s. However, when he was contacted by American physicists who were being persecuted because of their political conviction during McCarthyism, Rosenfeld helped out when his position allowed.[55] In late August 1951, Rosenfeld complained to Bohr that there were many more highly qualified applicants than fellowships available in Manchester; "I therefore had to turn down two Americans (Freistadt and Bohm), among others, who are forced to leave the United States because of political victimization. Meanwhile, Freistadt has got a scholarship in Dublin for one year; as for Bohm I do not know what has happened to him".[56] However, Rosenfeld's concern and sympathy for David Bohm's political victimization was soon replaced by grief and worries about Bohm's new research project. He first heard about Bohm's new project from Jean Louis Destouches, a philosophically minded French physicist and former student of de Broglie, in December 1951. Destouches mentioned the effects Bohm's work, still not published, had in the French milieu particularly among the young people, who he said were animated by Marxist determinism among other things.[57]

Bohm was a very promising theoretical physicist during his stay at the Radiation Laboratory in Berkeley as a student of Robert Oppenheimer during the Second World War, but his political commitment in the same period later ruined his future in America under McCarthyism.[58] In the 1930s and 1940s, Berkeley was a center of a

[55]The American physicist G. E. Brown had been a member of the Communist Party for a short while. He came to Birmingham in early 1950 to work with R. Peierls. When he attempted to have his passport renewed the American authorities kept it. He came to Manchester the year after. Brown (1974), p. ii. Brown (2001), p. 6.
[56]BSC, Rosenfeld to Bohr, 25 Aug 1951. Originally in Danish. Einstein had written Blackett on behalf of Bohm. Freire Jr. (2005), p. 4. H. Freistadt was also interested in philosophy of science and was a committed Marxist. Olwell (1999), p. 753. He later attacked Rosenfeld's interpretation of dialectical materialism. Freistadt (1957), pp. 26–29.
[57]RP, "Epistemologie: correspondence générale", J. L. Destouches to L. Rosenfeld, 19 Dec 1951.
[58]Kojevnikov (2002).

strong local left-wing political culture. Bohm got involved in the Federation of Architects, Engineers, Chemists and Technicians, a trade union organized and financed by the Communist Party, which Bohm joined as a member in 1942.[59] From the fall of 1942, Bohm was involved with a project concerned with isotope separation investigating uranium plasma. He asked several times to be transferred to Los Alamos, but was rejected, unofficially because of his political commitment. The activities and individual members of the Federation of Architects, Engineers, Chemists and Technicians were under close surveillance by the FBI and the Military Intelligence Division of the Manhattan Project. In August 1943, intelligence agents discovered a case of espionage in the trade union group and Bohm was also seen as a suspect, but only for a short while.

In 1947, Bohm obtained an assistant professorship at Princeton University. In the second half of the 1940s, Bohm was not particularly politically active, but socialism seems to have inspired him in his work. According to Alexei Kojevnikov, for example, Bohm's formation of the concept "collective movement" of electrons in plasma physics was inspired by socialism.[60] Another example may have been his textbook *Quantum Theory* (1951), building on his lectures at Princeton University in the late 1940s. This textbook was quite atypical seen in an American context. In it Bohm gave a historical presentation of quantum mechanics and provided a thorough introduction to experimental experiences and paradoxes which led to the development of quantum mechanics. Only after his treatment of this wealth of experience did Bohm proceed to introduce the mathematical formalism. By contrast, the more axiomatic American textbooks made it seem as if a genius had invented the Schrödinger equation out of the blue, according to Bohm.[61] Later Bohm stated that the textbook project was motivated in an attempt to understand Bohr, but that he felt he still failed in that.[62]

---

[59]Forstner (2008), p. 218.
[60]Kojevnikov (2002).
[61]Forstner (2008), p. 220.
[62]RP, "Epistemology: Measuring Process", Copy of letter from D. Bohm to A. Loinger, 10 Nov 1966.

In May and June 1949, Bohm was charged before the Committee on Un-American Activities, suspected in the supposed spy case in the Berkeley Radiation Laboratory during the war, even though there were no new proofs of spying. But the old files of the Military Intelligence Division of the Manhattan Project gained new importance during the McCarthy era and in the end, a former colleague of Bohm was identified as the spy. Bohm pleaded the Fifth Amendment right declining to testify against his former colleagues. As a result Bohm was indicted for contempt of Congress one and a half years after the hearing. After that Princeton University, which had otherwise acted in solidarity with Bohm, suspended him from all duties and forbade him to enter the university campus. In early June 1951 Bohm was acquitted, and the suspension was revoked. However, three weeks later his contract ended and was not renewed. As a result Bohm could no longer find a job in the academic sector in the US and he then attempted to find a job in Manchester with Rosenfeld and Blackett. Eventually Bohm accepted a post at the University of São Paulo in Brazil in October 1951. Shortly after his arrival in Brazil, Bohm's passport was retained by the American Consulate and he only recovered his American citizenship in 1986.

Soon after the publication of his textbook, Bohm started working out an alternative interpretation of quantum mechanics, a hidden variables interpretation. In the summer 1951 he submitted a paper entitled "A suggested interpretation of quantum mechanics in terms of "hidden" variables. I and II" to *Physical Review*, which was published in 1952.[63] He also submitted a shorter paper entitled "Causal and Continuous Interpretation of the Quantum Theory" to *Nature* sometime in 1951 or early 1952.[64] At this time, however, Rosenfeld acted as a consultant for *Nature* and due to his intervention the publication procedure was stopped and the paper rejected in March 1952.[65] It is clear that Rosenfeld regarded Bohm's new interpretation

[63]Bohm (1952).
[64]Archive de L'Académie des Sciences, Institute de France, Paris. D. Bohm, manuscript, A Causal and Continuous Interpretation of the Quantum Theory.
[65]RP, "Epistemology", *Nature* editors to Rosenfeld, 11 Mar 1952.

entirely in continuation of the Soviet criticism of the ideas of Bohr and Heisenberg.[66]

In his attack on the usual interpretation, Bohm took his starting point in Heisenberg's uncertainty relation, which he claimed was the backbone of the theory. Thus he did not take his starting point in the universal quantum of action as Bohr did and which is done in an historical account of the quantum theory, and which Bohm had in fact done in his textbook. Bohm wanted to find a description of each individual quantum system in precisely definable states that would change with time according to traditionally deterministic laws.[67] He began with the non-relativistic Schrödinger equation and rewrote the complex wave function by means of two real functions. The dynamical equation of motion could then be put into a form analogous to Newton's second law, where the force acting on microscopic particles was determined by the usual classical potential but in addition by a new quantum potential which introduced non-classical and non-local effects in the description. Bohm's theory meant that the electron was conceived as a particle with a well-defined trajectory, which again meant that it had a simultaneously well-defined position and momentum, even if these were the non-local hidden variables in the theory. In this way, he wanted to reintroduce determinism into the theory. In its motion the particle was guided by a field, which Bohm called a Psi-field, which obeyed the Schrödinger equation. The wave function also determined the probability and thus it had two different meanings in Bohm's theory.[68] Bohm's interpretation of the wave function turned out to be rather similar to de Broglies pilot wave interpretation from 1926 (see Chapter 1), and de Broglie in fact took up his old approach again after he learned about Bohm's work.

Bohm perceived the usual interpretation to be idealistic because of what could be said, or rather what could *not* be said, about the behavior of individual quantum systems, and hence matter, was dependent on an experimental context and ultimately on the presence

[66]Rosenfeld (1979t).
[67]Bohm (1952), pp. 166–167.
[68]*Ibid.*, pp. 169–171. Cushing (1998), pp. 331–338, 343.

of an observer. He did not express himself so militantly as the Party ideologists in Russia. However, with his new interpretation Bohm was seeking a materialistic or realistic, as well as causal, deterministic, and objective description of quantum systems. He distinguished sharply between the mathematical equations of the quantum theory and its interpretation. On the other hand, Bohm's interpretation was meant as an ontological or a realistic interpretation concerned with what could be called be-ables rather than only focussing on observables as in the orthodox interpretation. In the second part of the paper for the *Physical Review* Bohm accused the usual interpretation of being positivistic, because its protagonists were so cautious about avoiding "entities which cannot now be observed".[69] Bohm gave the typical example of Ernst Mach who rejected the atomistic hypothesis on the basis of a positivistic attitude but was later proved wrong; atoms were eventually found to exist and could be observed indirectly. Thus, Bohm echoed Lenin's criticism of Mach which suggests that he was either familiar with Lenin's book *Materialism and Empirio-Criticism* or with contemporary Soviet critique of neo-positivism in general. Retrospectively, Bohm claimed that he had been inspired explicitly by Blokhintsev's and Terletskii's critique of the usual interpretation in constructing his hidden variables interpretation from 1951.[70] It is not clear how he got access to the Russian articles since they appeared in German and French translation only in 1952 and 1953 and he did not read Russian. An English translation was blocked by Rosenfeld in February 1952 when Rosenfeld was contacted by the *Society for Cultural Relations between the Peoples of the British Commonwealth and the USSR* (SCR) to give his opinion about the publication in *Nature* of a translation of a paper by Ya.

[69]Bohm (1952), p. 188.

[70]*Ibid.* In a letter Bohm uttered in January 1952: I ask myself the question "Why in 25 years didn't someone in USSR find a materialist interpretation of quantum theory? It wasn't really very hard.... Yet in USSR there has been much criticism of quantum theory on ideological grounds, but it produced no results, because it may have scared people away from these problems rather than stimulate them". D. Bohm to M. Yevick, 7 Jan 1952, quoted from Forstner (2008), p. 222. Bohm (1957b) p. 110. Terletskii (1952). Blokhintsev (1953b). Blokhintsev (1952).

Iakov Frenkel. In addition, Rosenfeld reviewed papers by Blokhintsev and Terletskii which summarized the current Soviet point of view in quantum foundations.[71] Given his standpoint, it is not surprising that he dismissed the whole lot. Rosenfeld's criticism of Blokhintsev's preference of the field idea in the interpretation of quantum theory was published the year after.[72] In any case, Bohm was rather keen to associate himself and his new interpretation with Marxism–Leninism.

Bohm's new interpretation or theory also sprang from a local context. He had discussed the matter at length with Einstein who had resided at the Institute of Advanced Study in Princeton since 1933. Bohm used Einstein's criticism of the usual interpretation as support for his attempt at constructing an alternative theory. Bohm had also been provoked by von Neumann's proof of the impossibility of hidden variables, he later told Rosenfeld.[73] Moreover, it is interesting to note that the fact that Bohm took his starting point in Heisenberg's uncertainty relation may reflect a more general perception of the history of the orthodox interpretation in Princeton after the war. Thus, Eugene Wigner did the same when he decided to "review the standard view" in 1962: "The standard view is an outgrowth of Heisenberg's paper in which the uncertainty relation was formulated". And he continued: "The far-reaching implications of the consequences of Heisenberg's ideas were first fully appreciated, I believe, by von Neumann".[74] Thus, the quantum of action and Bohr's complementarity are completely absent from this account of the history. Both Wigner and von Neumann resided in Princeton after the war, Wigner at the University of Princeton and von Neumann at the Institute for Advanced Study. More importantly probably was that they had both been in Göttingen in the late 1920s when the standard interpretation of quantum theory emerged. On the other hand, none of them had spent much time, if any, in Copenhagen.[75] If there was a certain Princeton perception of the history of the emergence of the

[71]Frenkel (1950). Blokhintsev (1951). Terlezki (1951). RP, P. Yates to L. Rosenfeld, 7 and 19 Feb 1952.
[72]Rosenfeld (1979t).
[73]RP, "Measuring Process", D. Bohm to L. Rosenfeld, 13 Dec 1966.
[74]Wigner (1963), p. 7.
[75]Von Neumann was in Copenhagen only briefly in September 1938.

orthodox view in quantum mechanics, it may therefore reflect how this perception also prevailed in Göttingen in the late 1920s and early 1930s among physicists there who never made it to Copenhagen.[76] This seems to be further supported by Born's case; even though Born is considered as one of the central proponents of the Copenhagen Interpretation, he was not among the group of physicists who assembled each year in Copenhagen during the 1920s and the early 1930s. Upon reading Rosenfeld's small article in *Nature* in 1950 about "Early History of Quantum Mechanics", Born mentioned to Rosenfeld that the role Bohr had played in the development in the late 1920s was new to him, and that he only knew little of Bohr's important ideas from that time, that is, the complementarity interpretation of quantum mechanics.[77]

Rosenfeld's official response to Bohm's, de Broglie's, and the Russian physicists' ideas came in the shape of a contribution to a festschrift for de Broglie's sixtieth birthday in 1952. It was a polemical paper "L'évidence de la complémentarité", which was also translated into English with the title "Strife about Complementarity" and published in *Science Progress* in 1953.[78] In this paper, referring to de Broglie and Bohm, Rosenfeld sarcastically remarked, "[I]t is understandable that the pioneer who advances in an unknown territory does not find the best way at the outset; it is less understandable that a tourist loses his way again after this territory has been drawn and mapped in the twentieth century".[79] After reading Rosenfeld's paper, Bohm wrote to Rosenfeld that the term "tourist" could just as well be applied to Rosenfeld; Bohr would then be the pioneer who did "not find the best way at the outset". Thus, according to Bohm, it "worked both ways".[80] In fact, the double meaning of the

[76]The terms "Princeton school" and "Copenhagen school" referring to different viewpoints within the orthodox interpretation of quantum mechanics emerged with the discussions between Rosenfeld and Wigner in the 1960s (see epilogue). Freire Jr. (2007). Freire Jr. (2009), p. 285.
[77]Rosenfeld (1950). RP, "Correspondance particulière", M. Born to L. Rosenfeld, 10 May 1951.
[78]Rosenfeld (1953). Rosenfeld (1979t), p. 482.
[79]Rosenfeld (1953), pp. 55–56.
[80]RP, "Epistemology: Correspondance générale, Volume de Broglie (1952). Strife about complementarity (1953)", D. Bohm to L. Rosenfeld.

word "tourist" in relation to Bohm, who was by then in more or less enforced exile in Brazil, was resented by the Czech-born physicist Guido Beck, at the time residing in Brazil, and the British physicist E. H. S. Burhop. Rosenfeld apparently acknowledged their point and as a result, this passage of the paper was left out in its English translation.[81]

Rosenfeld later replied to Bohm's accusation that the usual interpretation had been founded on a positivistic basis, by dismissing Bohm's interpretation as metaphysical.[82] Both Bohm and Rosenfeld believed that science underwent a continuous dialectical development. Hence, there were no such things as closed or final theories, as Heisenberg would argue, for example.[83] In his later books, Bohm used this idea to emphasize the historical contingency of scientific theory. In 1957, Bohm suggested that "one must conceive of the law of nature as necessary only if one abstracts from *contingencies*", where "contingency is that which could be otherwise".[84] He continued "Of course, we may take an infinity of different views, but associated with each view there is always an opposite view. Thus, while we can always view any given process from any desired side (e.g. the causal side) by going to a suitable context, it is always possible to find another context in which we view it from the opposite side (in this case, that of contingency)".[85] In Bohm and Rosenfeld's correspondence at that time, Bohm did not insist that he had found absolute truth, with his hidden variables theory, only that he *might* have found truth, just as well as Niels Bohr *might* have. They could both be right, Bohm suggested to Rosenfeld in a letter in 1952, or rather he suggested "that either one of us may be in error", a statement which did not at all please Rosenfeld.[86] Rosenfeld did not believe in

[81]RP, "Epistemology: Correspondance générale, Volume de Broglie (1952). Strife about complementarity (1953)", E. H. S. Burhop to L. Rosenfeld, 5 May 1952. G. Beck to L. Rosenfeld, May 1952.
[82]Rosenfeld (1979h), pp. 497–498 [43–44].
[83]See for example, Bokulich (2004).
[84]Bohm (1957b), p. 2.
[85]*Ibid.*, p. 3.
[86]RP, "Epistemology: correspondence générale", Bohm to Rosenfeld [not dated]. Bohm Papers, Birkbeck College, Library, London, L. Rosenfeld to D. Bohm, 30 May 1952.

historical contingency and a more or less random development of scientific theory, which seemed partly to lie behind Bohm's motivation for proposing a new interpretation at this particular time.[87]

On the other hand, Rosenfeld admitted that Bohm's interpretation was "very cleverly contrived" and that "one would look in vain for any weakness in its formal construction".[88] In 1953, Rosenfeld spent three months lecturing in Brazil. On that occasion he met Bohm in São Paulo. After their meeting Bohm wrote to Aage Bohr, "Professor Rosenfeld visited Brazil recently, and we had a rather hot and extended discussion in São Paulo, following a seminar that he gave on the foundations of the quantum theory. However, I think that we both learned something from the seminar. Rosenfeld admitted to me afterwards that he could at least see that my point of view was a possible one, although he personally did not like it".[89] Rosenfeld's arguments against Bohm were based on empirical realism and his own interpretation of dialectical materialism in that light; the scientist learned from experience and his mind somehow adapted gradually to his experience. He categorically wrote to Bohm in May 1952: "The role of a physicist is to adapt his thinking to Nature, regardless of whether this process is simple or not. The main thing is not to accept any other guidance than that of Nature herself".[90] Thus, Rosenfeld opposed the idea of proposing theory change just for the sake of it. He insisted that "the conception of complementarity forces itself upon us with logical necessity... It arises from an effort to adapt our ideas to a novel experimental situation in the realm of atomic physics".[91] Therefore, according to Rosenfeld, the usual interpretation was materialistic and Bohr was in this sense an empiricist and materialist. Rosenfeld insisted that there was no new empirical evidence that could justify giving up complementarity and proposing a new interpretation of quantum mechanics in 1951. Since Bohm had

[87]See for example, Rosenfeld (1979x), p. 539 and Rosenfeld (1979d), p. 484.
[88]Rosenfeld (1979t), p. 475.
[89]Aage Bohr Papers, NBA, D. Bohm to Aa. Bohr, 13 Oct 1953. Quoted from Freire Jr. (2005), p. 13. RP, "Manchester (1947–1958), 3: Conferences & lectures (1953–1955)".
[90]Bohm Papers, Birkbeck College, Library, London, L. Rosenfeld to D. Bohm, 30 May 1952. See also Rosenfeld (1979d), p. 484.
[91]Rosenfeld (1979t), p. 466.

expressed difficulty in understanding Bohr's complementarity principle, Rosenfeld taught Bohm in the same letter, that *dialectics* was the way to comprehend complementarity: "The difficulty of access to complementarity which you mention is the result of the essentially metaphysical attitude which is inculcated to most people from their very childhood by the dominating influence of religion or idealistic philosophy on education. The remedy to this situation is surely not to avoid the issue but to shed off this metaphysics and learn to look at things dialectically".[92] From a Marxist philosophical perspective, the main difference between Rosenfeld and Bohm seems to be that Rosenfeld, a Marxist with a positivistic inclination and an epistemological standpoint, was able to combine dialectical materialism and complementarity. One may say that he emphasized the dialectical, while Bohm, a Marxist with a metaphysical inclination and an ontological position thought complementarity could only be inconsistent with dialectical materialism, because he emphasized materialism. In other words, Bohm let his alliance with Marxism–Leninism determine his view on quantum physics, whereas this relationship was inverted in Rosenfeld who let his alliance with the achievements of modern physics and Bohr's complementarity in particular determine his view on Marxism–Leninism.

In connection with the political development in Russia after Stalin's death, particularly the Russian invasion of Hungary in 1956 which crushed the revolution there, Bohm broke with communism and Marxism–Leninism altogether. In the same period Bohm changed his view on the interpretation of quantum mechanics. In a letter to Martin Strauss in the fall of 1956 Rosenfeld announced that he had just met Bohm in London, where "he told me that he had changed his standpoint considerably".[93] At a symposium on philosophy of science, "Observation and Interpretation", in London in April 1957, where Bohm and Rosenfeld both gave talks, Bohm still

[92]Bohm papers, Birkbeck College, Library, London, L. Rosenfeld to D. Bohm, 30 May 1952.
[93]RP, "Correspondance particulière", copy of letter from L. Rosenfeld to M. Strauss, 2 Nov 1956. Freire Jr. (2011), p. 294.

defended his hidden variables interpretation while Rosenfeld argued that Bohm had completely misunderstood Bohr's views.[94] The same year saw the publication of Bohm's book *Causality and Chance*, which Rosenfeld utterly disliked and therefore reviewed in his usual vitriolic style. Rosenfeld warned "the unwary student" that this book gave "a picture distorted beyond recognition by misinterpretation and misrepresentation".[95] De Broglie who had written the preface in Bohm's book replied to Rosenfeld's review. In 1928 de Broglie seemed to have given up his own pilot wave interpretation of Schrödinger's wave function because of the critique by Pauli (see Chapter 1). He now attempted to bolster his critique of the orthodox interpretation with the comment that "scientists as eminent as Planck, Einstein and Schrödinger have always expressed doubts on this subject. The idea of Professor Bohm... therefore seems perfectly defensible to me".[96] He maintained that new points of view which could provide a more intelligible image of the physical world than the obscure conception of complementarity were justified.

"Strife about Complementarity"

The fact that Bohm was a Marxist was to some extent significant both in determining who supported his approach and where he expected support.[97] All in all his interpretation did not receive much positive attention in the beginning. However, as mentioned above, Bohm's interpretation was welcomed and supported by a group of young communist physicists in France including Jean-Pierre Vigier, Evry Léon Schatzman, and G. Vassails, who were strongly motivated in their support of Bohm by Marxist thought.[98] Before he knew that de Broglie also supported Bohm, this French constellation provoked

[94]Bohm (1957a), pp. 33–40. Rosenfeld (1979h). See also Rosenfeld (1979d).
[95]Rosenfeld (1958).
[96]de Broglie (1958).
[97]RP, "Epistemology: Measuring process", Copy of letter from D. Bohm to A. Loinger, 10 Nov 1966. Freire Jr. (2005).
[98]Freire Jr. (2005), pp. 14–16. There is a good description in Cross (1991), pp. 747–751 of the French milieu.

Rosenfeld to write to Pauli with the usual scathing irony and sarcasm that pervades their correspondence:

> Is it not delightful that poor Bohr's only supporters in Paris should be this logical couple [de Broglie and Destouches], while all the youth is in arms against him "under the banner of Marxism"? Poor Marx too, I would add, since I belong, as you know, to the almost extinct species of *genuine* Marxists; the kind of theology dished up under this name today is just as repulsive to me as to you, perhaps even more so because I see it against the background of what Marx *really* meant.[99]

"Under the banner of Marxism" was a reference to the influential Soviet journal of philosophy of science *Pod znamenem marksizma*, in which discussions of the philosophy of quantum mechanics had taken place between 1922 and 1944.[100] Bohm's revival of de Broglie's old interpretation initiated a discussion of the topic and of Marxism between Rosenfeld and Pauli in connection with their contributions to the de Broglie Festschrift during March and April 1952. Rosenfeld expressed sincere concern about the development he witnessed: "We may well have a laugh at it between us... but I feel we have also a duty to help these people out of the mire if we can".[101] A few weeks later, Rosenfeld wrote to Frédéric Joliot-Curie in Paris to get him to intervene in the activities of the French group of young physicists devoted to the causal program.

> I think it my duty to inform you of a situation which I consider quite serious and which is close to you. It concerns your "foals" Vigier, Schatzman, Vassails and the whole lot,

---

[99]L. Rosenfeld to W. Pauli, 20 Mar 1952, in von Meyenn (1996), Vol. 4, Part 1, pp. 587–588, on p. 588. Emphasis in original.

[100]Graham (1987), p. 326. Joravsky (1961), pp. 78–80.

[101]L. Rosenfeld to W. Pauli, 20 Mar 1952 in von Meyenn (1996), Vol. 4, Part 1, pp. 587–588. The same message is conveyed in RP, "Epistemology: Correspondance générale 1955–1958", draft of letter from L. Rosenfeld to L. L. Whyte, 17 Mar 1958. See also W. Pauli to L. Rosenfeld, 1 Apr 1952, in von Meyenn (1996), Vol. 4, Part 1, pp. 590–593, on p. 592. L. Rosenfeld to W. Pauli, 6 Apr 1952, *ibid.*, pp. 598–600.

> all young intelligent people and all full of desire to do well. Unfortunately, at the moment, they are quite sick. They have gotten it into their heads that it is necessary persistently to shoot down complementarity and save determinism. The ill fortune is that they have not understood the problem and — what is even worse — that they have made no serious effort to understand it. I have done what I could to redeem them.... I have taken pains to do an explicit Marxist analysis of the question and clearly show the simultaneously dialectical and materialistic character of complementarity. As the only response, Schatzman sent me a polemical writing full of incorrect physics and quotations from Stalin which he maintains, through a casuistry that frightens me, oppose physical evidence. This reveals a profound crisis among these young people and it is high time (if it is not already too late) to straighten them out. They are under the spell of a scholasticism which borrows the external forms of Marxism, but is as opposed to its genuine spirit as is the blackest Catholicism. The best Soviet physicists are subjected to attacks from this scholasticism, which creates even more havoc in Moscow than in Paris. Surely it would be desirable that the French physicists show themselves capable of distinguishing the wheat from the chaff. At the moment, these young fanatics are the laughing stock of the theoreticians and they discredit Marxism and the Party to the great joy of reactionaries such as Destouches and de Broglie. I am ready to give you all possible support to redress the situation, on the basis of the ideas explained in my paper. But it is up to you, my dear Joliot, to take the initiative.[102]

Joliot-Curie usually did not take much interest in the philosophical subtleties of quantum physics, but from his reply he seems to have agreed with Rosenfeld in his critique of those youngsters and

[102] Archives Joliot-Curie, Fonds Frédéric Joliot. F120, L. Rosenfeld to F. Joliot-Curie, 6 Apr 1952. Originally in French.

on the need to take action.[103] Perhaps they both saw it as a give and take situation since at the same time Joliot-Curie was pressing Rosenfeld to approach Bohr on behalf of the World Federation of Scientific Workers to take the leadership in a conference bringing together prominent scientists from East and West (see previous chapter).

Rosenfeld also discussed the issue with his long-term friend Martin Strauss, whose criticism of Bohm Rosenfeld was familiar with. Strauss was one of only few scholars who remigrated to East Germany after the Second World War and now lived in East Berlin.[104]

> You cannot imagine with how much satisfaction I read your judgment of the Bohmian matter. Finally, I had to cry out, a reasonable Marxist who senses the declining mechanistic tendency of all such attempts. You should know that I have had animated discussions about this matter, first with the Paris group of communist physicists, then with the London group around Bernal, without succeeding to convince them that the true dialectics is expressed through complementarity, and that determinism represents an outdated metaphysical idea. Hopefully you have more success in that.[105]

In a long letter, Strauss replied that he would make sure Rosenfeld's "Strife" paper became the topic for a colloquium on foundations of quantum theory at the Humboldt University and invited Rosenfeld to East Berlin. Strauss agreed with Rosenfeld's criticism of Bohm, Schrödinger, and also the writings on this topic by the Hungarian physicist Lajos Jánossy. However, whereas in the 1930s Strauss had collaborated with the logical positivists just like the Danish philosopher Jørgen Jørgensen, now he seemed to have swung around completely and, loyally following the new Soviet ideological line, he accused Bohr and Rosenfeld for positivism. Furthermore,

[103] Archives Joliot-Curie, Fonds Frédéric Joliot, F120, F. Joliot-Curie to L. Rosenfeld, 21 Apr 1952. Pinault (2000), p. 508.
[104] For the cases of Fritz Lange and Klaus Fuchs, see Hoffmann (2009).
[105] RP, "Correspondance particulière", Copy of letter from L. Rosenfeld to M. Strauss, 24 Apr 1953. Originally in German.

according to Strauss, playing the Engels card against Lenin showed that Rosenfeld had not properly understood either Engels or Lenin; "all Marxists agreed about that".[106] Strauss claimed the new ideological line in the Soviet Union served to rid the quantum theory of idealistic quotes and erroneous interpretations. Strauss further claimed that what one found in Russia was an *open* debate contrary to what he believed had been the case with Bohr's "gentlemen's club". He sympathized with attempts to develop further what he conceived as the dehydrated quantum theory.[107] Rosenfeld and Strauss continued discussing these issues in their correspondence during the period 1953–1957 and Rosenfeld visited Strauss in East Berlin for ten days in June 1954.[108] They eventually found common ground. Through Strauss and Herneck Rosenfeld's attention was drawn to the Argentine philosopher of science Mario Bunge's writings. Bunge subscribed to Bohm's ideas and heavily criticized Rosenfeld's position in his articles.[109] The same did Hans Freistadt, whose criticism of Rosenfeld's interpretation of dialectical materialism was brought to Rosenfeld's attention by the German-born American philosopher Adolf Grünbaum. Grünbaum was critical of Bohr's generalization of complementarity into other domains of human knowledge.[110]

Rosenfeld's polemical papers and vitriolic reviews of Bohm as well as his reference to dialectical materialism in his publications puzzled non-Marxist scientists and philosophers of science in the West, and still does.[111] Furthermore, his efforts to defend Bohr's doctrine

[106]RP, "Correspondance particulière", M. Strauss to L. Rosenfeld, 30 May 1953.
[107]*Ibid.*
[108]RP, "Correspondance particulière: Strauss".
[109]RP, "Correspondance particulière", M. Strauss to L. Rosenfeld, 20 Oct 1956. Copy of letter from L. Rosenfeld to M. Strauss, 2 and 14 Nov 1956. RP, "History of Science 7: Herneck", F. Herneck to L. Rosenfeld, 4 Jun 1958. Bunge (1955). And from a Marxist perspective Bunge (1956).
[110]Freistadt (1957). RP, "Epistemology 1955–1958", A. Grünbaum to L. Rosenfeld, 20 Apr 1957. Copy of letter from L. Rosenfeld to H. Freistadt, 21 May 1957. Copy of letter from L. Rosenfeld to A. Grünbaum, 21 May 1957. Copy of letter from L. Rosenfeld to A. Grünbaum, 11 Dec 1957. Grünbaum (1957).
[111]See for example RP, "Correspondance particuliére", W. Yourgrau to L. Rosenfeld, 23 Dec 1958. Yourgrau soon completely supported Rosenfeld in his criticism, however. W. Yourgrau to L. Rosenfeld, 18 May 1959. Another example is RP, "Manchester 7: Thermodynamis, opticks, etc. Foundations of statistical mechanics", J. Koefoed

in a Marxist context met with opposition from the founders of quantum mechanics, Born, Heisenberg, and Pauli. Indeed, Rosenfeld ended his "Strife" paper with the sentence "I seem to quarrel with everybody".[112] Surely Pauli, Born, and Heisenberg did not like that the Copenhagen Interpretation was criticized from the perspective of Marxism, and they all criticized Bohm, but nor did they agree with Rosenfeld's blending of complementarity with dialectical materialism. Rosenfeld was confronted with their criticism in a friendly and collegiate atmosphere reflected, for instance, in Pauli's pertinent nick-name for Rosenfeld these years; in their correspondence from the early 1950s Pauli addressed him: "$\sqrt{Bohr \times Trotsky} =$ Rosenfeld".

Pauli's nickname for Rosenfeld.[113] Courtesy of The Niels Bohr Archive.

Rosenfeld and Born maintained a cordial relationship and corresponded on a regular basis. From their correspondence it appears that Born was growing increasingly concerned that his contributions to quantum mechanics were being forgotten. Heisenberg alone won the 1932 Nobel Prize for matrix mechanics whereas the contributions of Jordan and Born were overlooked.[114] It therefore pleased Born exceedingly when Rosenfeld emphasized Born's role in the formation of quantum mechanics in his *Nature* article "Early History of Quantum Mechanics".[115] According to Born, it was the first time his work

---

to L. Rosenfeld, 13 Feb 1957. RP, "Epistemology 1955–1958", Draft of letter from L. Rosenfeld to J. Koefoed, 20 Feb 1957. See also RP, "Epistemology: Correspondance générale 1955–1958", draft of letter from L. Rosenfeld to L. L. Whyte, 17 Mar 1958. Freire Jr. (2007).

[112]Rosenfeld (1979t), p. 482.

[113]RP, "Correspondance particuliére", W. Pauli to L. Rosenfeld, 28 Sep 1954, printed in von Meyenn (1999), Vol. 4, Part 2, p. 769. See also W. Pauli to W. Heisenberg, 13 May, 1954, *ibid.*, pp. 620–621.

[114]After receiving the prize, Heisenberg immediately wrote to Born to express dismay that they were not sharing the prize. Friedman (2001), p. 175. RP, "Correspondance particulière", M. Born to L. Rosenfeld, 21 Mar 1949.

[115]Rosenfeld (1950).

in the early quantum theory was acknowledged.[116] Four years later Born finally received the Nobel Prize "for his fundamental research in quantum mechanics, especially for his statistical interpretation of the wavefunction" and got the recognition he had longed for.[117] By giving the prize to Born at this particular time and for such an "abstract idea" as the statistical interpretation of the wavefunction, the Nobel Committee, whether intentionally or otherwise, was making a solid statement in the on-going quantum controversy.[118] Also Born's Nobel lecture may be regarded as a polemical statement in the on-going debate. In these years Born engaged in discussions about the interpretation of quantum mechanics with Schrödinger who criticized the Copenhagen Interpretation with renewed energy. Basically, Schrödinger wanted to reduce quantum theory to wave mechanics and thereby avoid the discontinuous transitions or quantum jumps of particles between stationary states.[119] A meeting between the two protagonists was set up in London in December 1952 at the British Society for the History and Philosophy of Science, in which, incidentally, Rosenfeld acted as Vice President at the time. Born hoped Rosenfeld would be present at meeting in order to assist Born "if it comes to real discussion".[120] However, Rosenfeld was not able to attend the meeting, and in the end, Schrödinger had to cancel due to an appendicitis operation. Still, both contributions were published in the society's journal.[121]

In his Nobel lecture Born outlined the history of the formation of the orthodox interpretation in the 1920s. He also discussed the objections based on the question of determinism and the question of reality given by "great scientists such as Einstein, Schrödinger, and De Broglie", but he kept silent about the recent opposition in the Soviet Union in line with his general avoidance of openly discussing

[116] RP, "Correspondance particulière", M. Born to L. Rosenfeld, 10 May 1951. Freire Jr. (2001), p. 249.
[117] http://nobelprize.org/nobel_prizes/physics/laureates/1954/ (accessed 2 May 2011). Friedman (2001), pp. 220–221.
[118] RP, "Correspondance particulière", M. Born to L. Rosenfeld, 27 Nov 1954.
[119] Schrödinger (1952).
[120] RP, "Correspondance particulière", M. Born to L. Rosenfeld, 12 Nov 1952.
[121] Moore (1989), p. 451. Schrödinger (1952). Born (1953).

Cold War political issues.[122] As mentioned in the previous chapter, Born joined the British organization Science for Peace in 1951 when he had been convinced that it was not under the influence of the communists. Born hated publicity, his health was poor, and he considered himself politically inexperienced. However, prompted by the American detonation of the Castle Bravo thermonuclear device on the Bikini Atoll on 1 March 1954, which turned out to be many times more powerful than first estimated and caused global ecological consequences, combined with receiving the Nobel Prize later the same year, Born gave considerable thought to "using my present popularity... to try and arouse the consciences of our colleagues over the production of ever more horrible bombs".[123] He got involved in creating the Russell–Einstein Manifesto, but also and simultaneously the Mainau Declaration. In trying to persuade Bohr to sign the Mainau Declaration, Born mentioned that he and Otto Hahn "were very reluctant to sign [the Russell–Einstein Manifesto], since there are so many communists among the initiators, and we believe that the only way to have any effect is to avoid this".[124] Born had witnessed what had happened to East Germany, the part of Germany where he had grown up, and he could only "pity the poor people who are compelled to live under" communism.[125] Born nevertheless remained among the signatories of the Russell–Einstein Manifesto, which in the end probably gained a larger impact than the Mainau Declaration partly because it also included communist signatories.

In 1953, Born characterized Rosenfeld's contribution to the de Broglie Festschrift, "L'évidence de la complémentarité", as "your tight-rope walk over the abyss of being either heretical to your St. Niels or to your St. Marx" and saw that Rosenfeld "will be, of course, always between two stools and will be attacked from both sides".[126] However, Born grew less understanding of Rosenfeld's position and in the mid 1950s eagerly confronted Rosenfeld about it in

---

[122]Born (1954), p. 8.
[123]Born (1971), pp. 229–231. Schirrmacher (2007).
[124]BSC, M. Born to N. Bohr, 20 Apr 1955.
[125]RP, "Correspondance particulière", M. Born to L. Rosenfeld, 28 Mar 1957.
[126]RP, "Correspondance particulière", M. Born to L. Rosenfeld, 21 Jan 1953.

their private correspondence; he still did not wish to go public about his criticism because he wished "to remain on good terms with the Russians and East-Germans ... until they themselves will abandon the more excessive and non-sensical Marxist–Engelist doctrines".[127] However, with his close friend Rosenfeld he felt safe to speak freely.

Born did not distinguish between variants of Marxist philosophy and Marxist politics. Since Rosenfeld defended Marx and Engels and dialectical materialism, Born put him in the same category as the communists. Prompted by the English translation of Rosenfeld's "Strife" paper, Born wrote a long piece in which he discussed dialectical materialism from the point of view of science, philosophy, and policy and sent it to Rosenfeld. He attacked Rosenfeld for declaring that the development of modern physics could be reconciled with dialectical materialism and failed to realize Rosenfeld's subtle critique of the communists' subordination of science and culture under *Zhdanovschina*. Only when Rosenfeld sent Born his review of Bernal's book *Science in History* did it occur to Born that Rosenfeld possessed an independent opinion in philosophy of science despite the fact that he adhered to dialectical materialism.[128]

Besides criticizing Rosenfeld for blending complementarity and dialectical materialism, Pauli and Born also expressed their distaste for Rosenfeld's criticism of Heisenberg's idealism.[129] Heisenberg for his part did not conceal his idealistic views just as he was quite open to discussing philosophical matters with Rosenfeld. He suggested,

[127] RP, "Correspondance particulière", M. Born to L. Rosenfeld, 14 Nov 1955. For Born's opinion about Rosenfeld's amalgamation of Marxism and complementarity, see also Born (1978), p. 234.

[128] M. Born, Dialectical materialism and modern physics, msc enclosed in RP, "Correspondance particulière", M. Born to L. Rosenfeld, 14 Nov 1955. The manuscript is published in Freire Jr. and Lehner (2010). RP, Supplement "Evolution of scientific thought", Copy of letter from L. Rosenfeld to M. Born, 8 Nov 1955. See also Freire Jr. (2001). W. Pauli to L. Rosenfeld, 1 Apr 1952, in von Meyenn (1996), Vol. 4, Part 1, pp. 590–593. RP, "Correspondance particulière", M. Born to L. Rosenfeld, 28 Mar 1957.

[129] "I am first concerned... with Heisenberg and with you. I have already corresponded with Heisenberg (whom you are free to consider *idealistic*, if you only consider *this* letter to be sufficiently *materialistic*).... To say it frankly, I am very much against your writing again on *complementarity* (it is urgent that Bohr do this himself again), simply because it is slowly getting dull. We should not keep repeating the same *Schmus*". W. Pauli to L. Rosenfeld, 21 May 1954, in von Meyenn (1999), Vol. 4, Part 2, pp. 639–640, on p. 640.

for example, that Engels' writings would not last for as long as those of Platon.[130] On the question of how Bohr reacted to Rosenfeld's blending of complementarity and dialectical materialism, Rosenfeld wrote to Born in 1955 that "I have had occasion of course to explain all this to Bohr who I think understands my attitude".[131] Moreover, Bohr agreed with Rosenfeld in so far as the problem with Marxist philosophers was that they let their ideology determine their view on new scientific acquisitions. Bohr had made that clear already during his visit in Russia in 1934 (see Chapter 3).

Thaw in East–West Relations

The decade from Stalin's death in 1953 till 1961 was a time of reform in which both the image and practice of Soviet science underwent dramatic transformations. It gradually freed itself from the rhetoric of ideological differences between Soviet science and bourgeois science. The changes came quietly from within the academic and scientific community, particularly due to pressure from the physicists. In 1954, the members of the Academy of Sciences began to express more openly the need for communication and international cooperation of science. The pressure from the Academy overlapped with the interests of the Soviet leaders, "who were eager to explore the possibilities of 'science diplomacy' as a means to open up dialogue with the West, lessen Soviet international isolation, and decrease tensions".[132]

Steps towards international collaboration were taken at the other side of the Atlantic too. At a meeting in the UN in December 1953, President Dwight D. Eisenhower launched a new "Atoms for Peace" policy

[130] See for example, Heisenberg (1955), p. 28. Heisenberg (1970), pp. 32–33. Rosenfeld, (1979b). Rosenfeld, (1979t), pp. 480–481, RP, "Correspondance particulière", W. Heisenberg to L. Rosenfeld, 16 Apr 1958.

[131] RP, Supplement "Evolution of scientific thought", Copy of letter from L. Rosenfeld to M. Born, 8 Nov 1955. Born replied, "You indicate that he [Bohr] shares your views. If I should see him again I shall ask him point-blank". RP, "Correspondance particulière", Born to Rosenfeld, 24 Oct 1955. And in another letter: "I think you are quite wrong in your judgment about Heisenberg's and other people's attitude and I also cannot believe that Bohr agrees with you". RP, "Correspondance particulière", Born to Rosenfeld, 14 Nov 1955.

[132] Ivanov (2002), p. 324.

in order to strengthen American world leadership and change the image of the US as a country concerned solely with destructive uses of nuclear technology. [133] As a result of Eisenhower's "Atoms for Peace" initiative, the UN General Assembly decided in early December 1954 "that an international technical conference of Governments should be held, under the auspices of the United Nations, to explore means of developing the peaceful uses of atomic energy through international cooperation".[134] The aim of the conference was to attain free discussion, exchange, and sharing of technical knowledge about atomic energy and "to ensure, notwithstanding the quite obvious and important political, economic and social implications of a conference of this nature, that it would be scientific in the most objective sense and free from all political bias".[135] The conference eventually took place in Geneva in August 1955. It followed the first post-war summit between Eisenhower and Krushchev, in the same city in July. The conference managed to assemble 1,428 delegates from 73 countries. It was further attended by approximately 1,350 observers from academic institutions and industrial enterprises and 905 representatives from the press. Most importantly, the Russians adopted the Atoms for Peace policy and sent a delegation of highly technical competent physicists and engineers to the conference. This was the first conference in the West in which Russian physicists took part since the late 1930s. Atomic energy was not Rosenfeld's area of research, and he was not present at the conference. However, the conference clearly seemed to bode a renewed hope for Bohr's idea of an open world by the declassification and internationalization of research on reactors and on peaceful applications of atomic energy that followed.[136]

In fact, Niels and Aage Bohr reacted promptly to the new openness in international relations. Already prior to the Geneva

[133] Hewlett and Holl (1989), p. 240.

[134] *Proceedings of the International Conference on the Peaceful Uses of Atomic Energy* (1956), p. ix. Hewlett and Holl (1989), pp. 232–235, 250. *Bulletin of the Atomic Scientists* (1955).

[135] *Proceedings of the International Conference on the Peaceful Uses of Atomic Energy* (1956), p. x.

[136] Lapp (1955). Smith (1955). Kojevnikov (2011).

conference, Aage Bohr was contacted by the Russian nuclear physicist Lev A. Sliv from the Joffe Institute in Leningrad concerning a copy of a special issue of a recent treatise. The ensuing correspondence between Sliv and Aage Bohr resulted in an invitation for Sliv to visit the institute in Copenhagen which he did in the spring of 1956. The visit came to constitute the initiation of relations between Danish and Russian physicists. In the following years, several Russian physicists visited Copenhagen and Danish physicists visited the USSR for extended periods. Hence the element of scientific diplomacy in Bohr's open world was partly realized. From the mid-1950s, the Institute for Theoretical Physics at Copenhagen constituted an important meeting place for the East–West cooperation among physicists. Niels Bohr's efforts in working for openness and international relations in science were carried on successfully and indefatigably by Aage Bohr until revolutions in 1989 overthrew the Soviet bloc.[137]

The thaw in the 1950s was due not least to Krushchev's introduction of the policy of de-Stalinization. The Castle Bravo test with its widespread radioactive fallout and contamination of human populations and habitats, made the world leaders realize that a war with hydrogen bombs would be the same as committing suicide. Kruschchev therefore rejected Stalin's doctrine of the inevitability of war between communism and capitalism and announced his new goal in terms of peaceful coexistence with other nations.[138] In April 1956 with the principle of cooperation with other nations now official Soviet policy, the Soviet Government let its most secret scientist, the leader of the Soviet thermonuclear project, Igor Vasilievitch Kurchatov, accompany Kruschchev on an official visit to the UK. Kurchatov was authorized by the Party and the Soviet Government to openly discuss the status of the top-secret Soviet research program on controlled thermonuclear reactions at the Atomic Energy Institute in Moscow at an international conference at the British Harwell Atomic Energy Research Establishment. Moreover, he was

[137]Aage Bohr Papers, "L. A. Sliv (Joffe Physico-Techn)". BGC, "Academy of Sciences, Moscow". Bjørnholm (1968).

[138]Gaddis (1997), pp. 208–209. Hewlett and Holl (1989), pp. 272–281.

authorized to make western scientists an offer of international cooperation in the field. Kurchatov's appearance at Harwell was a sensation; at the time such work was strictly classified in both the US and Britain. The event prompted the UK and the US to declassify and internationalize their respective efforts in this field starting 1957. Thus, Kurchatov's appearance had great political consequences and was another marker of the turning point in the relations between Soviet physicists and physicists in the West. The director of the research at Harwell, John D. Cockcroft, and other British physicists were invited by Kurchatov to visit his Atomic Energy Research Institute in Moscow in November 1958 and more exchanges between Soviet and British physicists followed the ensuing years.[139]

Kruschchev also encouraged communist parties outside Russia to adapt to local conditions. This proved an explosive message and combined with relinquishing the instruments of terror by which Stalin had built such a fearsome reputation, it severely destabilized Kruschchev's authority over communists outside Russia. Thus, by attempting reform Kruschchev was met with revolt in Poland and Hungary. The Hungarian leader Imre Nagy announced that Hungary would leave the Warsaw Pact to become a neutral state. In order to prevent this move and restore order and prevent the communist world from falling apart, Kruschchev commanded the Red Army to crush the Hungarian upheaval.[140]

Rosenfeld severely regretted Krushchev's reaction to the people's revolution in Hungary. A Hungarian physicist, Paul Roman, and his wife who had managed to escape out of Hungary in time before the border closed stayed with the Rosenfelds for a while.[141] Still, Rosenfeld kept his hope intact with respect to the new socialist movements in Poland, Hungary, China, and Yugoslavia, especially that these movements would eventually result in "intellectual revolution which will overcome the dogmatism" in the Party and the

[139]Ivanov (2002), pp. 324–334. Gaddis (1997), pp. 206–208. *Bulletin of the Atomic Scientists* (1956). Cockcroft (1963). *Time* (1956). Kojevnikov (2011).
[140]Gaddis (1997), pp. 208–211.
[141]BSC, L. Rosenfeld to N. Bohr, 14 Jan 1957.

Soviet leadership.[142] The events in Hungary did not crush this hope, as Rosenfeld made it clear to Max Born in 1957. Born seemed to have expected that Rosenfeld would abandone Marxism, because of the way the Russians behaved in Hungary, but Rosenfeld assured him otherwise: "May I finally cheer you up with regard to the present political situation. It is true that progress made last year has suffered a severe set-back, but that should not discourage us, because, however strong the reaction, it cannot put the clock back entirely. Something has happened which is irreversible and, after all, all the progress of mankind towards liberation from savagery and fanaticism has happened this way by a succession of upsurges followed by reactions. The main thing is that after each such upheaval there re[mains] a definitive advance on the road towards freedom and science. On this optimistic note I will conclude for the time being".[143] He thought it was not possible to stop entirely the liberation process from oppressive forces which had started among the Hungarian people and that eventually communist authorities would have to take their demands for freedom seriously. However, when the Prague Spring Reforms in Czechoslovakia were crushed by Warsaw Pact troops and tanks in 1968 Rosenfeld found it hard to maintain his optimism. As he wrote to the philosopher of physics Wolfgang Yourgrau "I kept away from Tiflis just because I could not stand the sight of Russians just now".[144] It was Fock who was hosting a physics and philosophy conference, The Fifth Meeting of the Society on General Relativity and Gravitation, in Tiflis.

[142]RP, "Correspondance particulière", Born to L. Rosenfeld, 28 Mar 1957. A. Rosenfeld in email correspondence with author, 13 Oct 2004, pointed out: "Certainly both my parents were initially disillusioned by the Soviet Union's lack of support in China, though after the cultural revolution there, they both despaired of communism there as well. After the Hungarian invasion, we had a refugee couple from Hungary to stay with us a short while (he was a youngish physicist), and yes that was another reason for criticism of Soviet policy". See also BPP, j. nr. 120, L. Rosenfeld to N. Bohr, 21 Jun 1950.

[143]RP, "Correspondance particulière", L. Rosenfeld to M. Born, 10 Apr 1957. Born to Rosenfeld, 21 Jan and 28 Mar 1957. See also Freire Jr. (2001).

[144]RP, "Correspondance particulière", L. Rosenfeld to W. Yourgrau, 20 Sep 1968. Yourgrau also refused to go to Tiflis on this occasion. W. Yourgrau to L. Rosenfeld, 7 Oct 1968.

Fock's Visit in Copenhagen

In February 1956, Rosenfeld was asked by Pergamon Press Ltd. to review Fock's recent book *Theory of Space, Time and Gravitation* (in Russian, 1955) with a view to an English translation.[145] Fock sent a personal copy to Rosenfeld, who read it, recommended it for translation, and returned some positive comments about it to Fock. On the same occasion Rosenfeld took the opportunity to address the problems, as he saw them, of de Broglie, Bohm, and Vigier's causal program in the interpretation of quantum mechanics.[146] Fock replied that he was in general agreement with Rosenfeld's criticism of "the cause and effect of the 'maladie Bohm-Vigier'," and he was also critical of Blokhintsev's ensemble interpretation of quantum mechanics.[147] In the 1930s, Fock had been one of the leading adherents of Bohr's interpretation of quantum mechanics, but during the early period of *Zhdanovschina* he retreated a bit in his defense of Bohr and complementarity. Fock maintained independently of Rosenfeld that Bohr's interpretation, including complementarity, (which he basically considered as synonymous with Heisenberg's uncertainty relation) did not violate dialectical materialism. On the other hand, he had a problem with what he conceived as a positivistic vocabulary in Bohr's writings, and Bohr's talk of "incontrollable interactions" during measurement processes. Indeed, Fock thought Bohr exaggerated the role of the measuring instrument when interpreting quantum mechanics: "one may reproach Bohr for not appreciating the necessity of making abstractions and for underestimating the fact that the aim and purpose of investigation are not the instrument readings, but the properties of the atomic object".[148] To this objection Bohr would probably repeat the quantum postulate, that is, due to the smallness of the quantum of action the quantum physicist had to include the

---

[145]RP, "Manchester folder 6 Rosbaud", P. Rosbaud to L. Rosenfeld, 22 Feb 1956.
[146]This we know from Fock's reply to Rosenfeld. RP, "Epistemology: Correspondance générale 1955–1958", V. A. Fock to L. Rosenfeld, 7 Apr 1956.
[147]RP, "Epistemology: Correspondance générale 1955–1958", V. A. Fock to L. Rosenfeld, 7 Apr 1956. Fock (1966), p. 412. See also Graham (1966) and Graham (1987) for Fock's criticism of Blokhintsev's interpretation.
[148]Fock (1957b), pp. 645–646. Fock's views are also reviewed in Graham (1966).

measurement instrument when accounting for the behavior of atomic objects. All the same, in the letter to Rosenfeld, Fock concluded that "all in all I think that we understand each other", and he hoped to see Rosenfeld soon in Leningrad or in Moscow.[149]

As mentioned above, the professor of nuclear physics Lev A. Sliv from Leningrad visited Copenhagen exactly at this time. As it happened, Sliv was a former pupil of Fock. These circumstances may have been behind the decision to invite Fock to Copenhagen, linking the renewed openness with renewed efforts of propagating the epistemological lesson of quantum theory once more among Russian physicists and philosophers. In any case, Sliv returned to Leningrad in the spring of 1956 with an invitation to Fock from Møller, and Niels Bohr sealed the invitation through a more official letter to the Soviet Academy of Sciences.[150] Fock gladly accepted the invitation which was arranged in more detail by Aage Bohr during his visit to the Soviet Union in September the same year.[151] Aage Bohr stayed in the USSR for a month with the aim of exploiting the new openness and establishing future possibilities for cooperation and exchanges between physicists in Denmark and Russia. Among other things he stayed with Fock and his wife in their country house. After Aage Bohr's return to Copenhagen in the autumn 1956, Fock wrote him a letter in December 1956 in which he expressed his sincere belief in the necessity of international relations between scientists and in unprejudiced scientific practice free from propaganda, clearly reflecting new times.[152]

During Fock's visit from 12 February till 16 March, 1957, he had many meetings with Bohr where they discussed the philosophical significance of quantum mechanics and the alleged positivistic language in Bohr's papers. Rosenfeld was not in Copenhagen at this

---

[149]RP, "Epistemology: Correspondance générale 1955–1958", V. A. Fock to Rosenfeld, 7 Apr 1956. Originally in French.

[150]BGC, "Academy of Sciences, Moscow", translation of A. N. Nesmejanov to N. Bohr, 9 Apr 1956. Copy of N. Bohr to A. N. Nesmejanov, 13 Jun 1956. Bjørnholm (1968), pp. 223–230, 224.

[151]MP, V. A. Fock to C. Møller, 8 May 1956. Copy of letter from C. Møller to V. A. Fock, 13 Jun 1956. Fock (1957a).

[152]Aage Bohr Papers, V. A. Fock to Aa. Bohr, 31 Dec 1956.

time, but Aage Bohr and Niels Bohr's assistant Aage Petersen were present at the conversations. Fock and the historian of Russian science, Loren Graham, later suggested that Fock on this occasion may have influenced Bohr to subsequently use a less positivistic vocabulary.[153] From Bohr (and Rosenfeld's) perspective, what was more important was that after their meeting Fock may have influenced the quantum epistemological discourse in the Soviet Union by his translations of Bohr's papers and by his own papers on the subject. After his visit in Copenhagen, Fock officially joined Rosenfeld's critique of the causal program by Bohm, Vigier, and de Broglie; defended Bohr against accusations of positivism; acknowledged that "the behavior of atomic objects is not separated from their interaction with the observation means"; and fully agreed with Rosenfeld that the basic principles of quantum mechanics were not at all in conflict with materialistic philosophy.[154] In the meantime, Rosenfeld was offered a position at the Nordic Institute for Theoretical Atomic Physics (NORDITA) in Copenhagen, probably partly as recognition of his efforts in communicating Bohr's interpretation of quantum theory in the quantum controversy. A cordial correspondence took place between Bohr, Rosenfeld, and Fock the following years. As a token of friendship, Fock provided NORDITA with his own copy of the oeuvre of Mikhail Lomonosov.[155] Rosenfeld and Fock were finally able to meet each other at the 9th International Conference on High Energy Physics in Kiev in July 1959, a conference organized by Blokhintsev. They valued highly their friendship; in May the following year when Fock travelled through Copenhagen, he asked Rosenfeld for a brief rendezvous in the airport.[156]

[153]Fock (1957a). Graham (1987), pp. 337–338. Fock (1960).
[154]Fock (1957b), pp. 647, 649. Fock (1958), pp. 179, 182.
[155]BSC, L. Rosenfeld to V. A. Fock, 28 Oct 1959. N. Bohr to V. A. Fock, 27 Nov 1959. V. A. Fock to N. Bohr, 18 Dec 1959. L. Rosenfeld to V. A. Fock, 30 Dec 1959. L. Rosenfeld to V. A. Fock, 30 Mar 1960. RP, "Copenhague: Correspondance générale 1950–1961", V. A. Fock to L. Rosenfeld, 18 Apr 1960. BSC, L. Rosenfeld to V. A. Fock, 25 Apr 1960. Aa. Petersen to V. A. Fock, 8 Jul 1960. Copy of letter from N. Bohr to V. A. Fock, 16 Feb 1961. V. A. Fock to N. Bohr, 19 Feb 1961.
[156]BSC, L. Rosenfeld to V. A. Fock, 28 Oct 1959. L. Rosenfeld to V. A. Fock, 25 Apr 1960.

Niels Bohr was finally able to visit the Soviet Union in 1961. During the preceding five years, Niels Bohr frequently received invitations to visit the Soviet Union, but he postponed a visit due to other duties and journeys.[157] He was accompanied by his wife Margrethe, his son Aage, and his daughter-in-law Marietta and the visit took place from 3 to 19 May. They all stayed at Landau's house, met with Fock and Tamm among others, and visited the Lebedev Institute, the Soviet Academy of Sciences, and the Kurchatov Institute of Atomic Energy, among other things. The visit may be seen as symbolizing a small triumph for Bohr's idea of an open world as well as for complementarity in a Russian context. Following Fock's promotion of Bohr's complementarity and his translations into Russian of many of Bohr's central papers, complementarity seems to have fared more easily in a Russian context during the 1960s.[158] The times were opportune for it, too. As a result of the intellectual thaw, the relationship between science and ideology was gradually reversed in a Russian context. During the first years of the 1960s, it was still too early to openly raise objections to the use of ideological criteria in science. Rosenfeld experienced the death agony of this ill-fated policy when in May 1961 he was asked by the editor of the Russian Academy of Sciences' journal to revise a paper in order to meet the ideological demands proper.[159] Rosenfeld was asked to delete his references to Ernst Mach since favorable comments on this philosopher were still not considered politically correct. Rosenfeld answered that this demand "has caused me extreme displeasure". He continued in his characteristically dramatic way:

> It is necessary to realize that scientists are very strongly attached to the great tradition of absolute freedom of expression and disregard of authority which they inherited from the Renaissance pioneers, and the vindication of which was

[157]For invitations to N. Bohr see BGC, "Academy of Sciences, Moscow", A. N. Nesmejanov to N. Bohr, 9 Apr 1956, 15 Mar 1957, undated April 1958, and D. I. Blokhintsev to N. Bohr, 2 Jun 1959.

[158]BSC, V. A. Fock-N. Bohr correspondence. Cross (1991), p. 751.

[159]Rosenfeld (1962a).

> only won after a hard fight and the martyrdom of Michel Servet, Giordano Bruno, Galileo and many others. So-called philosophers ought to remember that it is only because this tradition was firmly established even in the stronghold of capitalism that Marx and Engels were able to publish their great work. They should know that scientists resent censorship of any kind, and only feel contempt for people who have the arrogance of trying to impose such censorship upon others. This being said, you will realize that I am not prepared to remove a single word from my article. If you are not able to publish it as it is, I request you to return to me at once both the original and the translation.[160]

The paper was published *exactly* as Rosenfeld had written it.[161] By this time, scientists were still obliged to study dialectical materialism in Russia, but especially physicists began to argue that Marxist philosophers lacked scientific knowledge and often interpreted the achievements of modern science incorrectly. The result was that whereas quantum foundations the previous years was supposed to adapt to the demands of the official Marxist–Leninist ideology, Marxist philosophers now began to adapt their philosophy to the lessons and new concepts of modern science. As a result, the epistemological position in dialectical materialism gained ground on account of the ontological position. This development paired with Fock's writings and translations of Bohr further resulted in the accusation of philosophical idealism which had previously been raised against Bohr's complementarity disappeared from public discourse.[162]

---

[160]RP, "History of science 7: Quantum mechanics, Quantum theory of radiation", Copy of letter from L. Rosenfeld to A. Grigorian, 26 May 1961.

[161]RP, "History of science 7: Quantum mechanics, Quantum theory of radiation", A. Grigorian to L. Rosenfeld, 13 Jun 1961.

[162]Ivanov (2002), pp. 318, 328–329. Graham (1987), p. 59. It appears from a report by the Danish physicists, H. H. Jensen, H. G. Jensen, B. Madsen, O. Nathan, and M. Pihl, who visited the Soviet Union in 1962 in order to study the Russian system of higher education and research in physics, that, besides physics courses, physicists were taught political subjects such as the history of the Communist Party, political economy, dialectical materialism, and historical materialism. These subjects were evaluated by obligatory examinations. Aage Bohr Papers, "Samarbejde USSR/DK (Diverse)".

Rosenfeld sometimes compared what he called the "pseudo-Marxists'" use of the quantum formalism leaving out complementarity with the "cunning" Jesuit mathematicians' use of Copernicus' heliocentric cosmos in the sixteenth and seventeenth centuries. Contrary to Bohr, both Bohm and Blokhintsev emphasized a sharp distinction between the mathematical equations of the quantum theory and the interpretation of the theory. According to the Jesuits' catholic faith, they had to condemn a heliocentric cosmos that reduced the Earth to an ordinary planet, but they claimed to be able to use Copernicus' system nonetheless for calculational purposes, not taking into account what they conceived as an erroneous interpretation of the theory.[163] The day before he died, Bohr suggested a similar analogy when he compared the slow reception of the Copernican system in the 16th and 17th centuries with the reception of complementarity in an attempt to elucidate his criticism of philosophers, *viz.*, that rather than adapting their philosophy to the newest achievements of science, which was to be preferred, history showed that philosophers often attempted to adapt science to prevalent philosophical schools. Bohr was confident that his complementarity description would survive in the long run despite the initial opposition against it, just as had the Copernican system.[164]

Rosenfeld repeated the comparison between the Jesuits and the Russian philosophers in a review in 1969 in his journal *Nuclear Physics*.[165] The review concerned the English and French translations of Blokhintsev's book *The Philosophy of Quantum Mechanics* (1968), originally published in 1965. Blokhintsev's book is essentially a presentation of his ensemble interpretation. The novelty in the book was that he reluctantly sought reconciliation with Bohr's complementarity as a generalization of Heisenberg's uncertainty relation:

> Bohr himself formulated the principle in a somewhat different way, reflecting his philosophical concepts which were far from those of materialism. His formulation has been the origin

[163]BGC, L. Rosenfeld to N. Bohr, 14 Oct 1958.
[164]Kuhn *et al.* (1962), pp. 3–4.
[165]Frank (1974).

> of the far-reaching conclusion that the current mechanics of the atom cannot be compatible with materialism... It would seem generally better to speak of a principle of exclusiveness rather than complementarity: dynamic variables should be divided into mutually exclusive groups, which do not coexist in real ensembles. However, out of respect for Bohr and his tradition we shall retain the usual terminology.[166]

However feeble an attempt at reconciliation this was, it nevertheless seems to have satisfied Rosenfeld. "This publication", he stated in his review, "marks the happy ending of an extraordinary episode in the history of quantum theory".[167] The review was mainly Rosenfeld's sarcastic comment to the fact that *Zhdanovschina* had outlived its role in physical science so that now the Russian physicists were finally able to recognize complementarity, which he referred to as "the objective state of affairs". His main criticism of Blokhintsev had concerned Blokhintsev's denial of wave–particle duality and complementarity and his ideological motivation. In general, Rosenfeld mainly criticized hidden variables and idealistic interpretations of quantum physics; he did not mind statistical interpretations as long as complementarity and indeterminism were acknowledged.

I conclude with a brief epilogue about Rosenfeld's later years.

---

[166] Blokhintsev (1968), p. 22. See also Graham (1987), pp. 333–336.
[167] Rosenfeld (1969b).

Epilogue

Rosenfeld's vitriolic remarks in the continued quantum controversy are well-known and do not always leave him a flattering portrait.[1] He continued for the rest of his life to fiercely defend Bohr's complementarity interpretation and vigorously engage in discussions about the measurement problem. It was no longer communists under *Zhdanovschina* Rosenfeld criticized; the primary mission that lay before him was to make sure that the new generations of physicists adopted what he considered to be the correct view of quantum foundations. However, his style may have resulted in the opposite effect of what he evidently intended by this campaign, so that colleagues felt repelled by his attitude rather than encouraged to study Bohr. He showed no mercy with younger physicists who wanted to contribute to the philosophical foundations of quantum theory on the basis of misunderstandings of Bohr.[2] In that sense, Rosenfeld did not appear especially open-minded about the development of the interpretation of quantum mechanics and he was sensitive to that point, "I realize that I may give the impression of being 'intolerant' and 'dogmatic', in spite of the fact that I have been all these years advocating tolerance and free thinking", as he wrote to Lancelot Law Whyte in 1958.[3] Rosenfeld was definitely perceived by the scientific and philosophical communities as Bohr's representative or heir, and he was therefore constantly approached concerning questions about Bohr and the

[1] See for example, Pessoa Jr. *et al.* (2008), Osnaghi *et al.* (2009). Freire Jr. (2007).
[2] The Tausk affair is a good example, Pessoa Jr. *et al.* (2008). Other examples include RP, "Supplement Box 4, M", L. Rosenfeld to P. A. Moldauer, 9 Jun and 7 Jul 1971. About a preprint by H. D. Zeh, see RP, "Epistemology 1967–1968", L. Rosenfeld to H. H. D. Jensen, 14 Feb 1968. "Supplement V", L. Rosenfeld to V. Votruba, 24 Oct 1966.
[3] RP, "Epistemology: Correspondance générale 1955–1958", Draft of letter from L. Rosenfeld to L. L. Whyte, 17 Mar 1958.

philosophical foundations of quantum theory. Rosenfeld responded to all these approaches by indefatigably attempting to correct the misunderstandings about Bohr's views. In his preaching of the subject, he clearly enjoyed telling the standard (Copenhagen) story about the emergence of quantum theory and its interpretation. In a lecture note from the 1960s, he compared Bohr's role in the development of quantum mechanics in the 1920s with "King Arthur's Ride".[4]

However, Rosenfeld could not prevent new interpretations of quantum mechanics from appearing. New challenges to the Copenhagen Interpretation kept popping up, such as, "Everett's damned nonsense", as Rosenfeld called Hugh Everett's Many World Interpretation from 1956 in private correspondence.[5] In the 1960s, Bohm's work inspired John Stewart Bell, an Irish physicist at CERN, to develop a series of papers rethinking quantum mechanics in terms of hidden parameters.[6] This development shows that the significance of Bohm's interpretation was not limited to the episode of *Zhdanovschina.*[7] Famously, Bell derived an inequality which if violated stated that the existence of local hidden variables, such as suggested in the EPR paper, is not compatible with quantum mechanics. The Bell inequality was an explicit theoretical prediction that began to be tested in the laboratory by EPR-like experiments from around 1970.[8] Rosenfeld's negative reaction to Bell's work was predictable. In 1966, he frankly told Bell that "I need not tell you that I regard

---

[4]RP, Supplement: "Evolution of scientific thought", one sheet with notes for a lecture on development of non-relativistic quantum mechanics. Loinger is mentioned, it must therefore be from the 1960s or early 1970s.

[5]Osnaghi *et al.* (2009). RP, "Supplement B", L. Rosenfeld to J. S. Bell, 30 Nov 1971. Rosenfeld discussed Everett's interpretation with F. J. Belinfante in the early 1970s. RP, Supplement.

[6]Olwell (1999), pp. 755–756. Bell's first articles which appeared between 1964 and 1966 bear witness to Bohm's influence, Bell (1964), Bell (1966).

[7]In fact, Bohm took up his hidden variables interpretation again in the late 1970s with Basil Hiley. They combined the hidden variables interpretation with Bohm's ideas of implicate and explicate order. Bohm's interpretation seems to have proved quite viable and has only gained in popularity over the years. Freire Jr. (2011).

[8]D'Espagnat (1979). This series of experiments culminated with a famous experiment by the French physicist Alain Aspect in 1982 that demonstrated a violation of the Bell inequality. See also Bromberg (2006) and Bromberg (2008) as regards experiments related to the foundations of quantum mechanics.

your hunting hidden parameters as a waste of your talent; I don't know... whether you should be glad or sorry for that".[9] Rosenfeld came to consider Bell as "one of the very few heretics from whom I always expect to learn something" which, however, may have been an ironic remark.[10]

Rosenfeld kept the following quote by the Danish existentialist philosopher Søren Kierkegaard in his drawer:

> "I prefer to speak to children; since you (may)
> dare to hope that they can become rational beings;
> but those who have done so — Good Lord!"[11]

It fittingly reflects Rosenfeld's concern with conveying the correct Bohrian interpretation of quantum mechanics to the youth and his quarrels with Bohr's critics. However, the fact that he kept this particular quote does not mean that he was attracted to Kierkegaard's philosophy, on the contrary.[12] In the 1960s, Bohm had learned through Bohr's former assistant Aage Pedersen, who held a position at the Yeshiva University, New York, that Bohr should have been influenced by Kierkegaard. This gave rise to resumed discussions about Kierkegaard and the influence of religious and idealistic thought on Bohr's thinking in the correspondence between Bohm and Rosenfeld. Rosenfeld rejected that Bohr should have been influenced by Kierkegaard's philosophy. By the 1960s, however, Rosenfeld and Bohm adopted a much more agreeable and relaxed tone compared to the 1950s, which may also reflect the decreased political tension in the world.[13] Rosenfeld grew quite fond of Bohm's writings, particularly his book *The Special Theory of Relativity* (1965).[14]

---

[9]RP, "Measuring Process 1966", L. Rosenfeld to J. S. Bell, 2 Dec 1966.
[10]RP, "Supplement B", L. Rosenfeld to J. S. Bell, 30 Nov 1971.
[11]Kierkegaard (1994), Vol. 1, p. 23. Originally in Danish.
[12]Rosenfeld was of the opinion that Kierkegaard's "epistemological pretensions... amount to a wholesale rejection of scientific thinking". RP, "Measuring Process Nov 1966–Jan 1967", L. Rosenfeld to A. Loinger, 23 Jan 1967.
[13]See for example RP, "Measuring Process", D. Bohm to L. Rosenfeld, 13 Dec 1966.
[14]Bohm (1965). RP, "Epistemology (1965–1966)", L. Rosenfeld to B. van Rootselaar, 22 Nov 1965. See also RP, "Cambridge Colloquium: Complementarity & beyond (1967–1968)", L. Rosenfeld to D. Bohm, 18 Apr 1967, about a talk by Bohm on creativity. Freire Jr. (2011).

Bohm's later work took inspiration from Eastern mysticism in reaching a holistic view on quantum physics. By this time, however, Rosenfeld was no longer in the world to criticize it.

Due to his concern for the next generations of physicists adopting the orthodox view of quantum theory, Rosenfeld was particularly attentive to textbooks on quantum mechanics.[15] In 1957, he was asked by Academic Press of America to contribute with some introductory chapters to a textbook on quantum theory for postgraduate students. Rosenfeld declined the invitation because he was too busy at the time. However, because the book was to include a final chapter on "Hidden variables", Rosenfeld burst out to the editor, D. R. Bates:

> Could this possibly mean that you are smitten with Bohmitis? I hope rather that your intention to include a discussion of this point arises from the scruple of giving every idea its chance however wrong you may think it is. As you probably know, I have strong views about this as a result of my experience with students. Bohm's ideas, apart from being demonstrably wrong in many points, are entirely unscientific in as much as they are based on metaphysical speculation, not supported by any experimental evidence whatsoever. They are, quite rightly, ignored by the majority of physicists. To give them a place in a textbook on quantum theory would therefore present the students with a distorted picture of the true situation in physics, and with the worst possible example.[16]

When the book entitled *Quantum Theory* was finally published in two volumes in 1961, Rosenfeld's colleague in Copenhagen, Stefan Rozental, found it scandalous that it included a long article by Bohm on hidden variables, and no article expressing the orthodox view. He therefore asked Rosenfeld to review it so that the review would not

[15] See for example Rosenfeld (1979).

[16] RP, "Manchester 1947–1958 folder 6: Correspondence about publications (1949–1958)", copy of letter from L. Rosenfeld to D. R. Bates, 22 May 1957.

be too gentle! Rosenfeld accepted to review it, but to my knowledge a review by Rosenfeld never appeared.[17]

The growing community in the field of quantum foundations took exception to Rosenfeld's attitude. In 1970, the first Varenna Summer School about the foundations of quantum mechanics took place — organized under the auspices of the Italian Society of Physics by the French physicist Bernard d'Espagnat, who was a former student of Louis de Broglie. Rosenfeld was invited to speak about the measurability of quantum fields and gladly accepted.[18] However, upon receiving the program for the summer school he withdrew his participation. With the program d'Espagnat enclosed a "Letter to the Participants" in which he proposed the following agreement:

> we should not take as our goal the conversion of the heretic but rather a better understanding of his standpoint; ... we should not suggest that we consider as a stupid fool anybody in the audience (lest the stupid fools should in the end appear clearly to be ourselves!); ... we should try to cling to facts; and ... nevertheless we should be prepared to hear without indignation very non-conformist views which have no *immediate* bearing on facts.[19]

When introduced to these intentions for the summer school, Rosenfeld responded that he refused to involve himself in "an adventure" which, according to him, would only provide "bad service to the young generation".[20] The list of speakers that d'Espagnat had assembled included Wigner, J. N. Jauch, Abner Shimony, Giovanni Maria Prosperi, Bell, Bryce S. DeWitt, H. D. Zeh, de Broglie, and Bohm, among others. Rosenfeld foresaw that in this forum and on the background of the above mentioned agenda, his very person would clearly stand in the way of the message he wanted to convey. In order to avoid this, he sent the Copenhagen physicist Jørgen Kalckar in his

[17]Bates (1961–1962). RP, "Copenhague 4", S. Rozental to L. Rosenfeld, 15 Feb 1962.
[18]RP, Supplement box 2, B. d'Espagnat to L. Rosenfeld, 25 Aug 1969. Copy of letter from Rosenfeld to d'Espagnat, 29 Aug 1969.
[19]D'Espagnat (1971), p. xiii. Emphasis in original.
[20]RP, Supplement box 2, Rosenfeld to d'Espagnat, 23 Dec 1969.

place.[21] D'Espagnat regretted Rosenfeld's decision not to participate and denied that the direction the meeting had taken was pointing at Rosenfeld in any way. Despite this incident and their general disagreements about quantum foundations, Rosenfeld and d'Espagnat seem to have stayed on friendly terms marked by mutual respect. It appears from their correspondence that they at least managed to find common ground in getting frustrated about American editors! In 1971 Rosenfeld arranged a large symposium about statistical causality in Copenhagen in recognition of the 50th anniversary of Niels Bohr's institute. D'Espagnat was invited along with Jean Piaget, Ilya Prigogine, Wladyslaw Opechowski, Jauch, Evgenii Lifshitz, Oskar Klein, and many more of Rosenfeld's old friends and colleagues from around the world, including of course all the Copenhagen physicists.[22]

Rosenfeld not only defended Bohr's interpretation of quantum physics; he also developed his own independent view on the macroscopic behavior of large quantum systems and the measurement problem.[23] Although Bohr recognized that macroscopic entities such as measuring apparatuses are composed of atoms, and atoms should be described by means of quantum mechanics, he maintained that macroscopic things should be described by classical physics on the basis of considerations such as the following from 1948:

> [I]t may be remarked that the construction and the functioning of all apparatus like diaphragms and shutters, serving to define geometry and timing of the experimental arrangements, or photographic plates used for recording the localization of atomic objects, will depend on properties of materials which are themselves essentially determined by the quantum of action. Still, this circumstance is irrelevant for the

[21]RP, Supplement box 2, D'Espagnat to Rosenfeld, 11 Dec 1969. Rosenfeld to d'Espagnat, 23 Dec 1969. Kalckar (1971).

[22]RP, Supplement box 6, "Symposium on Statistical Causality April 1971", B. d'Espagnat to L. Rosenfeld, 8 Jan 1971. Supplement box 2, d'Espagnat to Rosenfeld, 17 Oct 1971. Rosenfeld to d'Espagnat, 19 and 28 Oct 1971.

[23]This has not previously been recognized. See for example Freire Jr. (2007) and Freire Jr. (2009), p. 285.

> study of simple atomic phenomena where, in the specification of the experimental conditions, we may to a very high degree of approximation disregard the molecular constitution of the measuring instruments. If only the instruments are sufficiently heavy compared with the atomic objects under investigation, we can in particular neglect the requirements of [the uncertainty] relation [of momentum and position] as regards the control of the localization in space and time of the single pieces of apparatus relative to each other.[24]

Thus, macroscopic instruments must be so constructed that their action ultimately depends solely on its macroscopic features, in which Planck's constant plays no part. In that way they can be treated classically. In the complex idealized measurement described in the Bohr–Rosenfeld paper of 1933, the atomic structure of the extended test charge could be neglected, which permitted a classical treatment of it as a measuring device in line with Bohr's thought.

In his famous book from 1932, *Mathematische Grundlagen der Quantenmechanik*, von Neumann formulated what has come to be known as the standard measurement problem in quantum theory.[25] It arises when the measurement is treated as a quantum mechanical interaction between an atomic object and a measuring apparatus, and the measuring apparatus is described as a *quantum* mechanical system. Quantum evolution is linear and therefore a so-called superposition of eigenstates of the quantity to be measured will — due to the correlation between the quantum system and the measurement apparatus when they interact — be reflected by the corresponding observable of the measuring apparatus, such as a pointer in a meter, ending up being described by a superposed state. However, superposed states for *macroscopic* things such as pointer needles are never observed. The so-called measurement problem therefore involves reconciling the lack of macroscopic superposition in our world with a quantum theory that predicts such superpositions. Von Neumann

---

[24]Bohr (1996d), pp. 333–334 [315–316].
[25]von Neumann (1932), pp. 222–237.

discussed the solution of the problem by including more elements of the surroundings in the description of the measuring process, ending with the retina in the eyes of the observer that performed the experiment. Still, the quantum mechanical description would not bring about the collapse of the superposed macrostate and all he ended up with was an infinite regression of reasoning. In order to link quantum theory with what we actually perceive during a measurement, von Neumann added the postulate of what has later been called "the reduction of the wave packet" to quantum theory. The reduction postulate accounts for the development of the superposed state of a quantum system into a single eigenstate which happens during a measurement. This development of the quantum state differs from the continuous, reversible, and deterministic development of the state given by the Schrödinger equation in being indeterministic, discontinuous, and irreversible. The question now became, when does the collapse happen in the measurement process, if it happens. Is it really a process that involves the observer and is triggered by the retina or neurological processes in the brain? Is it a process that the quantum system to be measured undergoes during the interaction with the measurement apparatus, or is it a physical process that takes place in the measurement apparatus?

Rosenfeld fully recognized von Neumann's mathematical treatment of the statistics of quantum systems in his 1932 book in terms of the density matrix formalism. However, he deplored the way von Neumann involved the observer in the measurement process. As we would expect, Rosenfeld therefore also criticized Eugene Wigner's continuation of the line of thought von Neumann had initiated by introducing the role of consciousness in the measurement process in the 1960s.[26] Rosenfeld instead preferred "to speak in terms of fully

[26]Wigner (1961). Wigner (1963). See also In Freire Jr. (2007) in which Rosenfeld's "bitter arguments" against Wigner's position are described. Correspondence in RP suggests, however, that Rosenfeld and Wigner could maintain a good personal and colleagual relationship despite their at times direct and blunt criticism of each other's philosophical positions. Rosenfeld proposed Wigner for the Nobel Prize for 1963. See for example RP, "Epistemology 2: Complementarity and Measurement in quantum mechanics", L. Rosenfeld to P. A. Moldauer, 7 Nov 1963, L. Rosenfeld to K. Gottfried 24 Feb and 8 Mar 1967, and RP, Supplement, L. Rosenfeld to A. Loinger, 13 Jun 1972.

automatic registration, so that we keep out the human observer and all the problems that are connected with that aspect of the problem", in accordance with his materialistic standpoint.[27] In 1955, Rosenfeld prudently suggested that quantum measurement could be viewed analogously to irreversible processes in statistical thermodynamics, although "strictly speaking, the consideration of this kind of irreversibility falls outside the scope of thermodynamics, since it is connected with (ideal) processes affecting single atomistic elements".[28]

In the 1960s Rosenfeld joined forces with three Italian physicists, Angelo Loinger, Adriana Daneri, and Prosperi in proposing a theory of the measuring apparatus. In this theory, the reduction of the superposed state and the macroscopic character of the recording process were explained by treating macroscopic measuring apparatuses as thermodynamic systems obeying so-called ergodic behavior.[29] The reduction of the quantum state of the observed atomic object was not brought about by its interaction with the macroscopic measuring device but by a thermodynamical process in the latter *after* the interaction had ceased and the measurement apparatus had recorded the result of the measurement.[30] Therefore, according to Rosenfeld, "the reduction rule is not an independent axiom but essentially a thermodynamic effect".[31] Rosenfeld regarded the work of the Italians as continuing the line of reasoning with respect to the consistency of the use of classical concepts in quantum theory initiated by Bohr, which also lay behind the Bohr–Rosenfeld papers: "The Italians have carried out a similar proof of consistency for the analysis of the process of measurement in quantum mechanics, which also consists of a set of idealizations, a set of equations involving these, and rules of interpretation inseparable from the other elements. In particular, what they have beautifully shown is that the idealization 'macroscopic body' enters into the argument through the property of such bodies expressed by

---

[27] Rosenfed (1979s), p. 556. See also Rosenfeld (1979x), p. 538.

[28] Rosenfeld (1979o), pp. 799–801 [p. 799]. See also RP, "Epistemology 1955–1958", copy of letter from L. Rosenfeld to A. Grünbaum, 14 Feb 1956.

[29] Daneri *et al.* (1962). Rosenfeld (1979x), p. 540. Rosenfeld (1979p).

[30] Rosenfeld (1979x), pp. 545–546.

[31] RP, Supplement, L. Rosenfeld to F. Belinfante, 24 Jul 1972.

the ergodic theorem".[32] In this way Bohr's requirement to treat measurement devices classically could still be uphold. Rosenfeld found consistency between the predictions of quantum theory about the reduction of the quantum state during measurement with the Italians' argument from statistical thermodynamics for the description of the behavior of the measuring apparatus.[33] With this interpretation, Rosenfeld maintained that the "misunderstandings which go back to deficiencies in von Neumann's axiomatic treatment, have... been completely removed".[34] Moreover, "from the epistemological point of view, this completion of the process is of fundamental significance", Rosenfeld maintained, in accordance with Bohr's views, because "it corresponds to a sharp separation of object and subject, without which no intelligible statement can be made by the subject about the object".[35] Rosenfeld continued this line of thought when in 1970 he began research into the epistemological aspects of the time evolution of large quantum systems in collaboration with a team in Brussels under the leadership of Ilya Prigogine and Claude George.[36] In using this work for a discussion of the measuring process, Rosenfeld treated the measuring device as a macroscopic *quantum* system. Thus, by this time Rosenfeld's interpretation of the measurement process therefore appears to have deviated from Bohr's, since Bohr had maintained that macroscopic things such as measuring devices should be described classically.[37]

Let me conclude with a few words about Rosenfeld's commitment to research in the history of physics, including the history of quantum physics in which he had himself been an actor. Thomas S. Kuhn admired Rosenfeld's historical work and since he knew that Rosenfeld was largely responsible for arousing Bohr's interest in the preservation of historical records, he involved Rosenfeld in the project Sources

[32]RP, "Measuring Process Nov 1966–Jan 1967", L. Rosenfeld to D. Bohm, 19 Jan 1967.
[33]Rosenfeld (1979x).
[34]*Ibid.*, p. 537. See also Rosenfeld (1979s).
[35]RP, "Epistemology: correspondance générale", L. Rosenfeld to I. Bloch, 7 Oct 1958. (emphasis in original). See also Rosenfeld (1979x), p. 538.
[36]Rosenfeld *et al.* (1979). RP, Supplement, Rosenfeld–Prigogine correspondence.
[37]RP, Supplement, L. Rosenfeld to J. Wheeler, 3 Jul 1973.

for History of Quantum Physics (1962–1964) from very early on.[38] The aim of the project was to ensure the preservation of oral and written records for the history of quantum physics and related fields. Rosenfeld was instrumental in establishing the facilities in Copenhagen for the project of interviewing physicists beginning in the fall of 1962. In addition, Rosenfeld mentioned to Kuhn in early 1962, "that there was a real likelihood of establishing an archive under Professor Bohr's auspices in Copenhagen".[39] About a year later, the Niels Bohr Archive was established on an informal basis by Aage Bohr who was then the Director of the Institute for Theoretical Physics. Niels Bohr had died in the meantime. During the 1960s, Rosenfeld started the project of publishing the Niels Bohr Collected Works. However, the archive was not only related to Bohr's work but to the development of modern physics in general, and in the same period Rosenfeld also pursued other collections for the archive such as George Hevesy's Papers and Hendrik Kramer's Papers.[40] Rosenfeld served as the leader of the archive until he died on 23 March 1974. He had then been ill for some time following a heart attack in February 1973.[41]

---

[38] Kuhn *et al.* (1967).

[39] RP, "History of Science 4: Sources for History of Quantum Physics (1961–1964)", T. S. Kuhn to L. Rosenfeld, 3 Jan 1962.

[40] RP, "History of Science 4: Sources for History of Quantum Physics (1961–1964)".

[41] Robert Cohen Archive, Boston University Library, Boston. Correspondence between R. S. Cohen and L. and Y. Rosenfeld.

Bibliography

The footnotes for each chapter give detailed information on the unpublished sources and information in abbreviated form on published sources. This section provides the basic data on the source material used.

Abbreviations

AHQP, Archive for the History of Quantum Physics.
AIP, American Institute of Physics, College Park, Maryland, USA.
BCW, *Niels Bohr Collected Works*, 12 Vols., (eds.) L. Rosenfeld, E. Rüdinger and F. Aaserud. (Elsevier, Amsterdam, 1972–2007).
BP, John Desmond Bernal Papers, Cambridge University Library, Cambridge.
BGC, Niels Bohr General Correspondence, NBA.
Bohr MSS, Niels Bohr Manuscripts.
BPP, Niels Bohr Political Papers, NBA.
BPC, Niels Bohr Private Correspondence, NBA.
BSC, Niels Bohr Scientific Correspondence, NBA.
CP, Subrahmanyan Chandrasekhar Papers, Special Collections Research Center, University of Chicago Library.
DCW, *The Collected Works of P. A. M. Dirac 1924–1948*, (ed.) R. H. Dalitz (Cambridge University Press, Cambridge, 1995).
MP, Christian Møller Papers, NBA.
NBA, The Niels Bohr Archive, Copenhagen.
RP, Léon Rosenfeld Papers, NBA.
SP, *Selected Papers of Léon Rosenfeld*, Boston Studies in the Philosophy of Science, Vol. 21, (eds.) R. S. Cohen and J. J. Stachel (D. Reidel, Dordrecht, 1979).

Archival Document Collections

I have used the resources of the following archives:

Albert Einstein Archives, Jewish National and University Library, Jerusalem.

Arbejdermuseet and Arbejderbevægelsens Bibliotek og Arkiv, Copenhagen.
Archive de L'Académie des Sciences, Institute de France and Fondation Louis de Broglie, Paris.
Archives Joliot-Curie, Fonds Frédéric Joliot-Curie, Paris.
David Joseph Bohm Papers, Birkbeck College, Library, London.
J. G. Crowther Archive, University of Sussex Library, Special Collections, Brighton.
Johannes (Jan) Martinus Burgers Papers, Internationaal Instituut voor Sociale Geschiedenis, Amsterdam.
John Desmond Bernal Papers, Cambridge University Library, Cambridge.
P. A. M. Dirac Papers, Paul A. M. Dirac Science Library, Florida State University, Tallahassee.
Robert S. Cohen Papers, Boston University Library, Boston.
Sir Rudolf Peierls Papers, The Bodleian Library, University of Oxford, Oxford.
Staatsbibliothek, Preussischer Kulturbesitz, Berlin.
Subrahmanyan Chandrasekhar Papers, Special Collections Research Center, University of Chicago Library.
The American Philosophical Society, Philadelphia.
The Niels Bohr Archive, Copenhagen.
The Niels Bohr Library, AIP.
U.S. Department of Justice, Federal Bureau of Investigation, Washington, D.C., USA.
Werner Heisenberg Nachlass, Werner–Heisenberg Institute, Munich.

Books, Articles, Interviews, and Theses

Aaserud, F. (1990). *Redirecting Science: Niels Bohr, Philanthropy and the Rise of Nuclear Physics* (Cambridge University Press, Cambridge).
Aaserud, F. (1999). The Scientist and the Statesmen: Niels Bohr's Political Crusade During World War II, *Historical Studies in the Physical and Biological Sciences*, **30**(1), pp. 1–47.
Aaserud, F. (2005). Introduction, *BCW*, Vol. 11, pp. 3–83.
Aaserud, F. (2007). Introduction to Part II, *BCW*, Vol. 12, pp. 97–130.
Anderson, P. (1976). *Considerations on Western Marxism* (NLB, London).
Anderson, P. (1980). *Arguments within English Marxism* (Perry Anderson, London).
Arblaster, P. (2006). *A History of the Low Countries* (Palgrave Macmillan, New York).
Bacciagaluppi, G. and Valentini, A. (2008). *Quantum Theory at the Crossroads: Reconsidering the 1927 Solvay Conference* (Cambridge University Press, Cambridge).
Badash, L. (1995). *Scientists and the Development of Nuclear Weapons: From Fission to the Limited Test Ban Treaty 1939–1963* (Humanities Press, New Jersey).

Baneke, D. (2008). *Synthetisch Denken: Natuurwetenschappers over hun rol in een Moderne Maatschappij, 1900–1940* (Uitgeverij Verloren, Utrecht).

Baratta, J. P. (2004). *The Politics of World Federation*, 2 Vols. (Praeger, London).

Bates, D. R. (1961–1962). *Quantum Theory*, 2 Vols. (Academic Press, New York).

Baudet, J. C. (2007). *Histoire des Sciences et de l'Industrie en Belgique* (Jourdan Editeur, Brussels).

Bell, J. S. (1964). On the Einstein–Podolsky–Rosen Paradox, *Physics*, **1**, pp. 195–200.

Bell, J. S. (1966). On the Problem of Hidden Variables in Quantum Theory, *Review of Modern Physics*, **38**, pp. 447–452.

Bell, J. S. (1982). On the Impossible Pilot Wave, *Foundations of Physics*, **12**(10), pp. 989–999.

Bernal, J. D. (1939). *The Social Function of Science* (Routledge, London).

Bernal, J. D. (1954). *Science in History* (Watts and Co, London).

Bertomeu-Sánchez, J. R. and Nieto-Galan, A. (eds.) (2006). *Chemistry, Medicine, and Crime: Mateu J. B. Orfila (1787–1853) and His Times* (Science History Publications, Sagamore Beach).

Beyerchen, A. D. (1977). *Scientists under Hitler: Politics and the Physics Community in the Third Reich* (Yale University Press, New Haven).

Beyler, R. H. (1994). From Positivism to Organicism: Pascual Jordan's Interpretations of Modern Physics in Cultural Context, Doctoral Dissertation, Harvard University, Cambridge, Mass.

Biquard, P. (1965). *Frédéric Joliot-Curie. The Man and His Theories*, Tr. G. Strachan (Souvenir Press, London).

Bjørnholm, S. (1968). Samarbejde med Sovjetunionen inden for atomfysiken, *Nordisk Forum: Tidsskrift for Universitets- og Forskningspolitikk*, **4**, pp. 223–230.

Blackett, P. M. S. (1948). *Military and Political Consequences of Atomic Energy* (Turnstile, London).

Blokhintsev, D. I. (1951). *Uspekhi Fizicheskikh Nauk*, **44**(1).

Blokhintsev, D. I. (1952). Critique de la conception idéaliste de la théorie quantique, *Questions Scientifique*, tome 1 (Physique), Paris: Les editions de la nouvelle critique, pp. 95–129.

Blokhintsev, D. I. (1953a). *Grundlagen der Quantenmechanik* (Deutscher Verlag der Wissenschaften, Berlin).

Blokhintsev, D. I. (1953b). Kritik der philosophischen Anschauungen der sogenannten "Kopenhagener Schule" in der Physik, *Sowjetwissenschaft, Naturwissenschaftliche Abteilung*, **6**(4), pp. 546–574.

Blokhintsev, D. I. (1968). *The Philosophy of Quantum Mechanics* (D. Reidel, Dordrecht).

Boeselager, W. F. (1975). *The Soviet Critique of Neopositivism: The History and Structure of the Critique of Logical Positivism and Related Doctrines by Soviet Philosophers in the Years 1947–1967* (Reidel, Dordrecht).

Bohm, D. (1952). A Suggested Interpretation of Quantum Mechanics in Terms of "Hidden" Variables. I and II, *Physical Review*, **85**(2), pp. 166–193.

Bohm, D. (1957a). A Proposed Explanation of Quantum Theory in Terms of Hidden Variables at a Sub-Quantum-Mechanical Level, in *Observation and Interpretation: A Symposium of Philosophers and Physicists*, Proceedings of the Ninth Symposium of the Colston Research Society held in the University of Bristol April 1–4, (ed.) S. Körner (Butterworths Scientific Publications, London), pp. 33–40.

Bohm, D. (1957b). *Causality and Chance in Modern Physics* (Routledge and Kegan Paul, London).

Bohm, D. (1965). *The Special Theory of Relativity* (W. A. Benjamin, Inc., London).

Bohr, Aa. (1974). Professor Léon Rosenfeld, *Berlingske Tidende*, 26 March.

Bohr, H. (2008). *Nogle Erindringer om Familien* (Bohr Family, Frederiksberg), NBA.

Bohr, N. (1929). *Atomteori og Naturbeskrivelse*, Festskrift udgivet af Københavns Universitet i Anledning af Universitets Aarsfest November 1929 (Bianco Luno, Copenhagen).

Bohr, N. (1932). *La Théorie Atomique et la Description des Phénomènes*, Tr. A. Legros and L. Rosenfeld (Gauthier-Villars, Paris).

Bohr, N. (1934). *Atomic Theory and the Descriptions of Nature* (Cambridge University Press, Cambridge).

Bohr, N. (1949). Atom Alderen: Skandinavisk-Britisk Udstilling, Charlottenborg 1. til 18. September 1949, *Een Verden*, pp. 5–6. Originally published in English, A Challenge to Civilization, *Science*, 102, 363–364 (1945). English version in *BCW*, Vol. 11, pp. 125–129.

Bohr, N. (1971). *Izbrannye nauchnye trudy*, T.2, (Moskva, 1971). The Russian spelling is Nils Bor.

Bohr, N., Kramers, H. A. and Slater, J. C. (1984). The Quantum Theory of Radiation, *BCW*, Vol. 5, pp. 101–118. Originally published in *Philosophical Magazine*, **47**, 785–802 (1924).

Bohr, N. (1985a). Faraday Lecture: Chemistry and the Quantum Theory of Atomic Constitution, *BCW*, Vol. 6, pp. 371–384. Originally published in *Journal of the Chemical Society, London*, (1932), pp. 349–384.

Bohr, N. (1985b). Maxwell and Modern Theoretical Physics, *BCW*, Vol. 6, pp. 359–360. Originally published in *Nature* (*Suppl.*), **128**, 691–692 (1931).

Bohr, N. (1985c). The Quantum Postulate and the Recent Development of Atomic Theory, *BCW*, Vol. 6, pp. 147–158. Originally published in *Nature*, (Suppl.), **121**, 580–590 (1928).

Bohr, N. (1986). Atomic Stability and Conservation Laws, *BCW*, Vol. 9, pp. 99–114. Originally published in *Atti del convegno di Fisica Nucleare della Fondazione Alessandro Volta*, Oct. 1931 (Reale Accademia d'Italia Rome, 1932), pp. 119–130.

Bohr, N. (1996a). Can Quantum Mechanical Description of Physical Reality be Considered Complete?, *BCW*, Vol. 7, pp. 292–298. Originally published in *Physical Review*, **48**, 696–702 (1935).

Bohr, N. (1996b). Discussion with Einstein on Epistemological Problems in Atomic Physics, *BCW*, Vol. 7, pp. 339–381. Originally published in *Albert Einstein: Philosopher — Scientist*, The Library of Living Philosophers,

Vol. 7, (ed.) P. A. Schilpp (Evanston, Illinois, 1949), pp. 201–241.

Bohr, N. (1996c). Field and Charge Measurements in Quantum Theory, unpublished manuscript (1937), printed in *BCW*, Vol. 7, pp. 195–209.

Bohr, N. (1996d). On the Notions of Causality and Complementarity, *BCW*, Vol. 7, pp. 325–337. Originally published in *Dialectica*, **2**, 312–319 (1948).

Bohr, N. (1999a). Analysis and Synthesis in Science, *BCW*, Vol. 10, pp. 63–64. Originally published in *Encyclopedia and United Science, Foundations of the Unity of Science*, Vol. 1, No. 1, *International Encyclopedia of Unified Science*, **1**, (1938), p. 28.

Bohr, N. (1999b). Biology and Atomic Physics, *BCW*, Vol. 10, pp. 52–62. Originally published in *Celebrazione del second centenario della nascita di Luigi Galvani*, Bologna 18–21 ottobre 1937-XV: I. Rendiconto generale, Tipografia Luigi Parma 1938, pp. 68–78.

Bohr, N. (1999c). Causality and Complementarity, Address at the Second International Congress for the Unity of Science in Copenhagen 21–26 June 1936. *BCW*, Vol. 10, pp. 39–48. Originally published in *Philosophy of Science*, **4**(3), 289–298 (1937). Published in German in *Erkenntnis*, **6**, 293–303 (1937).

Bohr, N. (1999d). Natural Philosophy and Human Cultures, *BCW*, Vol. 10, pp. 240–249. Originally published in *Congrès international des sciences anthropologiques et ethnologiques, compte rendu de la deuxième session*, Copenhague 1938 (Munksgaard, Copenhagen, 1939), pp. 86–95. Reprinted in *Nature*, **143**, 268–272 (1939) and in *Atomic Physics and Human Knowledge* (John Wiley & Sons, New York, 1958), pp. 23–31.

Bohr, N. (2005a). *Open Letter to the United Nations, 9th June, 1950, BCW*, Vol. 11, pp. 171–185. Originally published by (Schultz, Copenhagen, 1950).

Bohr, N. (2005b). Science and Civilization, *BCW*, Vol. 11, pp. 123–124. Originally published in *The Times*, 11 August 1945.

Bohr, N. (2007a). Hommage à Lord Rutherford, sept huit et neuf novembre MCMXLVII, *BCW*, Vol. 12, p. 277. Originally published in (La federation mondiale des travailleurs scientifiques, Paris, 1948) pp. 15–16.

Bohr, N. (2007b). Newton's Principles and Modern Atomic Mechanics, *BCW*, Vol. 12, pp. 219–225. Originally published in *The Royal Society Newton Tercentenary Celebrations 15–19 July 1946* (Cambridge University Press, Cambridge, 1947), pp. 56–61.

Bohr, N. and Rosenfeld, L. (1979a). On the Question of Measurability of Electromagnetic Field Quantities, *SP*, pp. 357–400. Originally published in German in *Det Kongelige Danske Videnskabernes Selskabs Mathematisk-fysiske Meddelelser*, **12**(8), pp. 3–65. *BCW*, Vol. 7, pp. 123–166.

Bohr, N. and Rosenfeld, L. (1979b). Field and charge measurements in quantum electrodynamics, *SP*, pp. 400–412. Originally published in *Physical Review*, **78**, 794–798 (1950). *BCW*, Vol. 7, pp. 211–216.

Bokulich, A. (2004). Open or Closed? Dirac, Heisenberg, and the Relation between Classical and Quantum Mechanics, *Studies in History and Philosophy of Modern Physics*, **35**, pp. 377–396.

Bokulich, P. and Bokulich, A. (2005). Niels Bohr's Generalization of Classical Mechanics, *Foundations of Physics*, **35**(3), pp. 347–371.

Born, M. (1928). Sommerfeld als Begründer einer Schule, *Naturwissenschaften*, **16**, pp. 1035–1036.

Born, M. (1953). The Interpretation of Quantum Mechanics, *The British Journal for the Philosophy of Science*, **4**(14), pp. 95–106. Reprinted in M. Born, *Physics in my Generation: A selection of papers* (Pergamon Press, London, 1956), pp. 140–150.

Born, M. (1954). The Statistical interpretation of quantum mechanics, *Nobel Lecture, December 11, 1954*, http://nobelprize.org/nobel_prizes/physics/laureates/1954/born-lecture.pdf (accessed 30 April, 2011).

Born, M. (ed.) (1971). *The Born–Einstein Letters: Correspondence between Albert Einstein and Max and Hedwig Born from 1916 to 1955 with commentaries by Max Born* (Macmillan, London).

Born, M. (1978). *My Life. Recollections of a Nobel Laureate* (Taylor and Francis Ltd, London).

Bramsen, I. (1995). The Long-Term Psychological Adjustment of World War II Survivors on the Netherlands, Proefschrift Rijksuniversiteit Leiden.

Brendel, C. (2001). *Anton Pannekoek: Denker der Revolution* (ça ira Verlag, Freiburg).

Breuer, T. (2001). Von Neumann, Gödel and Quantum Incompleteness, in *John von Neumann and the Foundations of Quantum Physics*, (eds.) M. Rédei and M. Stöltzner (Kluwer, Dordrecht), pp. 75–82.

Bromberg, J. (1977). Dirac's Quantum Electrodynamics and the Wave–Particle Equivalence, *Proceedings of the International School of Physics "Enrico Fermi": History of Twentieth Century Physics*, (ed.) C. Weiner (Academic Press, New York), pp. 147–157.

Bromberg, J. (2006). Device Physics vis-à-vis Fundamental Physics in Cold War America: The Case of Quantum Optics, *Isis*, **97**(2), pp. 237–259.

Bromberg, J. (2008). New Instruments and the Meaning of Quantum Mechanics, *Historical Studies in the Natural Sciences*, **38**(3), pp. 325–352.

Bronstein, M. P. (1934). *Uspekhi fizitcheskikh nauk*, **XIV**(4), pp. 516–520.

Brown, G. E. (1974). Léon Rosenfeld 14 August 1904–23 March 1974, *NORDITA publications*, 556/557.

Brown, G. E. (2001). Fly with Eagles, *Annual Review of Nuclear and Particle Science*, **51**, pp. 1–22.

Bulletin of the Atomic Scientists (1955). **11**(8), October.

Bulletin of the Atomic Scientists (1956). Igor Kurchatov interview, September. The interview also appeared in *Pravda* 10 May 1956.

Bunge, M. (1955). Strife about Complementarity (I) and (II), *The British Journal for the Philosophy of Science*, **6**(21), pp. 1–12 and **6**(22), pp. 141–154.

Bunge, M. (1956). Über philosophische Fragen der modernen Physik, *Deutsche Zeitschrift für Philosophie*, **4**, pp. 467–496.

Bustamante, M. C. (1997). Jacques Solomon (1908–1942): Profil d'un physicien théoricien dans la France des années trente, *Revue d'histoire des sciences et de leurs applications*, **50**(1–2), pp. 49–87.

Butcher, S. I. (2005). The Origins of the Russell–Einstein Manifesto, *Pugwash History Series*, **1**, pp. 5–35. http://www.pugwash.org/publication/phs/history9.pdf (accessed on 10 November 2010).

Camilleri, K. (2009a). Constructing the Myth of the Copenhagen Interpretation, *Perspectives on Science*, **17**(1), pp. 26–57.

Camilleri, K. (2009b). *Heisenberg and the Interpretation of Quantum Mechanics: The Physicist as Philosopher* (Cambridge University Press, Cambridge).

Cambresier, Y. and Rosenfeld, L. (1933). On the Dissociation of Molecules in the Atmospheres of the Stars of the Main Sequence, *Monthly Notices of the Royal Astronomical Society*, **93**, pp. 710–723.

Carnap, R., Frank, P., Jørgensen, J., Schlick, M., Neurath, O., Reichenbach, H., Rougier, L. and Stebbing, S. (1936). Zweiter internationaler Kongress für Einheit der Wissenschaft — Kopenhagen 21–26. Juni 1936: Das Kausalproblem, *Erkenntnis*, **6**, pp. 275–277.

Carson, C. (2010). *Heisenberg in the Atomic Age: Science and the Public Sphere* (Cambridge University Press, Cambridge).

Casimir, H. B. G. (1983). *Haphazard Reality: Half a Century of Science* (Harper & Row, New York).

Cassidy, D. C. (1992). *Uncertainty: The Life and Science of Werner Heisenberg* (W. H. Freeman and Company, New York).

Cassidy, D. C. (2005). *J. Robert Oppenheimer and the American Century* (Pi Press, New York).

Chilvers, C. A. J. (2003). The Dilemmas of Seditious Men: The Crowther-Hessen Correspondence in the 1930s, *British Journal for the History of Science*, **36**(4), pp. 417–435.

Chilvers, C. A. J. (2006). La signification historique de Boris Hessen, postscipt in B. Hessen, Les Racines Sociales et Èconomiques des Principia de Newton (Vuibert, Paris), pp. 179–206.

Chimisso, C. and Freudenthal, G. (2003). A Mind of Her Own: Hélène Metzger to Émile Meyerson, 1933, *Isis*, **94**(3), pp. 477–491.

Chimisso, C. (2008). *Writing the History of the Mind: Philosophy and Science in France, 1900 to 1960s* (Ashgate, Aldershot).

Christie, J. R. R. (1990). The Development of the Historiography of Science, in *Companion to the History of Modern Science*, (eds.) R. C. Olby, G. N. Cantor, J. R. R. Christie and M. J. S. Hodge (Routledge, London), pp. 5–22.

Cini, M. (1982). Cultural Traditions and Environmental Factors in the Development of Quantum Electrodynamics (1925–1933), *Fundamenta Scientiae*, **3**(3–4), pp. 229–253.

Cockcroft, J. (1963). British Research in Controlled Thermonuclear Fusion: Kurchatov Memorial Article, *Atomnaya Energiya, Plasma Physics, Journal of Nuclear Energy Part C*, **5**, pp. 388–391.

Cook, B. A. (2004). *Belgium: A History* (Peter Lang, New York).

Corinaldesi, E. (1953). Some Aspects of the Problem of Measurability in Quantum Electrodynamics, *Nuovo Cimento, Supplemento*, **10**(2), pp. 83–100.

Cross, A. (1991). The Crisis in Physics: Dialectical Materialism and Quantum Theory, *Social Studies of Science*, **21**, pp. 735–759.

Crowther, J. G. (1930). *Science in Soviet Russia* (Williams and Norgate, London).

Crowther, J. G. (1970). *Fifty Years with Science* (Barrie & Jenkins, London).

Cushing, J. T. (1998). *Philosophical Concepts in Physics: The Historical Relation between Philosophy and Scientific Theories* (Cambridge University Press, Cambridge).

Daneri, A., Loinger, A. and Prosperi, G. M. (1962). Quantum Theory of Measurement and Ergodicity Conditions, *Nuclear Physics*, **33**, pp. 297–319.

Darrigol, O. (1986). The Origin of Quantized Matter Waves, *Historical Studies in the Physical and Biological Sciences*, **16**(2), pp. 197–253.

Darrigol, O. (1991). Cohérence et complétude de la mécanique quantique: l'exemple de "Bohr-Rosenfeld", *Revue d'Histoire des Sciences*, **XLIV**(2), pp. 137–179.

Darrigol, O. (1992). *From c-Numbers to q-Numbers: The Classical Analogy in the History of Quantum Theory* (University of California Press, Berkeley).

de Broglie, L. (1958). Physics and Metaphysics, *Nature*, **181**, p. 1814.

Deery, P. (2002). The Dove Flies East: Whitehall, Warsaw and the 1950 World Peace Congress, *Australian Journal of Politics and History*, **48**(4), pp. 449–468.

D'Espagnat, B. (ed.) (1971). *Foundations of Quantum Mechanics, Proceedings of the International School of Physics "Enrico Fermi"*, Course IL, Varenna on Lake Como, Villa Monastero, 29th June–11th July 1970 (Academic Press, New York).

D'Espagnat, B. (1979). The Quantum Theory and Reality, *Scientific American*, **241**, pp. 128–140.

Dirac, P. A. M. (1927). The Quantum Theory of the Emission and Absorption of Radiation, *Proceedings of the Royal Society (London) A*, **114**, pp. 243–265. *DCW*, pp. 231–255.

Dirac, P. A. M. (1928). The Quantum Theory of the Electron, *Proceedings of the Royal Society (London)*, **117**, pp. 610–624. *DCW*, pp. 303–319.

Dirac, P. A. M. (1932). Relativistic Quantum Mechanics, *Proceedings of the Royal Society (London) A*, **136**, pp. 453–464. *DCW*, pp. 621–634.

Dirac, P. A. M., Fock, V. A. and Podolsky, B. (1932). On Electrodynamics, *Physikalische Zeitschrift der Sowjetunion*, **2**(6), pp. 468–479. *DCW*, pp. 635–648.

Dresden, M. (1987). *H. A. Kramers. Between Tradition and Revolution* (Springer, New York).

Eddington, A. S. (1930). *The Nature of the Physical World*, 6th impression (Cambridge University Press, Cambridge).

Edgerton, D. E. H. (1996). British Scientific Intellectuals and the Relations of Science, Technology, and War, in *National Military Establishments and the Advancement of Science and Technology*, (eds.) P. Forman and J. M. Sánchez-Ron (Kluwer, Dordrecht), pp. 1–35.

Einstein, A., Podolsky, B. and Rosen, N. (1935). Can Quantum Mechanical Description of Physical Reality be Considered Complete?, *Physical Review*, **47**, pp. 777–780.

Einstein, A. (1948). A Reply to the Soviet Scientists, *Bulletin of the Atomic Scientists*, **4**(2), p. 35.

Elzinga, A. (1996a). Introduction: Models of Internationalism, in *Internationalism and Science*, (eds.) A. Elzinga and C. Landström (Taylor Graham, London), pp. 3–20.

Elzinga, A. (1996b). Unesco and the Politics of Scientific Internationalism, in *Internationalism and Science*, (eds.) A. Elzinga and C. Landström (Taylor Graham, London), pp. 89–131.

Elzinga, A. and Landström, C. (eds.) (1996). *Internationalism and Science* (Taylor Graham, London).

Enebakk, V. (2009). Lilley Revisited: Or Science and Society in the Twentieth Century, *British Journal for the History of Science*, **42**(4), pp. 563–593.

Enz, C. P. (2002). *No Time to be Brief: A Scientific Biography of Wolfgang Pauli* (Oxford University Press, New York).

Erkenntnis (1936). **6**, pp. 275–450.

Europhysics News (1999). May/June, The School High in the Alps.

Farmelo, G. (2009). *The Strangest Man: The Hidden Life of Paul Dirac, Quantum Genius* (Faber and Faber, London).

Favrholdt, D. (1999a). General Introduction: Complementarity Beyond Physics, in *BCW*, Vol. 10, pp. xxiii–xlix.

Favrholdt, D. (1999b). Introduction: Complementarity in Biology and Related Fields, in *BCW*, Vol. 10, pp. 3–26.

Favrholdt, D. (2009). *Filosoffen Niels Bohr* (Informations Forlag, Copenhagen).

Faye, J. and Folse, H. J. (1998). Introduction, in *Causality and Complementarity: Supplementary Papers*, The Philosophical Writings of Niels Bohr, Vol. 4, (eds.) Faye and Folse (Ox Bow Press, Woodbridge, Connecticut), pp. 1–23.

Faye, J. (2010). Niels Bohr and the Vienna Circle, in *The Vienna Circle in the Nordic Countries: Networks and Transformations of Logical Empiricism*, (eds.) J. Manninen and F. Stadler (Springer, Dordrecht), pp. 33–45.

Fermi, E. (1929). Sopra l'electrodinamica quantistica, *Rendiconti della R. Accademia dei Lincei*, **9**, pp. 881–887.

Fermi, E. (1932). Quantum Theory of Radiation, *Reviews of Modern Physics*, **4**, pp. 87–132.

Fock, V. A. (1957a). Report "The journey to Copenhagen", originally published in Russian in *Vestnik Akademii Nauk SSSR*, **27**(7), pp. 54–57. English translation in BSC.

Fock, V. A. (1957b). On the Interpretation of Quantum Mechanics, *Czechoslovakian Journal of Physics*, **7**, pp. 643–656.

Fock, V. A. (1958). Über die Deutung der Quantenmechanik, German translation of Fock (1957b), in *Max-Planck-Festschrift*, (eds.) B. Kockel, W. Macke and A. Papapetrou (VEB Deutschen Verlag der Wissenschaften, Berlin, 1958), pp. 177–195.

Fock, V. A. (1960). Critique épistémologique de théories récentes, *La Pensée: Revue du rationalisme moderne*, **91**, pp. 8–15.

Fock, V. A. (1966). Quantum Mechanics and Dialectical Materialism: Comments, *Slavic Review*, **25**(3), pp. 411–413.

Fock, V. A. and Jordan, P. (1931). Neue Unbestimmtheitseigenschaften des electromagnetischen Feldes, *Zeitschrift für Physik*, **69**, pp. 206–209.

Forman, P. (1984). *Kausalität*, *Anschaulichkeit*, and *Individualität*, or How Cultural Values Prescribed the Character and the Lessons Ascribed to Quantum Mechanics, in *Society and Knowledge: Contemporary Perspectives in the Sociology of Knowledge*, (eds.) N. Stehr and V. Meja (Transaction Books, New Brunswick), pp. 333–347.

Forstner, C. (2008). The Early History of David Bohm's Quantum Mechanics Through the Perspective of Ludwik Fleck's Thought-Collectives, *Minerva*, **46**, pp. 215–229.

Frank, M. D. (1974). Léon Rosenfeld. Author, Editor and Friend, *NORDITA publications*, No. 556/557, pp. ix–xi.

Frank, P. (1936a). Philosophische Deutungen und Missdeutungen der Quantentheorie, *Erkenntnis*, **6**, pp. 303–317.

Frank, P. (1936b). Schlusswort, *Erkenntnis*, **6**, pp. 443–450.

Freire Jr., O. (1997). Quantum Controversy and Marxism, *Historia Scientiarum*, **7**(2), pp. 137–152.

Freire Jr., O. (2001). Science, Philosophy and Politics in the Fifties. On Max Born's unpublished paper entitled "Dialectical Materialism and Modern Physics", *Historia Scientiarum*, **10**(3), pp. 248–254.

Freire Jr., O. (2003). A Story Without an Ending: The Quantum Physics Controversy 1950–1970, *Science and Education*, **12**, pp. 573–586.

Freire Jr., O. (2005). Science and Exile: David Bohm, the Cold War, and a New Interpretation of Quantum Mechanics, *Historical Studies in the Physical and Biological Sciences*, **36**, pp. 1–34.

Freire Jr., O. (2007). Orthodoxy and Heterodoxy in the Research on the Foundations of Quantum Physics: E. P. Wigner's Case, in *Cognitive Justice in a Global World: Prudent Knowledge for a Decent Life*, (ed.) B. S. Santos (Lexington Books, Lanham, MD).

Freire Jr., O. (2009). Quantum Dissidents: Research on the Foundations of Quantum Theory circa 1970, *Studies in History and Philosophy of Modern Physics*, **40**, 280–289 (2009).

Freire Jr., O. (2011). Continuity and Change: Charting David Bohm's Evolving Ideas on Quantum Mechanics, in *Brazilian Studies in Philosophy and History of Science*, Boston Studies in the Philosophy of Science 290, (eds.) D. Krause and A. Videira (Springer), pp. 291–299.

Freire Jr., O. and Lehner, C. (2010). "Dialectical materialism and modern physics", an unpublished text by Max Born, *Notes and Records of the Royal Society*, **64**, pp. 155–162.

Freistadt, H. (1957). Dialectical Materialism: A Further Discussion, *Philosophy of Science*, **24**(1), pp. 25–40.

French, A. P. (1999). The Strange Case of Emil Rupp, *Physics in Perspective*, **1**, pp. 3–21.

Frenkel, Ya. I. (1950). On a Unified Field Theory, *Uspekhi Fizicheskikh Nauk*.

Freudenthal, G. (2010). Hélène Metzger 1889–1944, *Jewish Women's Archive* http://jwa.org/encyclopedia/article/metzger-helene (accessed 28 April 2011).

Friedman, R. M. (2001). *The Politics of Excellence: Behind the Nobel Prize in Science* (A. W. H. Freeman, New York).

Gaddis, J. L. (1997). *We Now Know: Rethinking Cold War History* (Clarendon Press, Oxford).

Gamow, G. (1935). Dialectics of Atomic Nuclei, *Journal of Jocular Physics*, Bohr Celebration Volume, Blegdamsvej 15, Copenhagen, pp. 2–4.

George, A. (1949). La puits de la vérité, chronique, *La Figaro*, October.

Gerber, J. (2005). Anton Pannekoek and the Quest for an Emancipatory Socialism, http://libcom.org/library/anton-pannekoek-and-the-quest-for-an-emancipatory-socialism (accessed 28 April 2011).

Geyl, P. (1947). Letter to the Editor, *Manchester Guardian*, 30 August.

Gorelik, G. (1995). *"Meine antisowjetische Tätigkeit...": Russische Physiker unter Stalin*, Tr. H. Rotter (Vieweg, Braunschweig/Wiesbaden).

Gorelik, G. (1997). The Top-Secret Life of Lev Landau, *Scientific American*, August, pp. 52–57.

Gorelik, G. E. (2005). Matvei Brönstein and Quantum Gravity: 70th Anniversary of the Unsolved Problem, *Physics-Uspehki*, **48**(10), pp. 1039–1053.

Gorelik, G. E. and Frenkel, V. Ya. (1994). *Matvei Petrovich Bronstein and Soviet Theoretical Physics in the Thirties*, Tr. V. M. Levina (Birkhäuser, Basel).

Graham, L. (1966). Quantum Mechanics and Dialectical Materialism, *Slavic Review*, **25**(3), pp. 381–410.

Graham, L. (1985). The Socio-Political Roots of Boris Hessen: Soviet Marxism and the History of Science, *Social Studies of Science*, **15**, pp. 705–722.

Graham, L. (1987). *Science, Philosophy, and Human Behavior in the Soviet Union* (Columbia University Press, New York).

Grandin, K. (2008). Intermediate Theoretical Physics, in *Aurora Torealis: Studies in the History of Science and Ideas in Honor of Tore Frängsmyr*, (eds.) M. Beretta, K. Grandin and S. Lindqvist (Science History Publications, Sagamore Beach), pp. 193–214.

Gray, J. J. (1994). Differential Equations and Groups in *Companion Encyclopedia of the History and Philosophy of the Mathematical Sciences*, 2 Vols., Vol. 1, (ed.) I. Grattan-Guinness (Johns Hopkins University Press, Baltimore), pp. 470–474.

Grünbaum, A. (1957). Complementarity in Quantum Physics and Its Philosophical Generalization, *The Journal of Philosophy*, **54**(23), pp. 713–727.

Hall, K. (2005). "Think less about Foundations": A Short Course on Landau and Lifshitz's *Course of Theoretical Physics*, in *Pedagogy and the Practice of Science: Historical and Contemporary Perspectives*, (ed.) D. Kaiser (The MIT Press, Cambridge, Massachusetts), pp. 253–286.

Hall, K. (2008). The Schooling of Lev Landau: The European Context of Postrevolutionary Soviet Theoretical Physics, *Osiris*, **23**, pp. 230–259.

Halleux, R., Vandersmissen, J., Despy-Meyer, A. and van Paemel, G. (eds.) (2001). *Histoire des sciences en Belgique 1815–2000*, 2 Vols. (La Renaissance du Livre, Brussels).

Hankins, T. L. (1979). In Defence of Biography: The Use of Biography in the History of Science, *History of Science*, XVII, pp. 1–16.

Hansen, K. G. (1984). Weber, Sophus Theodorus Holst, in *Dansk Biografisk Leksikon*, 3rd edn., Vol. 15 (Gyldendal, Copenhagen), p. 307.

Heijmans, H. G. (1994). *Wetenschap Tussen Universiteit en Industrie: De Experimentale Natuurkunde in Utrecht onder W. H. Julius en L. S. Ornstein 1896–1940* (Erasmus Publishing, Rotterdam).

Heilbron, J. L. (1985). The Earliest Missionaries of the Copenhagen Spirit, *Revue d'hisoire des sciences*, **38**(3–4), pp. 195–230.

Heilbron, J. L. (1986). *The Dilemmas of an Upright Man: Max Planck as Spokesman for German Science* (University of California Press, Berkeley).

Hein, P. (1932). Atomfysik og erkendelse, *Studenterbladet*, **1**(7), pp. 3–5; **1**(8), pp. 4–6; **1**(9), pp. 4–5; **1**(10), pp. 9–10.

Heisenberg, W. (1925). Über quantentheoretische Umdeutung kinematischer und mechanischer Bezeihungen, *Zeitschrift für Physik*, **33**, p. 879.

Heisenberg, W. (1927). Über den anschaulichen Inhalt der quantentheoretischen Kinematik und Mechanik, *Zeitschrift für Physik*, **43**, pp. 172–198.

Heisenberg, W. (1930). *The Physical Principles of the Quantum Theory*, Tr. C. Eckart and F. C. Hoyt (Dover, Chicago).

Heisenberg, W. (1931). Bemerkungen zur Strahlungstheorie, *Annalen der Physik*, **9**(3), pp. 338–346.

Heisenberg, W. (1934). Wandlungen der Grundlagen der exakten Naturwissenschaft in jüngster Zeit, *Angewandte Chemie*, **47**, pp. 697–702.

Heisenberg, W. (1955). The Development of the Interpretation of the Quantum Theory, in *Niels Bohr and the Development of Physics*, (eds.) W. Pauli, L. Rosenfeld, and V. Weisskopf (McGraw-Hill, New York), pp. 12–29.

Heisenberg, W. (1970). *Natural Law and the Structure of Matter* (Rebel, London).

Heisenberg, W. (1971). Quantum Mechanics and Kantian Philosophy (1930–1934), in *Physics and Beyond: Encounters and Conversations*, Tr. A. J. Pomerans (Harper and Row, New York), pp. 117–124.

Heisenberg, W. and Pauli, W. (1929). Zur Quantendynamik der Wellenfelder, *Zeitschrift für Physik*, **56**, pp. 1–61.

Heisenberg, W. and Pauli, W. (1930). Zur Quantentheorie der Wellenfelder. II, *Zeitschrift für Physik*, **59**, pp. 168–190.

Heitler, W. (1936). *The Quantum Theory of Radiation* (Clarendon, Oxford).

Heitler, W. (1954). *The Quantum Theory of Radiation*, 3rd edn. (Clarendon, Oxford).

Hermann, A., von Meyenn, K. and Weisskopf, V. F. (eds.) (1979–1985). *Wolfgang Pauli. Scientific Correspondence with Bohr, Einstein, Heisenberg, a.o.*, 2 Vols., Vols. 1 and 2 (Springer, Berlin).

Hermann, G. (1935). *Die naturphilosophischen Grundlagen der Quantenmechanik* (Verlag "Öffentliches Leben", Berlin).

Herzenberg, C. L. (2008). Grete Hermann: An Early Contributor to Quantum Theory, http://arxiv.org/abs/0812.3986 (accessed 28 April 2011).

Hessen, B. (1971). The Social and Economic Roots of Newton's 'Principia', in *Science at the Cross Roads: Papers presented to the International Congress of the History of Science and Technology held in London from 29th June to 3rd July 1931, by the Delegates of the U.S.S.R.*, (ed.) N. I. Bukharin, 2nd edn. (Frank Cass & Co, London), pp. 147–212. Originally published in 1931.

Hewlett, R. G. and Holl, J. M. (1989). *Atoms for Peace and War 1953–1961. Eisenhower and the Atomic Energy Commission* (Berkeley, University of California Press).

Hobsbawm, E. (1995). *Age of Extremes: The Short Twentieth Century 1914–1991*, 3rd edn. (Abacus, London).

Hoffmann, D. (1988). Zur Teilnahme deutscher Physiker an den Kopenhagener Physikerkonferenzen nach 1933 sowie am 2. Kongress für Einheit der Wissenschaft, Kopenhagen 1936, *NTM-Schriftenreihe Geschichte der Naturwissenschaft, Technik, und Medicin*, **25**(1), pp. 49–55.

Hoffmann, D. (2009). Fritz Lange, Klaus Fuchs, and the Remigration of Scientists to East Germany, *Physics in Perspective*, **11**, pp. 405–425.

Hook, E. B. (2002). *Prematurity in Scientific Discovery: On Resistance and Neglect* (University of California Press, Berkeley).

Hooyman, G. J. (1979). Leon Rosenfeld, *Fylakra*, 21, 23 January.

Horner, D. (1996). The Cold War and the Politics of Scientific Internationalism: The Post-War Formation and Development of the World Federation of Scientific Workers 1946–1956, in *Internationalism and Science*, (eds.) A. Elzinga and C. Landström (Taylor Graham, London), pp. 132–161.

Howard, D. (1990). "Nicht sein kann was nicht sein darf", or the Prehistory of EPR, 1909–1935: Einstein's Early Worries About the Quantum Mechanics of Composite Systems, in *Sixty-Two Years of Uncertainty: Historical, Philosophical, and Physical Inquiries into the Foundations of Quantum Mechanics*, (ed.) A. I. Miller (Plenum Press, New York), pp. 61–111.

Izvestia (1934). Interview with Professor Bohr, 12 May, *BCW*, Vol. 11, pp. 199–201. Originally in German.

Ivanov, K. (2002). Science after Stalin: Forging a New Image of Soviet Science, *Science in Context*, **15**(2), pp. 317–338.

Jacobsen, A. S. (1995). Nyere fortolkninger af kvantemekanikken og måleproblemet, Master's thesis, University of Aarhus.

Jacobsen, A. S. (2007). Léon Rosenfeld's Marxist Defense of Complementarity, *Historical Studies in the Physical and Biological Sciences*, 37, Supplement, pp. 3–34.

Jacobsen, A. S. (2008). The Complementarity between the Collective and the Individual: Rosenfeld and Cold War History of Science, *Minerva*, **46**(2), pp. 195–214.

Jacobsen, A. S. (2011). Crisis, Measurement Problems, and Controversy in Early Quantum Electrodynamics: The Failed Appropriation of Epistemology in the Second Quantum Generation, in *Quantum Mechanics and Weimar Culture: Revisiting the Forman Thesis, with Selected Papers by Paul Forman*, (eds.) A. Kojevnikov, C. Carson and H. Trischler (Imperial College Press, London), pp. 375–396.

Jahn, G. (1949). Award Ceremony Speech, http://nobelprize.org/nobel_prizes/peace/laureates/1949/press.html (accessed 28 April 2011).

Jammer, M. (1966). *The Conceptual Development of Quantum Mechanics* (McGraw-Hill, New York).

Jammer, M. (1974). *The Philosophy of Quantum Mechanics: The Interpretations of Quantum Mechanics in Historical Perspective* (Wiley, New York).

Jones, G. (1988). *Science, Politics and the Cold War* (Routledge, London).
Joravsky, D. (1961). *Soviet Marxism and Natural Science, 1917–1932* (Columbia University Press, New York).
Jordan, P. (1927). Zur Quantenmechanik der Gasentartung, *Zeitschrift für Physik*, **44**, pp. 473–480.
Jordan, P. (1936). *Anschauliche Quantentheorie* (Springer, Berlin).
Jordan, P. (1944). *Physics of the 20th Century* (New York).
Jordan, P. and Klein, O. (1927). Zum Mehrkörperproblem der Quantentheorie, *Zeitschrift für Physik*, **45**, pp. 751–765.
Jordan, P. and Wigner, E. (1928). Über das Paulische Äquivalenzverbot, *Zeitschrift für Physik*, **47**, pp. 631–651.
Jørgensen, J. (1956). *Indledning til logikken og metodelæren*, Ny udvidet udgave (Munksgaard, Copenhagen).
Josephson, P. R. (1991). *Physics and Politics in Revolutionary Russia* (University of California Press, Berkeley).
Jungk, R. (1956). *Heller als tausend Sonnen: Das Schicksal der Atomforscher* (Scherz & Goverts, Stuttgart). English translation by J. Cleugh (Harcourt Brace Jovanovich, New York, 1958).
Kaiser, D. (2005). *Drawing Theories Apart: The Dispersion of Feynman Diagrams in Postwar Physics* (The University of Chicago Press, Chicago).
Kaiser, D. (2007a). Comments on "Interpreting Quantum Mechanics: A Century of Debate", HSS Session, November, Washington.
Kaiser, D. (2007b). Turning Physicists into Quantum Mechanics, *Physics World*, May, pp. 28–33.
Kalckar, F. (1935). En Dag paa Bohrs Institut, *Journal of Jocular Physics*. Niels Bohr Celebration Number, October 7 (Institute of Theoretical Physics, Copenhagen).
Kalckar, J. (1967). Niels Bohr and His Youngest Disciples, in *Niels Bohr: His Life and Work as seen by His Friends and Colleagues*, (ed.) S. Rozental (North-Holland, Amsterdam), pp. 227–239.
Kalckar, J. (1971). Measurability Problems in the Quantum Theory of Fields, in *Proceedings of the International School of Physics "Enrico Fermi"*, Course IL, Varenna on Lake Como, Villa Monastero, 29th June–11th July 1970, (ed.) B. d'Espagnat (Academic Press, New York), pp. 127–168.
Kalckar, J. (1996). Introduction, in *BCW*, Vol. 7, pp. 3–51.
Khriplovich, I. (1992). The Eventful Life of Fritz Houtermans, *Physics Today*, July, pp. 29–37.
Kierkegaard, S. (1994). *Enten Eller*, 2 Vols. (Gyldendal, Copenhagen). Originally published in 1843.
Klein, O. (1927). Electrodynamics and Wave Mechanics from the Point of View of the Correspondence Principle, *Zeitschrift für Physik*, **41**, pp. 407–422.
Klein, O. (1933). *Einsteins Relativitetsteori i Allmäntillgänglig Form* (Natur och Kultur, Stockholm).
Klein, O. (1935a). On Political Quantization, *Journal of Jocular Physics*, Bohr Celebration Volume, Blegdamsvej 15, Copenhagen.

Klein, O. (1935b). *Orsak och Verkan i den nya Atomteoriens Belysning* (Natur och Kultur, Stockholm).

Klein, O. (1938). *Entretiens sur les idees fondamentales de la physique moderne*, Tr. L. Rosenfeld (Hermann, Paris).

Knudsen, H. (2005). Konsensus og konflikt: Organiseringen af den tekniske forskning I Danmark 1900–1960, PhD thesis, University of Aarhus.

Knudsen, H. (2010). *Videnskabens mand. Fysiologen, formidleren og forskningsaktivisten Poul Brandt Rehberg* (Aarhus Universitetsforlag, Aarhus).

Kojevnikov, A. B. (ed.) (1996). *Paul Dirac and Igor Tamm Correspondence Part 2: 1933–1936.* Commented by A. B. Kojevnikov, Max-Planck-Institut für Physik MPI-Ph/96-40, Werner-Heisenberg-Institut, Munich, Germany.

Kojevnikov, A. B. (2002a). The Great War, the Russian Civil War, and the Invention of Big Science, *Science in Context*, **15**(2), pp. 239–275.

Kojevnikov, A. (2002b). David Bohm and Collective Movement, *Historical Studies in the Physical and Biological Sciences*, **33**(1), pp. 161–192.

Kojevnikov, A. B. (2004). *Stalin's Great Science: The Times and Adventures of Soviet Physicists, History of Modern Physical Sciences*, Vol. 2 (Imperial College Press, London).

Kojevnikov, A. B. (2008). The Phenomenon of Soviet Science, *Osiris*, **23**, pp. 115–135.

Kojevnikov, A. B. (2011). Die Mobilmachung der sowjetischen Wissenschaft, in *Macht und Geist im Kalten Krieg*, (eds.) B. Greiner, T. B. Müller, and C. Weber (Hamburg), pp. 87–107.

Kragh, H. (1992). Relativistic Collisions: The Work of Christian Møller in the Early 1930s, *Archive for History of Exact Sciences*, **43**(4), pp. 299–328.

Kragh, H. (1999). *Quantum Generations: A History of Physics in the Twentieth Century* (Princeton University Press, Princeton).

Kragh, H. S. (2005). *Dirac: A Scientific Biography*, first paperback version (Cambridge University Press, Cambridge).

Kragh, P. J. E. (2003). Niels Bohr and the Soviet Union Between the Two World Wars: Resources at the Niels Bohr Archive, Master's thesis, University of Copenhagen.

Kuhn, T. S. (1963a). Interview with C. Møller, 29 July 1963, AHQP.

Kuhn, T. S. (1963b). Interview with Mrs. Bohr, Aage Bohr, and Léon Rosenfeld, 30 January 1963, AHQP.

Kuhn, T. S. (1978). *Black-Body Theory and the Quantum Discontinuity 1894–1912* (Oxford University Press, Oxford).

Kuhn, T. S., Petersen, Aa. and Rüdinger, E. (1962). Interview with Niels Bohr, 17 November 1962, AHQP.

Kuhn, T. S. and Heilbron, J. L. (1963a). Interview with Léon Rosenfeld, 1 July 1963, AHQP.

Kuhn, T. S. and Heilbron, J. L. (1963b). Interview with Léon Rosenfeld, 19 July 1963, AHQP.

Kuhn, T. S. and Heilbron, J. L. (1963c). Interview with Léon Rosenfeld, 22 July 1963, AHQP.

Kuhn, T. S., Heilbron, J. L., Forman, P. L. and Allen, L. (1967). *Sources for History of Quantum Physics. An Inventory and Report* (The American Philosophical Society, Philadelphia).

Kuzemsky, A. L. (2008). Works by D. I. Blokhintsev and the Development of Quantum Physics, *Physics of Particles and Nuclei*, **39**(2), pp. 137–172.

Kuznick, P. J. (1987). *Beyond the Laboratory: Scientists as Political Activists in 1930s America* (The University of Chicago Press, Chicago).

Lacki, L. (2000). The Early Axiomatizations of Quantum Mechanics: Jordan, von Neumann and the Continuation of Hilbert's Program, *Archive for the History of the Exact Sciences*, **54**, pp. 279–318.

Landau, L. (1955). On the Quantum Theory of Fields, in *Niels Bohr and the Development of Physics* (Pergamon Press, Oxford), pp. 52–69.

Landau, L. (1960). Fundamental Problems, in *Theoretical Physics in the Twentieth Century: A Memorial Volume to Wolfgang Pauli*, (eds.) M. Fierz and V. F. Weisskopf (Interscience, New York), pp. 245–248.

Landau, L. and Peierls, R. (1983). Extension of the Uncertainty Principle to Relativistic Quantum Theory, in *Quantum Theory and Measurement*, (ed.) J. A. Wheeler and W. H. Zurek (Princeton University Press, Princeton), pp. 465–476. Originally published in German in *Zeitschrift für Physik*, **69**, 56 (1931).

Landau, L., Abrikosov, A. A. and Halatnikov [=Khalatnikov], I. (1956). On the Quantum Theory of Fields, *Nuovo Cim. Suppl.*, **3**, p. 80.

Landström, C. (1996). Internationalism Between Two Wars, in *Internationalism and Science*, (eds.) A. Elzinga and C. Landström (Taylor Graham, London), pp. 46–77.

Lanouette, W. (1992). *Genius in the Shadows: A Biography of Leo Szilard, The Man Behind the Bomb* (The University of Chicago Press, Chicago).

Lapp, R. E. (1955). The Lesson of Geneva, *Bulletin of the Atomic Scientists*, **11**(8), pp. 275, 308.

Larsen, S. B. (1986). *Mod Strømmen: Den kommunistiske "højre"-og "venstre"-opposition i 30-ernes Danmark*, Selskabet til Forskning i Arbejderbevægelsens Historie (SFAH) Skriftserie nr. 17 (Copenhagen).

Lee, S. (2007–2009). *Sir Rudolf Peierls: Selected Private and Scientific Correspondence*, 2 Vols. (World Scientific, Singapore).

Lifshitz, E. M. (1989). Lev Davidovich Landau (1908–1968), in *Landau: The Physicist and the Man. Recollections of L.D. Landau*, (ed.) I. M. Khalatnikov (Pergamon Press, Oxford), pp. 7–27.

Lilley, S. (ed.) (1953). *Essays on the Social History of Science*, *Centaurus* **3**(1–2).

Lynning, K. H. and Jacobsen, A. S. (2011). Grasping the Spirit of Nature: *Anschauung* in Ørsted's Epistemology of Science and Beauty, *Studies in the History and Philosophy of Science*, **42**(1), pp. 45–57.

Machamer, P., Pera, M. and Baltas, A. (eds.) (2000). *Scientific Controversies: Philosophical and Historical Perspectives* (Oxford University Press, New York).

Maksimov, A. (1949). Lenins kamp mod den 'fysiske' idealisme, *Tiden: Tidsskrift for aktivt Demokrati*, **10**, pp. 253–268.

Masters, D. and Way, K. (eds.) (1946). *One World or None: A Report to the Public on the Full Meaning of the Atomic Bomb* (Whittlesey House, McGraw Hill, USA).

Mayer, A. K. (2000). Setting up a Discipline: Conflicting Agendas of the Cambridge History of Science Committee, 1936–1950, *Studies in the History and Philosophy of Science*, **31**, pp. 665–689.

Mayer, A. K. (2004). Setting up a Discipline, II: British History of Science and "the end of Ideology", 1931–1948, *Studies in the History and Philosophy of Science*, **35**, pp. 41–72.

McGucken, W. (1984). *Scientists Society and State: The Social Relations of Science Movement in Great Britain, 1931–1947* (Ohio State University Press, Ohio).

McLarty, C. (2005). Poor Taste as a Bright Character Trait: Emmy Noether and the Independent Social Democratic Party, *Science in Context*, **18**(3), pp. 1–22.

Mehra, J. and Rechenberg, H. (2001). *The Historical Development of Quantum Theory*, 6 Vols., Vol. 6: *The Completion of Quantum Mechanics 1926–1941*, Part 2: *The Conceptual Completion and the Extensions of Quantum Mechanics 1932–1941* (Springer, New York).

Mehrtens, H. (1987). Ludwig Bieberbach and "Deutsche Mathematik", in *Studies in the History of Mathematics*, *Studies in Mathematics*, Vol. 26, (ed.) E. R. Phillips (The Mathematical Association of America, USA), pp. 195–241.

Miller, A. I. (1994). *Early Quantum Electrodynamics: A Source Book* (Cambridge University Press, Cambridge).

Molenaar, L. (1994). "*Wij kunnen het niet langer aan de politici overlaten . . .*": *De geschiedenis van het Verbond van Wetenschappelijke Onderzoekers (VWO) 1946–1980* (Elmar, Delft).

Molenaar, L. (2003). *De rok van het universum: Marcel Minnaert astrofysicus (1893–1970)* (Uitgeverij Balans, Amsterdam).

Møller, C. (1974). Léon Rosenfeld 14 August 1904–23 March 1974, *Tale i Videnskabernes Selskabs møde den 18 Oktober 1974*, *Oversigt over Det Kgl. Danske Videnskabernes Selskabs Virksomhed*, pp. 1–8.

Møller, P. (1925). *En Dansk Students Eventyr* (Dansk Bogsamling, Copenhagen).

Moore, W. (1992). *Schrödinger: Life and Thought* (Cambridge University Press, Cambridge).

Mott, N. and Peierls, R. (1977). Werner Heisenberg 5 December 1901–1 February 1976, *Biographical Memoirs of Fellows of the Royal Society*, **23**, pp. 213–251.

Moyer, D. F. (1981a). Origins of Dirac's Electron, 1925–1928, *American Journal of Physics*, **49**(10), pp. 944–949.

Moyer, D. F. (1981b). Evaluations of Dirac's Electron, 1928–1932, *American Journal of Physics*, **49**(11), pp. 1055–1062.

Moyer, D. F. (1981c). Vindications of Dirac's Electron, 1932–1934, *American Journal of Physics*, **49**(12), pp. 1120–1125.

Neergaard, J. (1932). Studenterkommunisme — en børnesygdom! Leo Trotsky udtaler sig til Studenterbladet om studenter og politik, interview with Leo Trotsky by the social democratic student movement, *Studenterbladet* **1**(13), pp. 3–4, 8.

Neurath, O. (1935). Jordan, Quanthentheorie und Willensfreiheit, *Erkenntnis*, **5**, pp. 179–181.

New York Times (1935). Einstein Attacks Quantum Theory, 4 May.

Land og Folk (1950). Niels Bohrs åbne brev en alvorlig opfordring til at underskrive fredsappellen, 14 June.

Nielsen, K. H. (2008). Enacting the Social Relations of Science: Historical (Anti-)Boundary-Work of Danish Science Journalist Børge Michelsen, *Public Understanding of Science*, **17**, pp. 171–188.

Nørlund, I. (1949). Om hjærnespind, *Tiden: Tidsskrift for aktivt Demokrati*, **10**(8), pp. 314–316.

Nørlund, I. (1954). Videnskab i krise. Om "komplementaritetsteorierne" i den moderne fysik — og andre steder, *Dialog*, **4**(1), pp. 13–27.

Nørlund, I. (1991). *Den Sociale Samvittighed* (Gyldendal, Copenhagen).

Nye, M. J. (1975). Science and Socialism: The Case of Jean Perrin in the Third Republic, *French Historical Studies*, **9**(1), pp. 141–169.

Nye, M. J. (2004). *Blackett: Physics, War, and Politics in the Twentieth Century* (Harvard University Press, Cambridge, Mass.).

Nye, M. J. (2008). Re-Reading Bernal: History of Science at the Crossroads in 20th-Century Britain, in *Aurora Torealis: Studies in the History of Science and Ideas in Honor of Tore Frängsmyr*, (eds.) M. Beretta, K. Grandin, and S. Lindqvist (Science History Publications, Sagamore Beach), pp. 235–258.

Nygaard, B. (2005). Trotskijs Danmarksbesøg 1932: Intelligent Slubbert på Visit, *Socialistisk Information*, 193, January.

Olwell, R. (1999). Physical Isolation and Marginalization in Physics: David Bohm's Cold War Exile, *Isis*, **90**(4), pp. 738–756.

Oppenheimer, J. R. (1955). Physics in the Contemporary World, in idem., *The Open Mind* (Simon and Schuster, New York), pp. 81–102. Originally published in *Bulletin of the Atomic Scientists*, **4**(3), 65–68, 85–86 (1948).

O'Raifeartaigh, L. (1997). *The Dawning of Gauge Theory* (Princeton University Press, Princeton).

Oreskes, N. and Conway, E. M. (2010). *Merchants of Doubt: How a Handful of Scientists Obscured the Truth on Issues from Tobacco Smoke to Global Warming* (Bloomsbury Press, New York).

Osnaghi, S., Freitas, F. and Freire, Jr., O. (2009). The Origin of the Everettian Heresy, *Studies in History and Philosophy of Modern Physics*, **40**, pp. 97–123.

Pais, A. (1948). *Developments in the Theory of the Electron* (Princeton University Press, Princeton).

Pais, A. (1972). The Early History of the Theory of the Electron: 1897–1947, in *Aspects of Quantum Theory*, (eds.) A. Salam and E. P. Wigner (Cambridge University Press, Cambridge), pp. 79–93.

Pais, A. (1982). *Subtle is the Lord...: The Science and the Life of Albert Einstein* (Oxford University Press, Oxford).

Pais, A. (1986). *Inward Bound: Of Matter and Forces in the Physical World* (Oxford University Press, Oxford).

Pais, A. (1991). *Niels Bohr's Times: In Physics, Philosophy, and Polity* (Clarendon, Oxford).

Pais, A. (1997). *A Tale of Two Continents: A Physicist's Life in a Turbulent World* (Oxford University Press, Oxford).

Pannekoek, A. (1948). *Lenin as Philosopher: A Critical Examination of the Philosophical Basis of Leninism* (New Essays, New York). Originally published in Dutch in 1938.

Pauli, W. (1964). Review of P. Dirac, *The Principles of Quantum Mechanics*, in *Collected Scientific Papers by Wolfgang Pauli*, 2 Vols., Vol. 2, (eds.) R. Kronig and V. F. Weisskopf (Wiley Interscience, New York, 1964), pp. 1397–1398. Originally published in *Naturwissenschaften* **19**, (1931), p. 188.

Pauli, W. (1980). General Principles of Quantum Mechanics, Tr. P. Achuthan and K. Venkatesan (Springer, Berlin). Originally published in German in *Handbuch der Physik*, Vol. 24, part 1, 1933, (eds.) H. Geiger and K. Scheel.

Peierls, R. (1950). To a Just, Fair, and Steady Britain, *Sunday Express*, 5 Nov.

Peierls, R. (1963). Field Theory since Maxwell, in *Clerk Maxwell and Modern Science: Six Commemorative Lectures*, (ed.) C. Domb (Athlone, London), pp. 26–42.

Peierls, R. (1974). Obituary of Rosenfeld, *The Times*, 2 April.

Peierls, R. (1980). Preface, in A. Livanova, *Landau: A Great Physicist and Teacher* (Pergamon Press, Oxford).

Peierls, R. (1985a). *Bird of Passage: Recollections of a Physicist* (Princeton University Press, Princeton).

Peierls, R. (1985b). Some Recollections of Bohr, in *Niels Bohr: A Centenary Volume*, (eds.) A. P. French and P. J. Kennedy (Harvard University Press, Cambridge, Massachusetts), pp. 227–231.

Peierls, R. (1986). Introduction, *BCW*, Vol. 9, pp. 52–76.

Pessoa Jr., O., Freire Jr., O. and de Greiff, A. (2008). The Tausk Controversy on the Foundations of Quantum Mechanics: Physics, Philosophy, and Politics, *Physics in Perspective*, **10**, pp. 138–162.

Petersen, Aa. (1968). Quantum Physics and The Philosophical Tradition, Doctoral Dissertation, Belfer Graduate School of Science, Yeshiva University, New York.

Petitjean, P. (2008a). Introduction: Science, Politics, Philosophy and History, *Minerva*, **46**(2), pp. 175–180.

Petitjean, P. (2008b). The Joint Establishment of the World Federation of Scientific Workers and of UNESCO after World War II, *Minerva*, **46**(2), pp. 247–270.

Pihl, M. (1955). Om den dialektiske materialisme, *Dialog*, **5**(1), pp. 19–27.

Pinault, M. (2000). *Frédéric Joliot-Curie* (Editions Odile Jacob, Paris).

Presser, J. (1988). *Ashes in the Wind. The Destruction of Dutch Jewry*, tr. Arnold Pomerans (Wayne State University Press, Detroit).

Prigogine, I. (1974). Léon Rosenfeld et les fondements de la physique moderne, Extrait du *Bulletin de l'Académie royale de Belgique* (Classe des Sciences), Séance du samedi 6 juillet, pp. 841–854.

Proceedings of the International Conference on the Peaceful Uses of Atomic Energy (1956). Geneva 8–20 August 1955. (United Nations, New York).

Ramskov, K. (1995). Matematikeren Harald Bohr, PhD thesis, University of Aarhus.

Ravetz, J. R. (1981). Bernal's Marxist Vision of History, *Isis*, **72**, pp. 393–402.

Rechenberg, H. (2005). Kopenhagen 1941 und die Natur des deutschen Uranprojektes, in *Werner Heisenberg 1901–1976, Beiträge, Berichte Briefe. Festschrift zu seinem 100. Geburtstag*, (eds.) C. Kleint, H. Rechenberg and G. Wiemers (Sächsischen Akademie der Wissenschaften, Leipzig).

Rédei, M. and Stöltzner, M. (eds.) (2001). *John von Neumann and the Foundations of Quantum Physics* (Kluwer, Dordrecht).

Reisch, G. A. (2005). *How the Cold War Transformed Philosophy of Science: To the Icy Slopes of Logic* (Cambridge University Press, New York).

Rhodes, R. (1986). *The Making of the Atomic Bomb* (Simon and Schuster, London).

Richardson, A. (2008). Scientific Philosophy as a Topic for History of Science, *Isis*, **99**(1), pp. 88–96.

Rip, A. and Boeker, E. (1975). Scientists and Social Responsibility in the Netherlands, *Social Studies of Science*, **5**(4), pp. 457–484.

Robertson, P. (1979). *The Early Years: The Niels Bohr Institute 1921–1930* (Akademisk Forlag, Copenhagen).

Rodian, S. and Garrity, J. (1950). Perturbed Men: Foreign-Born Atom Experts Disturbed by Pontecorvo Case, *Sunday Express*, 30 October.

Rosenfeld, L. (1928). Le premier conflit entre la théorie ondulatoire et la théorie corpusculaire de la lumière, *Isis*, **11**(35), pp. 111–122.

Rosenfeld, L. (1929). Edmund Hoppe (1854–1928), *Isis*, **13**, pp. 45–50.

Rosenfeld, L. (1930a). Bemerkung über die Invarianz der kanonischen Vertauschungsrelationen, *Zeitschrift für Physik*, **63**, pp. 574–575.

Rosenfeld, L. (1930b). Zur Quantelung der Wellenfelder, *Annalen der Physik*, **5**(1), pp. 113–152. English translation and commentary of this paper in Salisbury (2009).

Rosenfeld, L. (1930c). Über die Gravitationswirkungen des Lichtes, *Zeitschrift für Physik*, **65**, pp. 589–599.

Rosenfeld, L. (1931a). Zur Kritik der Dirachsen Strahlungsteorie, *Zeitschrift für Physik*, **70**, pp. 454–462.

Rosenfeld, L. (1931b). Zur korrespondenzmässigen Behandlung der Linienbreite, *Zeitschrift für Physik*, **71**, pp. 273–278.

Rosenfeld, L. (1931c). Bemerkung zur korrespondenzmässigen Behandlung des relativistischen Mehrkörperproblems, *Zeitschrift für Physik*, **73**, pp. 253–259.

Rosenfeld, L. (1932a). La théorie quantique des champs, Conférences faites à l'Institut Henri-Poincaré en février 1931, *Annales de l'Institut Henri Poincaré*, **2**, pp. 25–91.

Rosenfeld, L. (1932b). Über die quantentheoretische Behandlung der Strahlungsprobleme, in *Convegno di Fisica Nucleare*, (ed.) O. M. Corbino (Reale Accademica d'Italia, Rome), pp. 131–135.

Rosenfeld, L. (1932c). Über eine mögliche Fassung des Diracschen Programmes zur Quantenelektrodynamik und deren formalen Zusammenhang mit der Heisenberg-Paulischen Theorie, *Zeitschrift für Physik*, **76**, pp. 729–734.

Rosenfeld, L. (1933). The Dissociation of Molecules in the Atmospheres of the Carbon Stars, *Monthly Notices of the Royal Astronomical Society*, **93**, pp. 724–729.

Rosenfeld, L. (1935). Kvanteteori og Feltfysik, *Fysisk Tidsskrift*, **33**, pp. 109–121.

Rosenfeld, L. (1936). Sur l'enseignement au Danemark, *Bulletin Association des Amis de l'Université de Liège*, **8**, pp. 135–149.

Rosenfeld, L. (1937a). C'est un animateur, un apôtre (Langevin 65 years old), *L'Humanité*, 22 January.

Rosenfeld, L. (1937b). Le dualisme entre ondes et corspuscules, *Archeion*, **19**, pp. 74–77.

Rosenfeld, L. (1938). Remarques sur la question des précurseurs, *Archeion*, **21**, pp. 74–77.

Rosenfeld, L. (1945a). Gesprek met Niels Bohr, Uitvinder van de atoombom, *De Baanbreker*, **1**(20), 17 November.

Rosenfeld, L. (1945b). Le grand physicien danois Niels Bohr nous parle de la bombe atomique, *L'Éclair*, **1**(25), 30 October.

Rosenfeld, L. (1945c). Niels Bohr, naar aanleiding van zijn 60e verjaardag, *De Vrije Katheder*, **5**(28), p. 268.

Rosenfeld, L. (1946a). *De ontsluiting van de atoomkern: Zes voordrachten, gehouden te Utrecht voor studenten aller faculteiten* (N.V. de Arbeiderspers, Amsterdam).

Rosenfeld, L. (1946b). Ierse indrukken, *De Vrije Katheder*, **6**(16), 23 August.

Rosenfeld, L. (1947). Letter to the Editor, *Manchester Guardian*, 22 August and 3 September.

Rosenfeld, L. (1948a). Dynamisch denken in de wetenschap, *De Vrije Katheder*, **7**(42), p. 668.

Rosenfeld, L. (1948b). J. K. Lubaski: Obituary Notice, *Acta Physica Polonica*, IX(Fasc. 2–4), pp. 63–64.

Rosenfeld, L. (1948c). *Nuclear Forces* (North-Holland, Amsterdam).

Rosenfeld, L. (1949). Some Impressions of University Life in Belgium and Holland, *Universities Quarterly*, **3**, pp. 593–599.

Rosenfeld, L. (1950). Early History of Quantum Mechanics, *Nature*, **4230**, November 25, pp. 883–884.

Rosenfeld, L. (1951). Review of *Beiträge zum neuzeitlichen Weldbild der Physik: Das Plancksche Wirkungsquantum*, by P. Jordan; *Dialektischer Materialismus und theoretische Physik*, by K. Zweiling (Akademie-Verlag, Berlin, 1950) in *Proceedings of the Physical Society*, **64**, Sec B, p. 324.

Rosenfeld, L. (1952). Review of W. Heisenberg, *Philosophic Problems of Nuclear Science*, in *Proceedings of the Physical Society*, London, **65**, p. 864.

Rosenfeld, L. (1953). L'évidence de la complémentarité, in *Louis de Broglie: Physicien et penseur*, (ed.) A. George (Paris), pp. 43–65. The De Broglie Festschrift appeared in a German translation (Claassen Verlag GmbH, Hamburg, 1955).

Rosenfeld, L. (1954). The Social Responsibility of the Scientist, published in French in *Comprendre*, No. 12, pp. 139–141. English version in RP, Supplement: "Science and society".

Rosenfeld, L. (1955). Dr. Julius Podolanski, *Nature*, **175**, p. 795.

Rosenfeld, L. (1955–1956). Review of A. R. Hall, *The Scientific Revolution 1500–1800* (1954), *Centaurus*, **4**, pp. 171–174.

Rosenfeld, L. (1958). Review of D. Bohm, *Causality and Chance in Modern Physics* (Routledge and Kegan Paul, London, 1957), in *Nature*, **181**, p. 658.

Rosenfeld, L. (1960). Heisenberg, Physics and Philosophy, review of W. Heisenberg, *Physics and Philosophy: The Revolution in Modern Science* (George Allen and Unwin, Ltd., London, 1959), in *Nature* **186**(4728), pp. 830–831.

Rosenfeld, L. (1962a). Einstein i razvitiye fiziko-matematicheskoi mysli, (sbornik statei) (Izdatel'stvo Akademischeskikh Nauk, Moscow, 1962), pp. 89–93. This paper appeared in English in *Zeitschrift für Physik*, **171**, 242–245 (1963). Rosenfeld (1979u).

Rosenfeld, L. (1962b). Miss Margrethe Have: In Memoriam, *Nuclear Physics*, **31**, p. 689.

Rosenfeld, L. (1963). Nogle minder om Niels Bohr, in *Niels Bohr: Et Mindeskrift* (Copenhagen), pp. 65–75, special issue of *Fysisk Tidsskrift*, **60**, (1962).

Rosenfeld, L. (1967). Niels Bohr in the Thirties: Consolidation and Extension of the Conception of Complementarity, in *Niels Bohr: His Life and Work as Seen by his Friends and Colleagues*, (ed.) S. Rozental (North-Holland, Amsterdam), pp. 114–136.

Rosenfeld, L. (1968). Some Concluding Remarks and Reminiscences, in *Fundamental Problems in Elementary Particle Physics, Proceedings of the Fourteenth Conference on Physics at the University of Brussels*, October 1967 (Interscience, London), pp. 231–234.

Rosenfeld, L. (1969a). Review of M. Jammer, *The Conceptual Development of Quantum Mechanics* (McGraw-Hill, New York), in *Nuclear Physics A*, **126**, p. 696.

Rosenfeld, L. (1969b). Review of D. I. Blokhintsev, *The Philosophy of Quantum Mechanics* (Reidel, Dordrecht, 1968), in *Nuclear Physics A*, **139**(3), pp. 698–699.

Rosenfeld, L. (1971). Quantum Theory in 1929: Recollections from the First Copenhagen Conference, in *Institute for Theoretical Physics — The Niels Bohr Institute 1921–71* (Rhodes, Copenhagen).

Rosenfeld, L. (1979a). A Voyage to Laplacia, *SP*, pp. 704–708. Originally published in *Journal of Jocular Physics*, 3 (Institute of Theoretical Physics, Copenhagen, 1955).

Rosenfeld, L. (1979b). Berkeley *Redivivus*, review of W. Heisenberg, *Natural Law and the Structure of Matter*, *SP*, pp. 686–687. Originally published in *Nature*, **228**, 479 (1970).

Rosenfeld, L. (1979c). Bibliography of the Writings of Léon Rosenfeld, in *SP*, pp. 911–921.

Rosenfeld, L. (1979d). Complementarity and Statistics I and II, *SP*, pp. 484–494. Originally published in *Det Kongelige Norske Videnskabernes Selskabs Forhandlinger*, **31**(9–10) (1958).

Rosenfeld, L. (1979e). Epistemology on a Scientific Basis, in *SP*, pp. 643–654. Originally published in Norwegian in *Physica Norvegica*, **5**, 319–326 (1971).

Rosenfeld, L. (1979f). Jacques Solomon (obituary), in *SP*, pp. 297–301. Originally published in *A la Mémoire de 15 savants français assassinés par les Allemands 1940–1945* (Comité à la mémoire des savants français victimes de la barbarie allemande 1940–1945, Paris, 1959), pp. 25–28.

Rosenfeld, L. (1979g). Men and Ideas in the History of Atomic Theory, *SP*, pp. 266–296. Originally published in *Archive for the History of Exact Sciences*, **7**, 69–90 (1971).

Rosenfeld, L. (1979h). Misunderstandings About the Foundations of Quantum Theory, in *SP*, pp. 495–502. Originally published in *Observation and Interpretation: A Symposium of Philosophers and Physicists, Proceedings of the Ninth Symposium of the Colston Research Society held in the University of Bristol 1st April–4th April*, (ed.) S. Körner (Butterworths Scientific Publications, London, 1957), pp. 41–45.

Rosenfeld, L. (1979i). My initiation, *SP*, pp. xxxi–xxxiv. Originally published in *Journal of Jocular Physics*, **2**(7) (Institute of Theoretical Physics, Copenhagen, October 1945).

Rosenfeld, L. (1979j). Niels Bohr: An Essay Dedicated to Him on the Occasion of His Sixtieth Birthday, October 7, 1945, in *SP*, pp. 313–326. Originally published by North-Holland, Amsterdam.

Rosenfeld, L. (1979k). Nuclear Reminiscences, *SP*, pp. 335–345. Originally published in *Cosmology, Fusion and Other Matters, George Gamow Memorial Volume*, (ed.) F. Reines (Colorado Associated Univeristy Press, 1972), pp. 289–299.

Rosenfeld, L. (1979l). On Quantization of Fields, *SP*, pp. 442–445. Originally published in *Nuclear Physics*, **40**, 353–356 (1963).

Rosenfeld, L. (1979m). On Quantum Electrodynamics, *SP*, pp. 413–441. Originally published in *Niels Bohr and the Development of Physics* (Pergamon Press, London, 1955), pp. 70–95.

Rosenfeld, L. (1979n). On the Energy-Momentum Tensor, in *SP*, pp. 711–735. Originally published in French in *Mémoires de l'Academie royale de Belgique*, **18**(6), 1–30 (1940).

Rosenfeld, L. (1979o). On the Foundations of Statistical Thermodynamics, in *SP*, pp. 762–807. Originally published in *Acta Physica Polonica*, **14**, 3–39 (1955).

Rosenfeld, L. (1979p). Questions of Irreversibility and Ergodicity, *SP*, pp. 808–829. Originally published in *Ergodic Theories*, (ed.) P. Caldirola, (*Rendiconti della Scuola Intern. D. Fisica "Enrico Fermi"*), corso 14, Varenna 1960 (Zanichelli, Bologna, 1962), pp. 1–20.

Rosenfeld, L. (1979q). Review of J. D. Bernal, *Science in History*, *SP*, pp. 7–15. Originally published in *Centaurus*, **4**, 285–296 (1956).

Rosenfeld, L. (1979r). Social and Individual Aspects of the Development of Science, *SP*, pp. 902–910. Originally published in Danish in *Fysisk Tidsskrift*, **69**(4–5), 97–106 (1971). English translation in *Problems of Theoretical Physics* (Igor Tamm Festschrift) (Nauka, Moscow, 1972).

Rosenfeld, L. (1979s). Statistical Causality in Atomic Theory (A General Introduction to Irreversibility), *SP*, pp. 547–570. Originally published in *The*

Interaction Between Science and Philosophy, (ed.) Y. Elkana (The van Leer Jerusalem Foundation Series) (Humanities Press, New York, 1974), pp. 469–480.

Rosenfeld, L. (1979t). Strife about Complementarity, *SP*, pp. 465–483. Originally published in *Science Progress*, **163**, 393–410 (1953).

Rosenfeld, L. (1979u). The Epistemological Conflict between Einstein and Bohr, *SP*, pp. 517–521. Originally published in *Zeitschrift für Physik*, **171**, 242–243 (1963). This paper is essentially the same as Professor Einstein's Dilemma, *The Listener*, **44**, 823 (1950). Published in French in *Revue de Métaphysique et de Morale*, **67**, 147–151 (1962), and subsequently published in Russian (Rosenfeld (1962a)), German, and English.

Rosenfeld, L. (1979v). The Evolution of the Idea of Causality, *SP*, pp. 446–464. Originally Rosenfeld's inaugural lecture in Utrecht and published in Dutch in 1942, *Inaugurale rede Universiteit Utrecht* (Amsterdam, 1942), pp. 1–27. Published in French the same year, *Mém. Soc. Sc. Liège*, **6**, 59–87 (1942). Published in Norwegian in 1944, *Fra Fysikkens Verden*, p. 49.

Rosenfeld, L. (1979w). The First Phase in the Evolution of the Quantum Theory, *SP*, pp. 193–234. Originally published in French in *Osiris*, **2**, 149–196 (1936).

Rosenfeld, L. (1979x). The Measuring Process in Quantum Mechanics, *SP*, pp. 536–546. Originally published in *Progress of Theoretical Physics, Supplement, Commemoration Issue for the 30th Anniversary of the Meson Theory by Dr. H. Yukawa* (1965), pp. 222–231.

Rosenfeld, L. (1979y). The Method of Physics, in *SP*, pp. 614–636. Originally published as a Unesco Report in 1968.

Rosenfeld, L. (1979z). The Organization of Scientific Research, *SP*, pp. 881–891. Originally published in French in *Hommage national à Paul Langevin et Jean Perrin* (Paris, 1948), pp. 50–60.

Rosenfeld, L. and Møller, C. (1943). Electromagnetic Properties of Nuclear Systems in Meson Theory, *Det Kongelige Danske Videnskabernes Selskabs Mathematisk-fysiske Meddelelser*, **20**(12), pp. 1–66.

Rosenfeld, L., George, C. and Prigogine, I. (1979). The Macroscopic Level of Quantum Mechanics, *SP*, pp. 571–598. Originally published partly in *Det Kongelige Danske Videnskabernes Selskabs Mathematisk-fysiske Meddelelser*, **38**(12) 1–44 (1972), and partly in *Nature*, **240**, 25–27 (1972).

Rosenfeld, L. and Solomon, J. (1931a). Sur la théorie quantique du rayonnement, *Journales de Physique*, **2**(7), pp. 139–147.

Rosenfeld, L. and Solomon, J. (1931b). Zur Theorie der Hohlraumstrahlung, *Die Naturwissenschaften*, **19**, p. 376.

Rosenfeld, L. and Witmer, E. E. (1928a). Über die Hohlraumstrahlung und die Lichtquantenhypothese, *Zeitschrift für Physik*, **47**, pp. 517–521.

Rosenfeld, L. and Witmer, E. E. (1928b). Über die Beugung der de Broglischen Wellen am Krystallgitter, *Die Naturwissenschaften*, **16**, p. 149.

Rosenfeld, L. and Witmer, E. E. (1928c). Über die Beugung der de Broglischen Wellen am Krystallgittern, *Zeitschrift für Physik*, **48**, pp. 530–540.

Rosenfeld, L. and Witmer, E. E. (1928d). Über den Brechungsindex der Elektronenwellen, *Zeitschrift für Physik*, **49**, pp. 534–540.

Rozental, S. (1967). The Forties and the Fifties, in *Niels Bohr: His Life and Work as Seen by His Friends and Colleagues*, (ed.) S. Rozental (North-Holland, Amsterdam), pp. 149–190.

Rozental, S. (1971). Interview with Léon Rosenfeld at Rosenfeld's summerhouse, Tisvilde, Denmark, NBA.

Rozental, S. (1975). Léon Rosenfeld 14 August 1904–23 March 1974, *Fysisk Tidsskrift*, **73**(3), pp. 99–105.

Salamandre (1940). Monsieur Rosenfeld, *l'Etudiant Libéral*, 20 February.

Salisbury, D. C. (2006). Peter Bergmann and the Invention of Constrained Hamiltonian Dynamics, http://arxiv.org/PS_cache/physics/pdf/0608/0608067v1.pdf (accessed 10 May 2011).

Salisbury, D. (2007). Rosenfeld, Bergmann, Dirac, and the Invention of Constrained Hamiltonian Dynamics, http://arxiv.org/PS_cache/physics/pdf/0701/0701299v1.pdf (accessed 10 May 2011).

Salisbury, D. (2009). Translation and Commentary of Leon Rosenfeld's "Zur Quantelung der Wellenfelder", see Rosenfeld (1930b), Preprint 381, Max Planck Institute for the History of Science, Berlin.

Schaffer, S. (1984). Newton at the Crossroads, *Radical Philosophy*, **37**, pp. 23–28.

Schiff, L. I. (1949). *Quantum Mechanics*, 1st edn., 4th impression (McGraw-Hill, New York).

Schirrmacher, A. (2007). Max Born und Politik. Auf der Suche nach Verantwortung in einem verlorenen Vaterland, unpublished ms. (Talk given on the occasion of Max Born's 125th birthday at a symposium at the Max Born Institute for Nonlinear Optics and the Max Planck Institute for the History of Science, Dec. 11/12).

Schrödinger, E. (1952). Are There Quantum Jumps? Part I and II, *The British Journal for the Philosophy of Science*, **3**(10), pp. 109–123 and **3**(11), pp. 233–242.

Schweber, S. S. (1986). The Empiricist Temper Regnant: Theoretical Physics in the United States 1920–1950, *Historical Studies in the Physical and Biological Sciences*, **17**(1), pp. 55–98.

Schweber, S. S. (1994). *QED and the Men Who Made It: Dyson, Feynman, Schwinger, and Tomonaga* (Princeton University Press, Princeton).

Schweber, S. S. (2000). *In the Shadow of the Bomb: Oppenheimer, Bethe, and the Moral Responsibility of the Scientist* (Princeton University Press, Princeton).

Schweber, S. S. (2002). Enrico Fermi and Quantum Electrodynamics: 1929–1932, *Physics Today*, June, pp. 31–36.

Schweber, S. S. (2003). Fermi and Quantum Electrodynamics (QED), *Proceedings of the International Conference "Enrico Fermi and the Universe of Physics", Rome, September 29–October 2, 2002*, (eds.) C. Bernardini, L. Bonolis, G. Ghisu, D. Savelli and L. Falera (ENEA, Rome), pp. 167–197.

Schwinger, J. (1958). *Selected Papers on Quantum Electrodynamics* (Dover, New York).

Scott, W. T. and Moleski, M. X., S. J. (2005). *Michael Polanyi: Scientist and Philosopher* (Oxford University Press, Oxford).

Segré, G. (2007). *Faust in Copenhagen: A Struggle for the Soul of Physics* (Jonathan Cape, London).

Serpe, J. (1980). Léon Rosenfeld (1904–1974), *Académie Royale de Belgique, Classe des sciences Bruxelles*, pp. 389–402.

Seth, S. (2010). *Crafting the Quantum: Arnold Sommerfeld and the Practice of Theory*, (1890–1926) (MIT Press, Cambridge Massachusetts).

Sheehan, H. (1993). *Marxism and the Philosophy of Science. A Critical History*, 2nd edn. (New Jersey, Humanities Press International, New Jersey).

Smith, A. K. (1955). Sidelights on Geneva, *Bulletin of the Atomic Scientists*, **11**(8), p. 276.

Smyth, H. D. (1945). *Atomic Energy for Military Purposes*, Princeton University Press, Princeton.

Social-Demokraten (1932). 24 Nov, http://www.marxist.dk/Home.php?Page=Artikel&ArtikelID=1748 (accessed 14 April 2011).

Söderqvist, T. (1998). *Hvilken kamp for at undslippe: En biografi om immunologen og nobelpristageren Niels Kaj Jerne* (Borgen, Copenhagen).

Söderqvist, T. (2006). What is the use of writing lives of recent scientists? In *The Historiography of Contemporary Science, Technology, and Medicine: Writing recent science*, (eds.) R. E. Doel and T. Söderqvist (Routledge, London), pp. 99–127.

Solomon, M. J. (1931). 1re Thèse: L'Électrodynamique et la Théorie des Quanta. 2e Thèse: Propositions donnés par la Faculté, Doctoral Dissertation, La Faculté des Sciences de l'Université de Paris.

Solomon, J. (1933). Remarques sur la théorie du rayonnement, *Journal de physique et le radium*, **4**(7), July, pp. 368–387.

Somsen, G. J. (2008). Value-Laden Science: Jan Burgers and Scientific Politics in the Netherlands, *Minerva*, **46**(2), pp. 231–245.

Stachel, J. (1999). The Early History of Quantum Gravity, in *Black Holes, Gravitational radiation and the Universe*, (eds.) B. R. Iyer and B. Bhawai (Kluwer, Dordrecht), pp. 528–532.

Stanley, M. (2007) *Practical Mystic: Religion, Science and A. S. Eddington* (The University of Chicago Press, Chicago).

Strauss, M. (1936). Komplementarität und Kausalität im Lichte der logischen Syntax, *Erkenntnis*, **6**, pp. 335–339.

Strauss, M. (1972). *Modern Physics and its Philosophy: Selected Papers in the Logic, History, and Philosophy of Science* (D. Reidel, Dordrecht).

Stuewer, R. H. (1985). Bringing the news of fission to America, *Physics Today*, October, pp. 48–56.

Swings, P. (1964). Deux contributions cruciales de Léon Rosenfeld en astrophysique moléculaire, *Nuclear Physics*, **57**, pp. 299–302.

Swings, P. (1974). In memoriam Léon Rosenfeld, *Académie Royale de Belgique, Bulletin de la Classe de Sciences Bruxelles*, **60**(5), pp. 656–662.

Terlezki, J. P. (1951). *Voprosy Filosofii*, **5**, p. 51.

Terlezki, J. P. (1952). Probleme der Entwicklung der Quantentheorie, *Sowjetwissenschaft: Naturwissenschaftliche Abteilung*, **5**(4), pp. 597–608.

The Manchester Guardian (1950). Professor Bohr, 14 June.

Thing, M. (1993). *Kommunismens Kultur: DKP og de intellektuelle 1918–1960*, 2 Vols. (Tiderne Skifter, Århus).

Tiden (1947–1948). Diskussion om Sovjetfilosofi i Moskva Juni 1947, *Tiden: Tidsskrift for aktivt Demokrati*, **9**, pp. 127–134.

Time (1956). Science: Soviet-Controlled Fusion, Monday 7 May 1956.

Trotsky, L. (1932a). In defence of October. A speech delivered in Copenhagen, Denmark, November 1932. http://www.marxists.org/archive/trotsky/1932/11/oct.htm (Accessed 15 April 2011), p. 21.

Trotsky, L. (1932b). *The History of the Russian Revolution*, http://www.marxists.org/archive/trotsky/1930/hrr/ (accessed 14 April 2011).

Ullrich, W. (2007). The British Government and the Second World Peace Congress: The Role of Perception in Cold War Policy Making, Master's thesis, The London School of Economics and Political Science.

van den Burg, F. (1983). *De Vrije Katheder 1945–1950: Een platform van communisten en niet-communisten* (Amsterdam).

van Dongen, J. (2007). Emil Rupp, Albert Einstein, and the Canal Ray Experiments on Wave–Particle Duality: Scientific Fraud and Theoretical Bias, *Historical Studies in the Physical and Biological Sciences*, **37** (Supplement), pp. 73–119.

van Walsum, S. (1995). *Ook al Voelt Men Zich Gewond: De Utrechtse Universiteit Tijdens de Duitse Bezetting 1940–1945* (Universiteit Utrecht, Utrecht).

Vavilov, S., Frumkin, A. N., Ioffe, A. F. and Semyonov, N. N. (1948). Open Letter to Dr. Einstein — From four Soviet Scientists, *Bulletin of the Atomic Scientists*, **4**(2), pp. 34, 37.

von Meyenn, K. (ed.) (1993–2005). *Wolfgang Pauli. Scientific Correspondence with Bohr, Einstein, Heisenberg, a.o.*, 2 Vols., Vols. 3 and 4 (Springer, Berlin).

von Neumann, J. (1927a). Mathematische Begründung der Quantenmechanik, *Nachrichten von der Gesellschaft der Wissenschaften zu Göttingen*, pp. 1–57.

von Neumann, J. (1927b). Wahrscheinlichkeitsteoretischer Aufbau der Quantenmechanik, *Nachrichten von der Gesellschaft der Wissenschaften zu Göttingen*, pp. 245–272.

von Neumann, J. (1932). *Mathematische Grundlagen der Quantenmechanik* (Springer, Berlin).

von Weizsäcker, C. F. (1985). A Reminiscence From 1932, in *Niels Bohr: A Centenary Volume*, (eds.) A. P. French and P. J. Kennedy (Harvard University Press, Cambridge Massachusetts), pp. 183–190.

Vucinich, A. (2001). *Einstein and Soviet Ideology* (Stanford University Press, Stanford).

Wali, K. C. (1991). *Chandra: A Biography of S. Chandrasekhar* (The University of Chicago Press, Chicago).

Walker, M. (1989). *German National Socialism and the Quest for Nuclear Power 1939–1949* (Cambridge University Press, Cambridge).

Walker, M. (1995). *Nazi Science: Myth, Truth, and the German Atomic Bomb* (Plenum, New York).

Wamberg, N. B. (1980). Hein, Piet, *Dansk Biografisk Leksikon*, 3rd edn., Vol. 6 (Gyldendal, Copenhagen), pp. 188–190.

Wang, J. (1999). *American Science in the Age of Anxiety: Scientists, Anticommunism, and the Cold War* (The University of North Carolina Press, Chapel Hill).

Warmbrunn, W. (1963). *The Dutch under German Occupation 1940–1945* (Stanford University Press, Stanford, California).

Warwick, A. (1992). Cambridge Mathematics and Cavendish Physics: Cunningham, Campbell and Einstein's Relativity 1905–1911 Part I: The Uses of Theory, *Studies in History and Philosophy of Science*, **23**(4), pp. 625–656.

Weart, S. R. (1977). Interview with Subrahmanyan Chandrasekhar, 17 May 1963, AIP. http://www.aip.org/history/ohilist/4551_1.html (accessed 3 May 2011).

Weart, S. R. (1979). *Scientists in Power* (Harvard University Press, Cambridge, Mass).

Weart, S. R. and Szilard, G. W. (eds.) (1978). *Leo Szilard: His Version of the Facts. Selected Recollections and Correspondence* (MIT Press, Cambridge Massachussetts).

Weisskopf, V. (1939). On the Self-Energy and the Electromagnetic Field of the Electron, *Physical Review*, **56**, pp. 72–85. Reprinted in Schwinger (1958), pp. 68–81.

Wentzel, G. (1973). Quantum Theory of Fields (until 1947), in *The Physicists's Conception of Nature*, (ed.) J. Mehra (Reidel, Dordrecht), pp. 380–403.

Werskey, G. (1988). *The Visible College: A Collective Biography of British Scientists and Socialists of the 1930s* (Free Association Books, London).

Werskey, G. (2007a). The Visible College Revisited: Second Opinions on the Red Scientists of the 1930s, *Minerva*, **45**, pp. 305–319.

Werskey, G. (2007b). The Marxist Critique of Capitalist Science: A History in Three Movements? *Science as Culture* http://www.human-nature.com/science-as-culture/werskey.html (Accessed 3 September 2009).

Wigner, E. P. (1961). Remarks on the Mind-Body Question, in *The Scientist Speculates*, (ed.) I. J. Good (W. Heinemann, London), pp. 284–302. Reprinted in *Quantum Theory and Measurement*, (ed.) J. A. Wheeler and W. H. Zurek (Princeton University Press, Princeton), pp. 168–181.

Wigner, E. P. (1963). The Problem of Measurement, *American Journal of Physics*, **31**, pp. 6–15.

Williams, L. P. (1957). Review of J. D. Bernal, *Science in History*, *Isis*, **48**, pp. 471–473.

Zhdanov, A. A. (1947). Vystuplenie na diskussiii po knige G. F. Aleksandrova "Istoriia zapadnoevropeiskoi filosofii", 24 iiunia.

Zilsel, E. (1935). P. Jordans Versuch, den Vitalismus quantenmechanisch zu retten, *Erkenntnis*, **5**, pp. 56–64.

Index

Abrikosov, A. A., 88
Abyssinian War, 122
Adler, Ellen, 56
Advisory Committee on Atomic Energy, 233
Anderson, Carl D., 35, 81
Anderson, Sir John, 195, 241
Anti-communism, 186, 188, 191, 232, 234–235
Aspect, Alain, 304
Atomic bomb, 173, 183, 186–188, 194–199, 208, 240, 241, 251
    British, 173, 195, 233
    German, 169, 176
    Russian, 190, 235
    see also Manhattan Project and hydrogen bomb
Atomic Energy Commission of the United Nations, 187–188
Atomic Scientists' Association, 191
Atoms for Peace, 290, 291

Baker, John, 226
Baruch, Bernard Mannes, 187, 241
Baruch Plan, 187–188
Bates, D. R., 306
Bauer, Edmund, 17, 142
Beck, Guido, 117, 278
Belgian American Educational Foundation, 144, 155
Belgium, 6, 13, 14, 46, 75, 150, 154, 238
    Flemish part, 12, 115–116
    German occupation of, 153, 161–162, 165–166, 174–175, 177, 179, 181
    history of science in, 13
    political situation in, 12, 106, 115–116, 119, 188, 203
    science and engineering in, 13, 14, 54, 154–158, 160, 237–239
    Walloon part, 12, 115, 238
Bell, John Stewart, 304–305, 307
Bergman, Peter, 38
Berlin crisis 1948, 190
Bernal, John Desmond, 112, 120, 143, 205, 207, 210, 211, 212, 215, 217, 220, 223, 224, 226, 227, 228, 229–232, 248, 250, 284, 289
    Science in History, 224, 229–232, 289
    The Social Function of Science, 120, 212, 215–216
Bernalism, 212, 227–228
Bethe, Hans, 11
Bieberbach, Ludwig, 123
Biquard, Pierre, 119, 231, 247
Blackett, P. M. S., xii, 35, 81, 110, 112, 120, 143, 190, 205, 213–215, 217, 224, 232–233, 238, 239, 271, 273
Bloch, Felix, 51
Blokhintsev, D. I., 261, 262, 275, 276, 295, 297, 298, 300–301
Blum, Léon, 119
Bohm, David, 9, 141, 255, 271–282, 284, 285, 286, 295, 297, 300, 304, 305–306, 307
Bohr, Aage, 86, 173, 249–250, 279, 291–292, 296, 297, 298, 313
Bohr, Hans, 114, 148
Bohr, Harald, 56, 123, 181, 196

Bohr, Margrethe, xi, 56, 59, 108, 109, 111, 114, 161, 169, 170, 172, 176, 241, 242–243, 269, 298
Bohr, Marietta, 298
Bohr, Niels, v, vi, xi, 1–5, 11, 14, 17, 18, 21, 25, 26, 27, 31, 32, 35, 41, 79, 81, 95, 97, 105, 122, 123, 162–163, 169, 170, 172, 175, 181, 182, 193, 205, 214, 216, 223, 226, 240, 255, 271, 277, 312–313
  and the atomic bomb, 153, 172–173, 186, 194–195, 197, 208, 240–242
  and Einstein, 20, 26, 53, 65–66, 90–94, 112, 124, 255, 257
  and fission, 144–149
  and philosophy, 52, 124–134, 256, 258, 260–262, 267, 268–270, 279, 284, 285, 290, 296–297, 299–300, 305
  collaboration and relationship with Rosenfeld, see Rosenfeld
  controversy with Landau and Peierls, 46, 49, 61, 69–75, 82–89, 111, 124
  disagreement with Pauli, 72–74, 84–85
  interpretation of quantized fields of, 63, 67, 70, 77, 83–84, 88
  interpretation of quantum mechanics of, 20, 25, 32, 50, 53, 63–64, 97, 112, 124, 128, 131, 132, 133, 138, 140–141, 171, 186, 256, 258, 265, 269, 270, 274, 276, 277, 278, 280, 281, 282, 295, 297, 300–301, 303–304, 305
  leadership of, 31, 51–52, 54, 60, 72, 85, 117, 129, 167–168, 171, 212, 227, 243, 247, 248–251, 260, 269, 284, 285, 288, 296, 304
  liquid drop model of, 144
  meeting with Heisenberg 1941, 169, 209
  on quantum measurement, 22–23, 61, 63–68, 71, 74–76, 81–87, 89, 93–94, 139, 257–258, 295–296, 303, 308–309, 311–312
  Open Letter to the United Nations of, 195, 209, 242–246, 248–249
  principle of the Open World of, 8, 196–197, 240–244, 247, 248–251, 291–292, 298
  political engagement of, 109, 121, 129, 143, 176, 185, 193, 195–196, 208–209, 212–213, 233–234, 240–244, 246, 249–252, 284, 288, 291–292
  Soviet Union, 1934 visit to, 7, 87–88, 95, 108–114, 124–125, 290
  Soviet Union, 1937 visit to, 114
  Soviet Union, 1961 visit to, 298
  working method, 2, 51, 56–60, 260
Bohr-Rosenfeld paper
  1933, 1, 5, 6, 49, 61–62, 67, 68, 77, 81–91, 112, 309, 311
  1950, 1, 257, 311
Born, Max, 8, 20, 52, 231, 265
  and the interpretation of quantum mechanics, 61, 277, 287
  and matrix mechanics, 20
  as Rosenfeld's employer, 6, 18, 26–28
  friendship with Rosenfeld, 286–290, 294
  Nobel Prize of, 286–287
  opposition to Rosenfeld's Marxism, 9, 286, 288–290
  probabilistic interpretation of wave function of, 20–21, 51
  political views of, 27, 224, 231, 250, 251, 288–289, 294
Boscovich, Ruggiero, 59
Breakthrough Movement, 121

Bridgman, P. W., 251
Brillouin, Léon, 15
Britain's War Cabinet, 120
British Academic Assistance Council (AAC), 116–117
British Association for the Advancement of Science, 120, 214, 247
British Association of Scientific Workers, 120
British Cultural Committee for Peace, 221
British Society for the History and Philosophy of Science, 287
Bronstein, Matvei Petrovich, xi, 11, 48, 107, 112, 113, 143
Brown, G. E., 271
Bruno, Giordano, 299
Brussels, 18, 20, 25, 48, 91, 114, 117, 238, 239, 312
Bukharin, Nikolai, 109, 111
Bulletin of Atomic Scientists, 232, 233
Bunge, Mario, 285
Burau, Florent-Joseph, 238
Burgers, Johannes Martinus (Jan), 210
Burhop, E. H. S., 235, 278
Bush, Vannevar, 209

Carnap, Rudolf, 58, 130
Cartan, Élie Joseph, 15
Casimir, Hendrik B. G., 54, 56, 159, 173, 174, 180, 182, 214
Castle Bravo Nuclear Test, 251, 288, 292
Causality, 21, 23, 84, 121, 129–131, 160, 171, 281, 308
Centaurus, 207, 231
Central Intelligence Agency (CIA), 235–236
CERN, 239
Chadwick, James, 173, 233
Chandrasekhar, Subrahmanyan, 14, 54, 82, 93, 103–104, 110–111, 119, 136–137, 150, 181, 182, 216, 238
Childe, V. G., 207
Churchill, Winston, 189, 195
Cockcroft, John D., 199, 233
Cohen, Robert S., vi, 208
Collège de France, 15, 16
Cominform (Communist Information Bureau), 189, 190, 218
Comintern (Communist International), 99, 118, 151, 189
Commission for the History of the Social Relations of Science, 206–207
Complementarity, v, 1, 2, 3, 7, 20, 51, 56, 61, 75, 90, 93, 94, 96, 97, 131, 171, 223, 256, 257, 258, 259, 267, 269, 270, 276, 277, 301
  and dialectical materialism, v, 5, 8, 96, 140–141, 186, 260, 263, 277, 279–286, 289–290, 295, 298, 299, 300
  and wave-particle duality, 63, 138, 140–141, 301
  popularization of, 96, 97, 121, 124, 131–133, 138–140, 256, 270
Compton scattering, 22
Conant, James B., 209
Congress of Cultural Freedom Committee on Science and Freedom, 226
Conservatoire des Arts et Métiers, 15
Constrained Hamilton dynamics, 1, 38
Copenhagen Conference
  1929, xi, 31–32, 50
  1931, 72
  1932, xi, 51, 75, 77, 80, 98, 111
  1933, xii, 117–118
  1936, 114, 129
  1937, 141
Copenhagen Interpretation, 2, 25, 61, 258, 259, 262, 268, 277, 286, 287, 304
Copenhagen School, 227
Copenhagen spirit, 50, 51, 60, 149, 205, 265

Copernicus, Nicolaus, 300
Corinaldesi, Ernesto, 87
Correspondence principle, 21, 50, 62, 70
Cortesao, Armando, 206, 207, 227
Cosyns, Max, 239
Crowther, J. G., 108, 111–113, 120, 129, 205, 209–210, 211, 213, 214, 217, 219, 220–221, 226, 227, 231
Curie, Marie, 16
Czechoslovakia, 212
    Coup d'état 1948, 190, 235
    Prague Spring, 294

Daneri, Adriana, 311
Danin, Daniil, 90, 216, 224, 256, 268, 269
Danish Organization for the Protection of Scientific Work, 212
Danish Committee for the Support of Refugee Intellectual Workers, 117
Darwin, Charles Robert, 32
Darwin, Charles Galton, 233
Deborin, Abram Moiseevich, 261
de Broglie, Louis, 6, 11, 15–17, 20–21, 23–25, 31, 65, 259, 262, 263, 271, 274, 277, 282, 283, 287, 288, 295, 297, 307
    de Broglie wavelength, 16
    pilot wave theory of, 20–21, 25, 274, 277, 281–282
de Donder, Théophile, 18
Degrelle, Léon, 115–116
Dehalu, Marcel, 15
Delbrück, Max, 51, 54, 79, 117, 129, 130, 216
Denmark, 5, 31, 55, 97, 117, 121, 135, 170, 194, 207, 213, 241, 296
    German occupation of, 150, 161, 162, 166, 168–169, 172, 178, 210
    political situation in, 99–101, 106, 168, 172, 188, 193, 196, 203
Demkov, Yuri, v
d'Espagant, Bernard, 307–308
Destouches, Jean Louis, 271, 282, 283
Determinism, 23–24, 75, 127, 128, 140, 161, 260, 261, 271, 274, 275, 283, 284, 287
Devons, Samuel, 239–240
De Vrije Katheder, 192–193, 203, 204
DeWitt, Bryce S., 307
DeWitt-Morette, Cécile, vii, 236–237
Dialectical materialism, v, 5, 8, 95, 96, 123, 140, 156–157, 201, 204, 205, 206, 216, 218, 224, 228, 260, 261, 263, 266, 269, 270, 271, 278, 279, 280, 283, 285, 286, 289, 290, 295, 299
Dirac, P. A. M., xi, 11, 25, 38, 60, 63, 104
    and the Soviet Union, 27, 108, 110
    critique of Heisenberg-Pauli theory, 42, 76–77
    "hole" theory of, 34–35, 43, 73, 78, 81
    in Göttingen, 27, 28
    quantum mechanics of, 24, 25
    quantum electrodynamics of, 32–35, 36, 37, 38, 39, 42, 43, 49, 61, 62, 68, 73, 75–81, 83, 84

École Normale Superieure, 15
Eddington, Sir Arthur, 127–128, 136–138
"Eddington-Jeans" idealism, 127–128, 136–138, 256
Ehrenfest, Paul, 27, 32, 51, 52, 53, 60, 80, 97, 117
Ehrenfest, Tatyana Alexeyevna Afanasyeva, 180
Ehrenfest, Tatyana Pavlovna, 27
Einstein, Albert, 11, 16, 20, 27, 29, 52, 53, 91, 112, 117, 124, 132, 137, 139, 159, 271, 288
    and World Government, 196, 233, 234
    EPR paper, 6–7, 20, 26, 50, 90–94, 140

opposition to quantum mechanics of, 25–26, 65, 255, 257, 259, 276, 281, 287
Photon box experiment, xi, 65–66
unified field theory of, 26
Eisenhower, Dwight D., 290, 291
Emmet, Dorothy, 224
Engels, Friedrich, 16, 125, 204, 228, 259, 260, 261, 263, 265–266, 285, 289, 290, 299
Erenburg, Ilja, 218
Evans, M. G., 224
Everett, Hugh, 304

Fadeyev, Alexander, 218
Farge, Yves, 208, 218
Farrington, Benjamin, 206, 207, 230, 231, 232
Fascism, 95, 96, 114–124, 188, 192, 198, 199, 200, 203, 212, 218, 221, 222, 235
Faust parody, 51, 55, 136
Federal Bureau of Investigation (FBI), 235–236, 272
Federation of American Scientists, 232, 237
Fermi, Enrico, 11, 36, 37, 39, 42, 43
Feynman, Richard, 257
First World War, 12, 46, 114, 115
Fission, 144–149
Fock, Vladimir A., v, 38, 67, 80, 112, 113, 114, 143, 255, 269, 294, 295–299
Fokker, Adriaan Daniel, 166, 173
Fond National de la Recherche Scientifique (FNRS), 35, 154
Forbes, R. J., 207
Franck, James, 117
Frank, Philipp, 126–132, 133, 136, 142
Francqui, Emile, 237
Francqui Foundation, 155
Francqui Prize, 1, 15, 45, 237–238
Freistadt, Hans, 271, 285
French Resistance, 17
Frenkel, Yakov I., 107, 113, 114, 276
Fries, Jakob Friedrich, 131
Frisch, Otto Robert, 117, 144–149
Frolich, H., 214
Frumkin, A. N., 233
Fuchs, Klaus, 191, 236

Galilei, Galileo, 299
Gamow, George, xi, 11, 27, 29, 45, 46–48, 54, 66, 69, 70, 72, 86, 97, 107–108, 114, 123, 150
George, Claude, 312
Geneva Conference on the Peaceful Uses of Atomic Energy 1955, 251, 291
Geneva Protocol, 122
Georgia-Augusta University or University of Göttingen, 18
Gestapo, 17, 138, 143, 165, 169, 177–178, 183
Goethe, Johann Wolfgang von, 51, 135
Gordon, Walter, 113, 117
Göttingen, xi, 6, 18, 20–32, 95, 112, 131, 134
Greek Civil War, 189, 241, 244
Groves, Leslie, 173
Grünbaum, Adolf, 127, 258, 285

Hafnium, 53
Hahn, Otto, 144, 147, 148, 176, 250, 288
Haldane, J. B. S., 120, 217, 224
Halifax, Lord, 195
Hall, Arthur Rupert, 232
Hartree, D. R., 213
Have, Margrethe, 54, 67
Hegel, Friedrich, 267
Hein, Piet, xi, 66, 96–98, 101–102, 129, 130, 234
Heisenberg, Werner, 7, 11, 15, 33, 45, 52, 60, 63, 85, 86, 91, 92, 123, 132, 153, 159, 176, 180–181, 216, 229, 278
and the interpretation of quantum mechanics, 50, 61, 71, 134–135, 258, 268, 270, 274, 276

and the Landau-Peierls paper, 67–71
critique of Dirac, 75, 77, 79
gamma ray microscope experiment of, 21–24, 65
Copenhagen, 1941 visit to, 169, 209
Copenhagen, 1944, visit to, 176
Holland, 1943 visit to, 153, 172–174
matrix mechanics of, 20, 70, 71, 76, 258, 286
opposition to Rosenfeld's Marxism, 9, 286, 289–290
quantum field theory of, 35–44, 49, 61, 62, 67, 69–72, 76–84
relationship with Bohr, 23, 54, 56, 69, 73
relationship with Rosenfeld, 135, 168, 175, 182, 289–290
Heitler, Walter, xii, 27, 32, 34, 42, 43, 141, 159
Hempel, Carl Gustav, 129
Henriot, M., 238
Hermann, Grete, 131, 134–135
Herneck, Friedrich, 26, 116, 117, 125, 204, 211, 230, 231, 261, 264, 285
Hessen, Boris M., 110, 143
Hevesy, George, 130, 170, 313
Hidden parameters, 84, 135, 255, 262, 273–276, 278, 281, 301, 304–305, 306
Hilbert, David, 18, 30
Hitler, Adolf 7, 55, 116, 123, 135, 150–151, 161, 188
Hobart Ellis Jr., R., 217
Høffding, Harald, 125
Holland, see The Netherlands
Hoppe, Edmund, 18
Houtermans, Charlotte, 143
Houtermans, Frederick (Fritz) G., 27, 105, 143
Humblet, Jean, 50
Hulthén, Lamek, 166, 167, 171
Huxley, Julien, 218
Huygens, Christiaan, 205
Hydrogen bomb, 187, 191, 251, see also Castle Bravo Nuclear Test

Idealized measurement, 21–23, 63–68, 74–77, 82–94, 257, 309, 311
Infeld, Leopold, 251
Institut Henri Poincaré, 17, 69, 75
Institute for Theoretical Physics, 6, 7, 14–15, 17, 49–59, 62, 63, 67, 69, 96–98, 111, 117, 122–123, 129–130, 144–146, 149, 153, 155, 169, 170, 175–176, 227, 269, 292, 308, 313
International Academy for the History of Science, 138, 206, 207
International Education Board, 25
Isis, 13, 18, 26, 207
Ivanenko, Dmitry, 48

James, William, 267
Jánossy, Lajos, 284
Jauch, J. N., 307, 308
Jazz Band, 48
Jeans, James, 127–128, 136–138
Jensen, H. G., 299
Jensen, H. H., 299
Joffe, Abram F., 108, 111, 233
Joliot-Curie, Frédéric, 17, 119, 143, 211, 212, 215, 222, 240, 247–251, 282, 283–284
Joliot-Curie, Iréne, 17, 119, 143, 218
Jordan, Pascual, 11, 26, 29, 32, 42, 118, 123, 128, 129–131, 133–135, 241, 262, 265, 286
quantization of waves, 33–34, 36, 37, 42, 43, 44
as Rosenfeld's teacher, 18, 21, 23–24, 29, 31, 134
Jørgensen, Jørgen, 58, 128–130, 269–270, 284
Journal of Jocular Physics, 51, 122–123
Joule, James Prescott, 229
Jungk, Robert, 209

Kahn, Boris, 172

Kalckar, Fritz, 54
Kalckar, Jørgen, vii, 193, 307
Kannegisser, Genia, 46, 48, 105
Kant, Immanuel, 131, 134, 135, 222
Kapitza, Pyotr, 114, 143
Kemmer, N., 200, 214
Kharkov Conference 1934, xi, xii, 112–113
Khatlatnikov, I. M., 88
Kierkegaard, Søren, 267, 305
Klein, Felix, 123
Klein, Oskar, 7, 11, 16, 24, 34, 39, 40, 54, 56, 57, 62, 68, 69, 70, 72, 73, 79, 122–123, 137–138, 167–168, 308
Koefoed, Jørgen, 263
Korean War, 191, 235, 246, 251
Kragh, Helge, v
Kramers, Hendrik Anthony, 11, 54, 56, 160, 162, 164, 166, 168, 170, 171, 173, 175, 180–181, 313
Kronig, Ralph de Laer, 173
Krushchev, Nikita, 255, 291, 292, 293
Kruyt, H. R., 164
Kuhn, Thomas, 312–313
Kurchatov, Igor Vasilievitch, 292, 293, 298

Landau, Lev Davidovich, xi, 6, 11, 45, 46–48, 52–53, 54, 61–63, 67, 86, 90, 107, 111, 112, 113, 114, 124, 142, 143, 261, 298
Landau-Peierls paper, 46, 49, 68–75, 77, 82–83, 85–89, 90, 91, 111, 124
Langevin, Hélène, 17
Langevin, Paul, 6, 15, 17, 18, 119, 142, 226
Lemaître, George, 238
Lenin, Vladimir I., 47, 99, 202, 203, 220, 228, 260–162, 264, 266, 275, 285
  Materialism and Empirio-Criticism, 260–262, 265, 268, 275
Leray, Jean, 17
Levi, Hilde, 55, 105, 117, 170
Levy, Hyman, 218
Liège, vii, 1, 13, 30, 59, 103–104, 159, 166, 181
Lifshitz, Evgenii, 90, 308
Lilley, Samuel, 207
Lindemann, F., 159
Logical Positivists, 126–133, 134, 135, 136, 139
Loinger, Angelo, 272, 281, 304, 305, 310, 311
Lomonosov, Mikhail, 297
Lubański, Jozeph Kazimir, 166, 168
Lysenko, T. D., 207, 223, 232

Mach, Ernst, 26, 132, 261, 275, 298
Madsen, B., 299
Mainau Declaration, 250, 251, 288
Maksimov, A. A., 268–269
Manchester, vii, xii, 1, 8, 13, 14, 87, 135, 166, 203, 213, 215, 224, 229, 239, 240, 249, 271, 273
Manhattan Project, 8, 110, 173, 191, 194, 272, 273
Marshak, R. E., 236, 237
Marshall, George Catlett, 241
Marshall Plan, 186, 189, 190, 235
Marvel, Josiah, 213, 241
Marx, Karl, 32, 47, 125, 202, 259, 263, 264, 265, 267, 282, 288, 289, 299
Marxism-Leninism, 9, 186, 230, 264, 265, 276, 280, 299
Mason, S. F., 207
Matter wave, 16, 33, 34, 39–40, 42, 63–64
MAUD Committee, 233
McCarthyism, 235, 271, 273
Measurement problem, 65, 303, 308–311
Meitner, Lise, 75, 144–148, 176
Metzger, Hélène, 133, 138–140
Milne, A. A., 136
Minnaert, Marcel, 210, 211
Møller, Christian, 7, 11, 15, 49, 54–55, 61, 62, 86, 146–150, 154, 159, 162–172, 175–177, 179–182, 194, 212, 213, 214, 216, 250, 269, 296

Møller, Poul Martin, 126
Morgan, Claude, 208, 231, 241
Muller, H. J., 251
Mussolini, Benito, 116, 122, 150

Nagy, Imre, 293
Nathan, O., 299
Nazism, 95, 96, 104, 114–124, 128, 129, 135, 142, 143, 150–151, 161–169, 171–183, 192, 229, 241
Needham, Dorothy Moyle, 224
Needham, Joseph, 120, 205, 206, 207, 224
Nelson, Leonard, 131
Neurath, Otto, 58, 126–130, 142
Newman, Max, 224
Nexø, Martin Andersen, 217
Niels Bohr Archive, v, vi, viii, 3, 313
Niels Bohr Institute, v, viii, see also Institute for Theoretical Physics
Nishina, Yoshio, 170
Noether, Emmy, 28, 38, 45, 131
Nordic Institute for Theoretical Atomic Physics (NORDITA), 8, 252, 255, 297
Nørlund, Ib, 167, 168–169, 269
Nuclear Physics, 1, 300

Occhialini, G. P. S., 35, 81
Opechowski, Wladyslaw, 166, 199, 202, 205, 215, 218, 219, 220–221, 227, 259, 260, 308
Oppenheimer, Robert, 36, 41, 173, 187, 209, 234, 271
Ornstein, Leonard Salomon, 159–161, 164
Orr, Lord Boyd, 225, 234
Ørsted, Hans Christian, 60

Pais, Abraham, 58, 160, 162, 164, 166, 167–168, 171–172, 175, 180–181
Pannekoek, Anton, 207, 210, 211, 264
Paper tiger, 58
Paris, 6, 8, 9, 15–18, 26, 30, 32, 46, 69, 95, 117, 119, 132, 142, 167, 183, 213, 222, 226, 282, 283, 284
Pauli, Wolfgang, 9, 11, 12, 32, 33, 45, 52, 57, 86, 122, 142–143, 282
  and the Landau-Peierls paper, 6, 61, 67–68, 69, 71–74, 85
  as Mephistopheles in Faust, 51–52, 55
  critique of Bohr, 72–73
  critique of de Broglie, 281, 282
  critique of Dirac, 75–81
  critique of Einstein, 26, 92
  critique of Rosenfeld, 9, 282, 286, 289
  exclusion principle, 34, 35
  mentor for Rosenfeld, 6, 38–40, 45
  nickname for Rosenfeld, xii, 286
  quantum field theory of, 26, 35–44, 49, 61, 62, 69, 72, 76–84
  Relationship with Bohr, 54, 56, 69, 72–74, 84–85
Pauling, Linus, 251
Petersen, Aage, 270, 297
Peierls, Rudolf, xi, 6, 11, 45–48, 54, 61, 67, 69, 86, 105–107, 111, 124, 142–143, 165, 172, 178, 191, 200, 214, 233, 271, see also Landau-Peierls paper
Pelseneer, Jean, 13, 226
Perrin, Jean, 17, 119, 143, 226
Piaget, Jean, 224, 308
Picard, Émile, 15
Picasso, Pablo, 218, 222
Pierre, Ada, 162
Pierre, Jeanne Marie-Laure Mathilde, 12, 13, 14, 164, 165, 174, 177, 181
Pihl, Mogens, 269, 270, 299
Placzek, George, 54
Planck, Max, 11, 137, 281, 309
Platon, 290
Plesset, M. S., 113
Podolanski, Julius, 166, 168, 172

Podolsky, Boris, 6, 80
Pohl, Robert, 29
Polanyi, Michael, 224, 226, 231, 235
Politzer, George, 16
Pomeranchuk, I. Ya., 88
Pontevorco, Bruno, 191
Popper, Karl, 129–130
Popular Front, 96, 114, 116, 118–119, 142, 150–151, 199–201, 217
Powell, C. F., 251
Prenant, Eugene-Marcel, 218
Prigogine, Ilya, 58, 224, 308, 312
Proca, Alexandre, 17, 167
Prosperi, Giovanni Maria, 307, 311

Quantum dissident, 2, 5
Quantum electrodynamics, 17, 32–45, 61–89, 163, 241, 257
Quantum field theory, 32–45, 49, 61–89, 112, 150, 159, 167, 171, 262
Quantum gravity, 16, 38, 39, 41, 112

Rehberg, Poul Brandt, 234
Rexism, 115–116
Rey, Abel, 15
Roman, Paul, 293
Roosevelt, Franklin D., 195, 208, 209, 242
Rosen, Nathan, 7
Rosenberg, Ethel, 191
Rosenberg, Julius, 191
Rosenblum, Salomon, 46
Rosenfeld, Andrée, viii, 13, 14, 104, 105, 165, 177, 180
Rosenfeld, Jean L. J., viii, 104, 164–165, 169, 177, 180
Rosenfeld, Léon (1872–1918), 12–13
Rosenfeld, Léon Jacques Henri Constant, v–vii, 1–2, 5
and Bohr, collaboration with, 1–2, 6, 7, 26, 31, 49–50, 54, 55–60, 74–75, 77, 81–94, 95, 105, 108, 163, 167, 212, 214, 241, 246, communicator of Bohr's views, 1–2, 4, 8, 9, 18, 94, 96, 132–134, 137–141, 255–256, 259–261, 265, 275–286, 288–290, 295–297, 300–301, critique of, 8, 185, 193, 195–196, 214, 240, 243–246, filial relationship with, 60–61, friendship with, v-vi, 4, 162–163, 167, 170, 171, 181, 185, 192–193, 205, 208, 233–234, 240–250, 252–253
and history of science, 1, 3, 13, 15, 26, 43–44, 58–59, 96, 111–112, 156–157, 202, 203–208, 226, 227–232, 286, 287, 289, 300–301, 312–313.
and Marxism, 2–3, 7, 32, 47, 95, 156, 186, 198–199, 200–207, 210–212, 215, 218–222, 224, 226–232, 255–256, 259–261, 263–267, 270, 278–286, 288–290, 295, 300
and a position in Brussels, 238–240
appointment in Copenhagen of, 252–253, 255, 297
appointment in Manchester of, 213–214
appointment in Utrecht of, 153, 158–163
as teacher, 50, 155–158, 160, 166–168, 171, 195
attitude to Soviet Union of, 47, 95, 105–106, 107–108, 110, 185, 206, 218–223, 233, 240, 243–244, 256, 259–261, 263–265, 267, 294, 298–301
attitude to United States of, 149–150, 185, 216, 219, 234–237, 246, 252, 271
childhood and education of, 6, 12–15
origin of interest in Russia of, 12
Jewish origin of, 7, 12, 164–165

on university reform and higher education, 154–158, 161, 200–202, 210
quantum field theory of, vi, 1, 5, 6, 11, 37–45, 49, 54, 59, 61–62, 69, 74, 77–87
political commitment of, vii, 1, 3, 4, 8, 16–18, 95–96, 98–99, 104, 123, 156, 182–183, 185–186, 192–252, 256, 259–267, 270–271, 277–280, 282–290, 293–295, 298–301, 303, 305, 311
relationship with Heisenberg, 134–135, 168, 172–175, 182, 289–290
Soviet Union, 1934 visit to, 7, 87–88, 95, 108–113
style in physics, 5, 26–32, 45, 58–60, 89–90, 159
United States, 1939 visit to, 144–151
view on religion of, 12, 203–204, 305
Rosenfeld, Yvonne nee Cambresier, xi, xii, 14, 103–105, 113, 150, 161, 162, 166, 169, 170, 172, 177, 180, 181, 182, 238, 241, 242, 245
Ross, Alf, 129, 270
Rotblat, Joseph, 250, 251
Rozental, Stefan, 2, 123, 153, 167, 169, 269, 306, 307
Rubin, Edgar, 128, 129
Rumer, Iurii, 141–143
Rupp, Emil, 29
Russell, Bertrand, 250, 251
Russell-Einstein Manifesto, 250–251, 288
Russia (see Soviet Union)
Russian Revolution, 47, 99–101, 114, 121

Sarton, George, 13, 150
Schatzman, Evry Léon, 281, 282, 283
Schelling, F. W. J., 126
Schiff, Leonard I., 43
Schlick, Moritz, 129
Schönberg, Mario, 218
Schrödinger, Erwin, 11, 15, 16, 20, 65, 140, 159, 259, 281, 284, 287
Schrödinger equation, 34, 70, 76, 272, 274, 310
Schrödinger's wave function, 20, 24, 25, 34, 62, 76, 274, 281
Schultzer, Bent, 270
Schwinger, Julian, 43, 257
Science for Peace, 224–226, 239, 249, 288
Second International Congress of the History of Science and Technology, 111, 120
Second International Unity of Science Congress, xii, 129–132, 269
Second World War, 1, 7, 8, 14, 41, 54, 90, 95, 112, 121, 134, 136, 141, 151, 154–183, 186, 188, 191, 192, 194, 195, 197, 198, 203, 208, 209, 210, 216, 224, 233, 234, 258, 269, 271, 273, 284
Semyonov, N. N., 233
Servet, Michel, 299
Serpe, Jean, 50, 166
Seventh Annual Rochester Conference in High Energy Nuclear Physics, 237
Seyss-Inquart, Arthur, 161
Shimony, Abner, 307
Shryock, Richard H., 207
Singer, Charles, 224, 231
Singer, Dorothea Waley, 207, 231, 267
Sliv, Lev A., 292, 296
Smyth, Henry D., 195
Smyth Report, 195
Social Relations of Science Movement, 120–121, 205, 210–211
Social responsibility of the scientist, 8, 120–121, 183, 187, 199, 225
Socialism, 3, 8, 15, 16, 17, 32, 95–96, 101, 106, 111, 119, 192–201, 212, 226, 267, 272

Society for the Freedom of Science (SFS), 207, 223, 226, 227
Solomon, Jacques, 11, 16–17, 29, 82, 95, 112, 183
Solvay, Ernest, 18
Solvay Conference, 31
 1927, xi, 18–20, 61, 90
 1930, 20, 63
 1933, 48, 81, 82, 85, 86, 114
Sommerfeld, Arnold, 11, 52, 159
Sorbonne, 15
Soviet Union, 27, 28, 47–48, 90, 95, 96, 99–100, 105–114, 116, 119, 120, 121, 142–144, 151, 185, 186, 188, 189–190, 199, 218, 221, 222, 234, 243–244, 246, 255, 260, 262, 267, 268, 285, 287, 290–292, 294, 296–299
 Cultural Revolution in, 47
 First Five Year Plan, 47, 99, 105
Specht, Minna, 134
Spinoza, Baruch or Benedict de, 125
Stachel, John J., vi, vii, 237
Stalin, Joseph, 2, 8, 47, 99, 101, 106, 143, 151, 186, 189–190, 203, 223, 255, 280, 283, 290, 292, 293
Stalinism, 239, 292
Stauning, Thorvald, 101
Stevin, Simon, 205
Stockholm Appeal, 240–242, 243, 245
Strassmann, Fritz, 144
Strauss, David Herrmann Martin, 117, 129–132, 134–135, 139, 140, 142, 180, 252, 280, 284–285
Swings, Polydore, 14, 15, 45, 103, 237, 239
Szilard, Leo, 29, 116, 117

Talmudist, 53
Tamm, Igor E., 27, 112, 113, 114, 298
Tansley, Sir Arthur G., 226
Tas, Eva, 202
Taton, R., 207
Taylor, A. J., 218, 219
Terletskii, J. P., 261, 262, 275, 276
Thaw in East-West-relations, 252, 255, 290–294, 298
The Communist Party, 95, 106, 118–119, 151, 185, 188, 190, 236, 283, 293
 American, 271, 272
 Belgian, 115, 188, 203
 Danish, 168–169, 188
 Dutch, 192, 200
 French, 16, 188, 211
 Greek, 189
 Russian, 5, 90, 99, 190, 222, 240, 255, 258, 259, 261, 263, 264, 268, 275, 292, 299
The Great Break, 47
The Great Depression, 106, 114, 120
The League of Nations, 122
The Manchester Guardian, 112, 192, 203, 214, 242
The Netherlands, 14, 105, 120, 121, 142, 153, 158–161, 199, 200, 203, 221
 German occupation of, 161–183, 210
 political situation in, 192, 203–04, 208
Tisza, L., 113
Tomonaga, Sin-Itiro, 257
Trotsky, Leon, xi, 98–103, 199, 200, 201, 202, 203, 286
Trotskyism, 144, 199
Truman, Harry S., 191
Truman Doctrine, 189

Uhlenbeck, George E., 158–159, 162, 172
Uncertainty relation, 22–23, 63, 65, 67, 68, 73, 74, 82, 84, 89, 90, 91, 258, 274, 276, 295, 300, 309
UNESCO, 206, 217
United Nations, 187–188, 195, 196, 209, 233, 241, 242, 248, 250, 251–252, 290–291
United States, 95, 142, 144, 155, 159, 173, 185–191, 195, 199, 209, 213, 216, 218–219, 221, 232–237, 241, 242, 244, 246, 250–252, 291, 293

Unity of Science Movement, see Logical Positivists
University Foundation, 15, 154
University of Brussels, 238–240
University of Copenhagen, 56, 128, 170, 269–270
University of Liège, 13, 14–15, 50, 59, 103, 116, 155, 159, 160, 166
University of Manchester, 213, 214, 239
University of Utrecht, 133, 159–162, 164, 166
    Physical Laboratory, 159–161
Utrecht, vii, ix, 1, 8, 13, 50, 155, 159–162
    German occupation of, 7, 153, 161–183

van der Waerden, B. L., 134
Vassails, G., 281, 282
Vavilov, N. I., 111
Vavilov, Sergei, 201, 233
Verbond van Wetenschappelijke Onderzoekers (VWO), 8, 210
Vessiot, Ernest, 15
Vienna Circle, see Logical Positivists
Vigier, Jean-Pierre, 281, 282, 295, 297
Volta, Alessandro, 20, 56
von Laue, Max, 143
von Neumann, John, 30, 65, 85, 135, 262, 276, 309, 310, 312

Waller, Ivar, 30, 36, 41
Wave-particle duality, 16, 20–21, 27, 33, 63, 65, 90, 140, 301
Weber, Sophus T. Holst, 170
Weisskopf, Victor, 41, 54, 59, 117, 143
Weizsäcker, Carl Friedrich von, 134–135, 143, 169
Wentzel, Gregor, 37, 167
Weyl, Hermann, 38
Wheeler, Janette, 59
Wheeler, John, 59, 145, 214, 312
Whyte, Lancelot Law, 282, 286, 303
Wigner, Eugene, 30, 42, 234, 276–277
Williams, E. J., 113
Willems, Jean, 116, 155, 239
Witmer, E. E., 29
World Congress of Intellectuals, Wroclaw, 217–223, resolution of, 218–221
World Federation of Scientific Workers (WFSW), xii, 8, 211–214, 222, 225, 226–229, 247–249, 284
World Federal Government, 196, 233–234
World Peace Council or Congress (WPC), 8, 222, 225, 240, 247, 251
World Peace Movement, 217, 221, 229, 231, 250, 251

Yourgrau, Wolfgang, 216, 285, 294
Yukawa, H., 251

Zaslawski, Dawid, 218
Zeh, H. D., 303, 307
Zhdanov, Andrei, 190, 222, 262, 268
Zhdanovschina, 223, 257–270, 275–276, 284–285, 289, 295, 298, 301
Zurich, 6, 35–36, 40, 45–47, 67, 74, 95, 167, 201
Zweiling, Klaus, 265